Peeling Potatoes or Grinding Lenses

Peeling Potatoes or Grinding Lenses

Spinoza and Young Wittgenstein
Converse on Immanence and Its Logic

ARISTIDES BALTAS

University of Pittsburgh Press

Published by the University of Pittsburgh Press, Pittsburgh, Pa., 15260

Copyright © 2012, University of Pittsburgh Press

Manufactured in the United States of America

Printed on acid-free paper

10 9 8 7 6 5 4 3 2 1

Library of Congress Cataloging-in-Publication Data

Baltas, Aristeides.

 Peeling potatoes or grinding lenses : Spinoza and young Wittgenstein converse on immanence
and its logic / Aristides Baltas.

    p. cm.

 Includes bibliographical references (p.   ) and index.

 ISBN 978-0-8229-4416-4 (hardcover : alk. paper)

 1. Spinoza, Benedictus de, 1632–1677. Ethica. 2. Wittgenstein, Ludwig, 1889–1951. Tractatus logico-
philosophicus. 3. Immanence (Philosophy) 4. Language and languages—Philosophy. 5. Logic,
Symbolic and mathematical. 6. God. I. Title.

 B3974.B35 2012

 192—dc23   2011039618

I can work best now while peeling potatoes. . . .
It is for me what lens-grinding was for Spinoza.

—LUDWIG WITTGENSTEIN, *Geheime Tagebücher 1914–1916*

Contents

Preface

No one can draw out of things, books included, more than he already knows. A man has no ears for that to which experience has given him no access.

—Friedrich Nietzsche, *Ecce Homo*

The National Technical University of Athens, where I started studying and where I continue teaching, has always been liberal with leaves of absence and helpful in many other ways. I am grateful to those who people and have peopled it—students, colleagues, secretaries, chairpersons, rectors—for unfailing support and encouragement.

The first draft of what is presented here was written at one go under the generous auspices of the University of Pittsburgh. The Center for Philosophy of Science and the Department of History and Philosophy of Science at the University of Pittsburgh provided sabbatical surroundings most propitious for the undertaking during the winter and spring terms of 2006–2007. Everybody involved—colleagues, students, administration—should be thanked most wholeheartedly. By and large, the writing as such took place in the cheery apartment of Giovanni and Lucia Camardi, where smoking—indispensable for punctuating ideas and their expression—was (almost) allowed. Evangelia suffered most from punctuation. Not because of smoking—she fully shares the good habit with me—but because rounding off an insecure paragraph invariably led me to make her stop whatever she was doing to answer questions such as, What do you make of this? Does that make sense to you? If, after my explanations, the never indulgent answer was encouraging, I felt I could go on; if not, I knew I had to rework the idea. For this reason and many others, this book is as much hers as it is mine.

Pittsburgh has become a second home. Peter Machamer, Barbara Diven Machamer, Tara and Michael, Peter's and Barbara's families, Merrilee Salmon, Ted McGuire and Barbara Tuchanska, Jim and Pat Lennox, John and Frances Earman, John Norton and Eve Picker, Sandy Mitchell and Joel Smith, Jim and Deb Bogen, Jerry Massey, and John and Andrea McDowell, together with Evgenia Mylonaki, Markos Valaris, Andreas Karitzis, Nikos Vestarhis, Thanos Raftopoulos, and Brian Hepburn—all these friends and colleagues formed the extended family whose hospitality fostered discussions on philosophy and politics, literature and film, horses and cats, Greece and the United States, personal and collective conundrums, and the mysteries of American football (the Pittsburgh Steelers won the Super Bowl that year), all taking place effortlessly over large pots of coffee or while eating, drinking, and being generally happy. The benign spirit of Marcello Pera, then speaker of the Senate of the Italian Republic, was hovering above Pittsburgh all that time, materializing by the occasional email and once by a quite interesting—if totally unexpected—book. Coming to know and to relate closely with Beatriz, Alejandro, and Laura Bruzual was an additional blessing. The visit of Ria Markoulakis added its own pinch of salt.

During my stay in Pittsburgh, the ideas presented here were given the chance of being tested in different places. Jim Conant at the University of Chicago, Gary Downey at Virginia Tech, Alice Crary at the New School of Social Research, Bruce Robbins at Columbia, and Stathis Gourgouris at UCLA issued generous invitations. In addition to the crucial contribution of these colleagues and friends, the questions and critical remarks of John Haugeland, Dick Burian, Joe Pitt, Jim Klagge, Raymond Petridis-Tzombanos, Akeel Bilgrami, Gil Anidjar, Neni Panourgiá, and Norton Wise, who were members of various audiences, helped me decisively to understand what I was doing. Alice Crary had the goodwill and patience to read my first chapters closely. An apparently innocent question of hers obliged me to rethink the whole approach and drastically change some of its basic ingredients. I am more grateful to her than I can say. Liana Theodoratou, liberally seconded by Dimitris Theodaratos, Eleftheria Astrinaki, and Niki Kekos, presided not only over classic and modern Greek poetry, language, and literature at NYU but over New York City generally: she took care not just of talks and presentations but also of accommodations, dinners, lunches, brunches, sightseeing, and theatrical performances. Giorgos Perrakis participated in some of these activities.

The ideas in the book took a long time to gestate. While a graduate student studying physics (and "the Revolution") in Paris from 1968 to 1973 (together with Platon Andreadis), I had become interested in Althusser, for, among other things, his work made me realize (perversely) that there is a question concerning the passage from classical mechanics to relativity theory and that what he had to say about Marx's contribution could illuminate the issue. Later I discovered that my idea might not have been that eccentric after all, for Althusser was relying heavily on the tradition of *épistémologie historique,* marked by Alexandre Koyré, Gaston Bachelard, Georges Canguilhem, and Jean Cavaillès, with the work of Dominique Lecourt clarifying the connections.

Having thus started to become comfortable with Althusser, I was somewhat shocked to learn that my favorite author felt obliged to present a "self-criticism" (1974a) wherein he "confessed" that he had been following Spinoza. My efforts to understand this turn foundered then on the impossibility of understanding why Althusser was invoking that incomprehensible philosopher. The clouds started to disperse after I read Macherey's book *Hegel ou Spinoza* (1979) and particularly when, during my first sabbatical at Pittsburgh in 1984–1985 (voted then the "most livable" city of the United States), I audited a seminar on Duns Scotus and Spinoza conducted by Peter Machamer, Ted McGuire, and Peter King.

Back in Greece, ideas took shape within an informal and geographically dispersed Spinoza "circle" that included, in one or another capacity, Gerassimos Vokos, Vana Grigoropoulou, Giorgos Fourtounis, Akis Gavriilidis, Aris Stylianou, Tasos Betzelos, Lara Skourla, and later Sofia Kousiantza. Other studies that helped shape the arguments in this book include Negri 1982; Wolfson 1983; Curley 1988 (as a rejoinder to Bennett 1984); Della Rocca 1996; Loyd 1994 and 1996; Nadler 1999, 2004, and 2006; Mandelbaum and Freeman 1975; Montag 1999; and Smith 2003—these works set in conjunction with or in tension to the "structural Spinozism" (going more or less together with the "Spinozistic structuralism" of Althusser, Lacan, and others, the characterizations being due to Fourtounis 2005) of Martial Guéroult, Alexandre Matheron, Gilles Deleuze, Pierre-François Moreau, Pierre Macherey, and Etienne Balibar.[1] Curley 1969 and Mason 1999—the latter discovered by chance in an inconspicuous but impressively rich bookstore in Chicago—gave me confidence that connecting Spinoza with young Wittgenstein might hold some water after all. A wider context emerged from works by Mason (2000 and 2007), Bloch (1993), Garver (1994), Norris (1991), Lock (1992), and Revault d'Allones (1993).

Very soon after I started teaching (physics, not "the Revolution") in Greece in 1976, Kostas Gavroglu introduced me to Kuhn, Lakatos, Feyerabend, and Popper. In sustained discussions with him, as well as with Pantelis Nicolacopoulos, Aris Koutoungos, Yorgos Goudaroulis, Dionysis Anapolitanos, Pantelis Basakos, and Theofilos Veikos, I came to realize that philosophy of science—or rather, what I took the thing to be—should have always lain closer to my heart than did "puzzle solving" in physics. My idiosyncratic exposure to Althusser allowed me to formulate my first big question: Do Althusser and Kuhn say fundamentally the same thing or not? And if not, what is the difference? The adventure leading to the pages at hand had started. For present purposes, I should say that a significant episode of that adventure has been trying to understand Kuhnian radical theory change as Wittgensteinian grammatical change, with Althusser's *idéologie pratique*, Bachelard's *obstacles épistémologiques*, Lakatos's "hidden lemmas," and Wittgenstein's "hinges" contributing their pieces. In this respect, the support of Wal Suchting, Gideon Freudenthal, Peter Machamer, Marcello Pera, Tom Nickles, Paul Humphreys, Jean Paul van Bendegem, Tom Kuhn, Jim Conant, and Léna Soler has been determinative.

Wittgenstein was forced upon me by Vasso Kindi. Her PhD thesis (later Kindi 1995), the first I was conducting, focused on the relations between Kuhn and Wittgenstein, thereby obliging me to come to terms with an author apparently far from my interests. A symposium in Rethymno organized by Myrto Dragona-Monachou and a Sunday morning study group in my house—including Vasso Kindi, Kostis Kovaios, Aris Koutoungos, Pantazis Tselemanis, and Stelios Virvidakis—undertook to streamline Wittgenstein's later philosophy and to help clarify it. The study group, however, was not interested in the *Tractatus,* and my own efforts to penetrate it struck an impenetrable wall.

The breakthrough occurred when I closely read McDonough's *Argument of the Tractatus* (1986). Despite my interpretative disagreements with it, to appear later, McDonough's presentation of the grandeur of Wittgenstein's vision as connected to the flimsiness of his starting point inspired a feeling of awe: from that moment on, I knew I had to appropriate the *Tractatus*. Crary and Read's collection *The New Wittgenstein* (2000) came at the opportune moment; the ideas I had started exploring found a firm basis to stand on. The work of Cora Diamond, Jim Conant, and Tom Ricketts (1985, 1996) and the collections of Tait (1997), Floyd and Shieh (1999), Reck (2002), and Crary (2007b), together with the long discussions I imposed on Jim Conant at various times and places, shaped the general coordinates of my understanding. Friedlander (2001) and Ostrow (2002) formed companionable interlocutors.[2] I disagree, though, with Ostrow's qualifying the internal movement of the *Tractatus* as "dialectic,"[3] not only because the term can have many interpretations, as Ostrow acknowledges, or because it is usually invoked for hiding rather than clarifying a difficulty, but mainly because the flow of the *Tractatus* exhibits no *Aufhebung*: at its close we are given not a synthesis of contradictory poles at a higher plane but a total erasure leaving nothing behind.

The thought of connecting the *Tractatus* with the *Ethics* popped up (I do not recall when, how, or why) as a vague idea that Wittgenstein's notorious identification of solipsism with realism should share something deep with Spinoza's "body-mind identity" thesis. I tried exploring this idea with the help of colleagues and students—Lara Skourla suffered most, closely followed by Spyros Petrounakos and Dimitris Papagiannakos—and in a couple of papers written in Greek. Eventually I was offered the opportunity of presenting the general contours of the point I had reached at a Wittgenstein symposium organized by Vasso Kindi in Delphi, with many renowned Wittgenstein scholars attending. I received mostly puzzlement and blank looks, with the few exceptions crucial for sustaining my self-confidence. A phrase of Stanley Cavell's to the effect that I should persevere with the task, a nod of approbation from Cora Diamond, and a phrase of John McDowell's some time later to the effect that my line of approach to the *Tractatus* was on the right track did the job. The same paper was presented in Pittsburgh at the invitation of Nancy Condee, Ronald Judy, and the cultural studies program. The reception was warmer overall, even as positive remarks from Jerry Massey and Peter Machamer pointed at the necessity for a more elaborated account.

It was such elaboration I had in mind when I started writing the present book. It soon appeared, however, that I had embarked on something of altogether different proportions. Though I was never sure whether the larger undertaking to which I had been led would reach closure, I persevered. Only the reader can assess the outcome, but I feel obliged to add that all this would have been impossible if my philosophical education during my first stay in Pittsburgh (thanks to the intervention of Dionysis Anapolitanos, the generosity of Nick Rescher and Adolf Grünbaum, and the companionship of Liana Theodoratou) and that in Princeton some years later (thanks to Alexander Nehamas, Dimitri Gondicas, and Liana Theodoratou) had not been what it was. Alexander Nehamas, at both Pittsburgh and Princeton, introduced me not only to Nietzsche, Plato, and the intricacies of literary criticism but also to the whole philosophical tradition of *epimeleia heautou,* or care of the self. Nehamas's works (especially 1985 and 1998) have been constant resources. In addition, the courses or seminars of Peter Hempel, John Haugeland, Joe Camp, Peter Machamer, Ted McGuire, Peter King, and John Beverley at Pittsburgh and those of Paul Benacerraf, Dan Sperber, and Stathis Gourgouris at Princeton, together with my tendency to ruthlessly exploit colleagues and, particularly, students,[4] made up for my total lack of formal philosophical training. Without the background provided by such high-quality teaching, the present book could not have been conceived, let alone written.

Subsequent drafts of the book received extended comments from Aristidis Arageorgis, Kostas Loukos, Peter Machamer, Evgenia Mylonaki, Kostas Pagondiotis, Spyros Petrounakos, and Markos Valaris. I thank them all for their time and generosity. Merrilee Salmon's intervention was decisive not only for encouraging me beyond measure and helping me finish the penultimate draft but also for prompting me to bring the work to the University of Pittsburgh Press. Merrilee, as well as the press's anonymous readers, helped me pump out unnecessary air from the final version. Thodoris Dimitrakos did a superb job with the index. Efi Kyprianidou helped crucially with references. Cynthia Miller was most encouraging from the first moment the manuscript landed in her hands. The staff of the University of Pittsburgh Press has been most helpful.

Time passes. It only remains to express my deepest regret for the absence of teachers, and I dare say friends, whose criticism would have been truly invaluable. I am referring to Theofilos Veikos, Wal Suchting, Michael Sprinker, Wes Salmon, Kosmas Psychopedis, Pantelis Nicolacopoulos, Tom Kuhn, Tamara Horowitz, Peter Hempel, John Haugeland, Yorgos Goudaroulis, Michael Frede, Angelos Elefantis, and Jacques Derrida.

This book is dedicated to the memory of Beatriz Bruzual, philosopher in life.

Note on References and List of Abbreviations

With respect to Spinoza I will capitalize "God" and its synonyms, as well as "Attribute." For citations to the *Tractatus* and the *Ethics,* I follow the standard conventions, as I do with respect to other works by Spinoza or Wittgenstein. Citations to the *Ethics* incorporate the standard symbols: "def" for definitions, "a" for axioms, "l" for lemmas, "p" for propositions, "c" for corollaries, "d" for demonstrations, "s" for scholia, "ex" for explications, "post" for postulates, "Pr" for prefaces, and "Ap" for appendixes. Thus E II p16c2 refers to corollary 2 of proposition 16 of part II of the *Ethics.* I base quotations on Shirley's translation of the *Ethics* (Spinoza 1982), though I also consulted those of Curley and of Parkinson (Spinoza 1985a and 2003, respectively). I take into account the French translation of Bernard Pautrat in his bilingual edition and the Greek translation of Evangelos Vandarakis (Spinoza 1988 and 2009, respectively). For the *Tractatus,* I base quotations on Ogden's translation (with the German *en face*); I also consulted the translation of Pears and McGuinness, the French translation of Gilles-Gaston Granger, and the Greek one by Thanassis Kitsopoulos (Wittgenstein 1986, 1974, 1993, and 1971, respectively).

Abbreviations for Texts

The following abbreviations are used for citations to works.

SPINOZA

E: *The Ethics*
L: *Letters*
ST: *Short Treatise on God, Man, and His Well-Being*
TEI: *Treatise on the Emendation of the Intellect*

WITTGENSTEIN

PI: *Philosophical Investigations*
TLP: *Tractatus Logico-Philosophicus*

Peeling Potatoes or Grinding Lenses

Coordinates of a Conversation

Even though the divergencies are admittedly tremendous, they
are due more to the difference in time, culture, and science.
—Friedrich Nietzsche, "Postcard to Overbeck"

THROUGHOUT THE FOLLOWING pages I argue that Wittgenstein's *Tractatus Logico-Philosphicus* (TLP) and Spinoza's *Ethics* (E) both pursue the same end. We can profitably take each as aiming to establish that there cannot be any position outside the world, thought, and language, that there can be no overarching standpoint from which anyone or anything can encompass the world, thought, and language as wholes, can act on them, regiment them, know them, or make meaningful pronouncements on them. Borrowing Spinoza's terminology, I call this philosophical perspective the *perspective of radical immanence.*

Spinoza and Wittgenstein worked in very different times. In Spinoza's time, it was almost impossible *not* to conceive God as occupying a position of overarching transcendence. Even Descartes, who had opened the philosophical vistas of modernity and arguably constitutes the principal influence on Spinoza, continued to conceive of God's relation to the world in this way. Accordingly, Spinoza's attempt to demonstrate the impossibility of such an overarching position has God as its main object: the burden of his work consists in trying to prove with the full rigor of geometry—and hence apodictically—that God can neither be nor be conceived as a supreme overarching entity who created the world and resides eternally outside it, overseeing its course with perfect freedom and benevolence.

In Wittgenstein's time God was of no prime concern to philosophers. If anything provided the intellectual focus for the environment in which Wittgenstein worked, it was logic in its relation to language. All sources attest that in his earlier years, Wittgenstein was consumed with such issues. But even if philosophical involvement with logic is not identical to philosophical involvement with God, there

1

is a parallel: logic, as it was being radically rethought in the early twentieth century, mainly by Frege and Russell—whom Wittgenstein himself labeled his principal philosophical teachers—was being surreptitiously taken out of the world, as it were, to become elevated to the overarching position I have described. The young Wittgenstein, then, whether despite or because of his active participation in these developments, vehemently rejected such elevation and tried to demonstrate with the full rigor of logic—and hence apodictically—that logic is *immanent* in language, in thought, and in a sense, in the world.

We might say, therefore, that independent of the enormous differences in historical circumstances and philosophical context and of the ways in which God and logic were conceived and discussed in each period, the principal philosophical influences on (and thus the main philosophical opponents of) both these authors had the same relationship to the world (and thought and language) in mind when thinking about God or logic, respectively. It was this position relative to the world (and thought and language) that concentrated the philosophical wrath of both Spinoza and Wittgenstein. Hence both set out to demonstrate, each in his own way, the impossibility of such a position. This is what I mean when I say that both Wittgenstein and Spinoza espouse the same philosophical perspective of radical immanence, and this is what I will cash out by reading the *Ethics* and the *Tractatus* in conjunction.

The perspective of radical immanence as I define it has been espoused by others, too. Many philosophers share, in some way or other, the insight that there can be no position overarching the world (and thought and language). With respect to this insight, however, Spinoza and Wittgenstein are distinguished by two things. First, both focus their work on this insight as such and concentrate their efforts on working out what the perspective it defines amounts to and what it involves for the whole of philosophy. Spinoza and Wittgenstein make the perspective of radical immanence appear most perspicuously in all its presuppositions and in all its consequences. Second, both of them reason implacably, with a kind of single-mindedness rarely found in the history of philosophy, mustering for the purpose the highest standards of rigor their respective periods afforded. We might say that both distill the perspective of radical immanence and present only this concentrate and its implications.

These characteristics make the *Ethics* and the *Tractatus* exemplary texts not just with respect to the perspective of radical immanence but for the whole history of philosophy. Consequently, any serious effort to read the two works in conjunction will offer insights into related discussions: trying to clarify the issues associated with the perspective of radical immanence in the cases of Spinoza and Wittgenstein will help us understand how other philosophers treat what amounts to the same perspective. In addition, trying to grasp the perspective at issue as Spinoza and Wittgenstein handle it might help us understand how this perspective is important for the philosophical endeavor in its entirety. Glimpses of such wider understanding will emerge in what follows.

Basically the same characteristics make the two texts notoriously difficult. It is therefore no coincidence that each has given rise to a vast array of divergent interpretations. Such difficulties are perhaps responsible for the continuing deferral of a canonical reading of either work that would lead to a wide consensus, despite valiant efforts to the contrary. Concomitantly, the same difficulties seem to deter efforts to read the two works systematically in conjunction despite their arresting similarities, which many others have noted. Wittgenstein himself came to accept Moore's mediation to that effect and thus came to owe to Spinoza the very title of the only book he published in his lifetime (Monk 1990). The absence of canonical readings, as well as the absence of systematic comparative readings, leaves room, so to speak, for the effort I undertake here.

• • •

The difficulties of these texts are due to more than just the denseness of the writers' subjects or the relentlessness of their reasoning and the austerity of their styles. They arise as well—and this decisively—from the focus on the perspective of radical immanence itself. We may get a foretaste of this kind of difficulty if we ask an apparently innocent question: from what standpoint is the proposition defining the perspective issued? On what does it rest, and what are the overall conditions of its enunciation? But once we ask this question, we cannot fail to realize that any statement of the perspective of radical immanence looks bizarre, with nonsense seeming to lurk threateningly in the background.

For one thing, if the proposition is true, if there can be no position outside the world (and thought and language), then there can be no position from which to issue this proposition—talking, as it does, of the world (and thought and language) from the outside. If the proposition is true, it cannot be meaningfully stated; it self-destructs because what it says precludes what it must presuppose to say what it says. And even if we disregard abstract issues of meaning, the proposition cannot be supported or vitiated by bringing in appropriate justificatory grounds. For again, from what standpoint and under what conditions of enunciation can such dialogue take place? Inside the world (and thought and language), outside it, or on some mysterious ground in between? And with regard to the last alternative, we might further ask what the nature of this in-between can possibly be and hence to what kind of argument either side of the dialogue can appeal. We are landing dangerously close to nonsense as our thoughts on the matter start disintegrating.

Given this, the only remaining option seems to be that the proposition in question is simply not true.[1] Things appear now as quite straightforward. If the proposition defining the perspective of radical immanence is not true, if it is not true that there can be no position outside the world, thought, and language from which one can talk about them as wholes, then it is true that there *can be* such a position.[2] Stating that such a position is available is stating it from the very position in question; it is stating it by simultaneously occupying the position outside the world (and thought and language) that the proposition states is possible. The proposition is thus consistent with the conditions of its enunciation and therefore apparently self-consistent. The net result seems to be that only a philosophical perspective countering that of radical immanence at its defining core can be self-consistent or, at least, can properly take into account the conditions of its own enunciation. An additional bonus is that espousing such a counterperspective should make the one espousing it quite self-satisfied: simply by assuming that there can be a position outside the world (and thought and language), one comes to occupy no less than the position traditionally assigned to God.

This apparently impeccable piece of reasoning can be only anathema for Wittgenstein and Spinoza, for it seems to leave as the only viable option exactly what both set out to demolish. But it is easy to see how they would have responded: they would have maintained that the argument just sketched assumes the possibility of a position outside the world (and thought and language), and in taking this for granted, it surreptitiously elevates this possibility above the world (and thought and language), thereby begging the question. Hence, despite appearances, it too short-circuits itself; despite appearances to the contrary, it boils down to nonsense.

Nonetheless, it is hard to see how either Spinoza or Wittgenstein could deny that the definition of the perspective of radical immanence self-destructs, thus hovering dangerously close to nonsense by its own lights. To someone pinpointing this, they might have answered that under conditions to be carefully specified, one can make good sense at least of the definition's intent. It seems to follow that both our authors would be maintaining, more generally, that propositions appearing to make sense might in fact be nonsense while, conversely, propositions appearing to be nonsensical might, at least under certain conditions, let one go through them to reach the intent in making them.

If this is the kind of answer Spinoza and Wittgenstein would have given to the previously stated objections, then both should have something to say about nonsense, something powerful enough to take care of these objections. Of course, nonsense notoriously constitutes one of the key elements of the *Tractatus*, and the way Wittgenstein handles it (among other things, by retrospectively inflicting the charge on his own work) continues to tax Wittgenstein scholars. In Spinoza's case, the subject of nonsense, or in his terms, of "confusion," is less pronounced but appears in the *Ethics* nevertheless. In E II p40s Spinoza explicitly characterizes what he calls universal and transcendental terms as "confusing" even as he goes on using such terms unrestrainedly. Savan (1958) has specifically discussed how Spinoza's work relates to self-inflicted confusion in ways I will discuss later on.[3]

The doubtful meaning status (to say it politely) that the definition of the perspective of radical immanence enjoys leads one to ask how this perspective can be supported, to investigate the strategy that Spinoza and Wittgenstein have to deploy to establish this perspective and defend it. Given the previous considerations, it seems impossible to argue straightforwardly for this perspective in open philosophical battle with the perspective countering it, namely, the assertion or silent presupposition that any position outside the world, thought, or language is available. And this seems impossible for two interconnected reasons.

First, such a straightforward way of arguing would leave the perspective of radical immanence open to the charge of nonsense, thereby making the philosophical battle follow circles like the one previously discussed. Second, if we disregard the indictment, we notice that the formulation of the perspective of radical immanence rules out the possibility of any neutral ground on which the battle in question could be fought: to maintain that there is no position outside the world (and thought and language) is simultaneously to maintain that there is no room for any philosophical approach that would allow the possibility of such a position. There is no place for a battle to occur and thus no space an opponent might occupy to wage a battle. The perspective of radical immanence seems to foreclose any opposition. More strongly put, the perspective takes itself to be the only possible philosophical perspective; consequently, there cannot be philosophical perspectives in the first place, so that the perspective of radical immanence cannot take itself as being *a* philosophical perspective. If philosophy could be reduced to the fight between these two mutually exclusive perspectives, and if the perspective of radical immanence indeed emerges victorious, then the whole of philosophy is done away with in the sense that no room remains for any philosophical dispute. The claim that these conditions are fulfilled so that the conclusion goes through is certainly exorbitant, and the task of establishing it is correspondingly enormous.

Be that as it may, however, the philosophical battle has to be fought, and this fight cannot be limited to accusations of nonsense and appeals to abstract possibilities.[4] But as the previously adduced considerations testify, the perspective of radical immanence finds itself at a relative disadvantage: the proposition defining it carries its self-destructive character on its face, while its contradictory does not. The former thus carries the burden of establishing its exorbitant claim even as massive parts of the philosophical tradition tend to side with its opponent. It is imperative, then, that the perspective of radical immanence find some terrain on which to engage the battle and fight it to the end. Thankfully, there seems one option left: establishing the perspective by working from within the opposing perspective, seeking to undermine it and destroy it from the inside.

As I construe it, working from within comprises two features. First, the strategy should involve provisionally *accepting* the possibility of a world-, thought-, or language-transcending standpoint. This applies to concrete philosophical views that may appear in various forms and guises and that might concern any philosophical subject whatsoever. Second, one using this strategy should advance some particular philosophical content that does not appear to differ in kind from the philosophical views against which he or she is arguing. At this stage of its deployment, then, anyone employing such a strategy should not hesitate in advancing philosophical content that appears to sanction a position outside the world, thought, or language. Setting up and advancing such content constitutes what we may call "the first movement" of the strategy.

The strategy involves granting legitimacy to the philosophical views opposing the perspective of radical immanence, which means that the particular debates in which someone using it engages take the standard form of philosophical debate, utilizing the standard philosophical tools of demonstrations, arguments, examples and counterexamples, comments, scholia, and so on. But their use can be only provisional, for the propositions and views they are used to establish must in turn be rejected, since they, too, sanction the possibility of the perspective of radical immanence. That is, the views for which one argues in the first movement can ultimately have no more validity than the views against which they were initially set. We may call the movement leading to such self-annihilation the "second movement" of the strategy.

In TLP 6.54, the penultimate proposition of the *Tractatus,* Wittgenstein all but openly admits the existence of such a second movement, and the strategy behind the *Ethics* enfolds a second movement as well, albeit less explicitly. Further, the two movements unfurl simultaneously in the context of the unique strategy they conjointly make up, while they need not be textually separated in some clear-cut fashion.

For this strategy to succeed, the content advanced should possess the requisite philosophical power; that is, it should be capable of convincing open-minded opponents that it indeed manages to undermine not only the views it addresses in the first movement but also those advanced in doing so. The strategy is thus not designed to gain some advantage: winning the battle brings both sides to exactly the same plane, one wherein all issues engaged have become elucidated. Thus the strategy ought not be seen as a devious or deceitful one, for though following it involves making claims only to later reject them, the ultimate point is to provide genuine elucidation for everybody concerned.

To win this battle by following the strategy described is therefore to close off the philosophical enterprise altogether, for the successful deployment of it shows as untenable both the denial of the perspective of radical immanence and the philosophical content advanced for winning the dispute. The loss of all contenders on the philosophical stage and hence any stage on which a philosophical debate can transpire touches all philosophical subjects, leaving nothing more for philosophy to address. The issues have become elucidated, philosophical shadows have disappeared, and a new light has been shed on the whole intellectual landscape.

Both Spinoza and Wittgenstein seem to endorse the gist of this conclusion. Wittgenstein expressly holds that he has solved "in essentials the problems of philosophy"—in a way, moreover, that is "unassailable and definitive" (TLP Pr ¶8)—while Spinoza holds equally expressly that the philosophical theory he composed and completed to his satisfaction (E V p42s) is the only true one offering "adequate knowledge of the essence of things" (E II p40s2). Barring the inessential for Wittgenstein and the possible applications of the true philosophical theory to subjects of more practical concern for Spinoza (such as those he envisages in his *Thelogico-Political Treatise* and his *Political Treatise*), there is nothing left with which philosophy might occupy itself. Philosophy as traditionally practiced is finished, and philosophical peace—or philosophical silence—has come to reign everlastingly, at least for those who earnestly engaged either treatise and followed it scrupulously to the end.

The strategy I am discussing cannot be effectively deployed in the ethereal medium where philosophy is usually taken to reside and its disembodied arguments to confront one another. Wittgenstein considers philosophy as an activity rather than a theory (TLP 4.112), while Spinoza, although ostensibly viewing philosophy as theory, nevertheless construes thinking in general as an activity of the mind (E II d3ex), a claim to which Wittgenstein would not object. Spinoza specifies the nature of that activity by expressly ushering the body into the picture: mind and body are one ("psychophysical parallelism" or "mind-body identity"), and hence philosophical activity is simultaneously bodily activity.

Since philosophical activity takes place in language, the ways in which Spinoza and Wittgenstein approach language will affect how they appraise the nature of philosophical activity and hence their strategies for pursuing such an appraisal. Remarkably, both take language as indissolubly linked to the body. Wittgenstein says that language "is part of the human organism" (TLP 4.002), while Spinoza maintains that "the essence of words is constituted solely by corporeal motions" of the human body (E II p49s). For both, the linguistic expression of philosophical activity cannot avoid involving the body fundamentally. It follows that engaging in philosophical activity and deploying a philosophical strategy amounts to doing something to somebody, a relationship that is not only mind to mind but also body to body, even if this doing involves only elucidating and convincing and even if this somebody is only oneself. According to all biographical accounts, both Spinoza and Wittgenstein were consumed by doing philosophy, with their own bodies bearing witness to the fact.

Hence the strategy in question is not merely to state this or that; it is also to *do* this or that. Spinoza and Wittgenstein do things with philosophical words and suffer things from philosophical words. Their strategy cannot be understood unless one takes into account the performative axis. We will see later how Wittgenstein uses nonsense as a performative instrument in the guise of what I will call "telling nonsense" and how Spinoza goes along in much the same way, if perhaps more hesitantly. We might say, therefore, that if the whole of philosophy is to be dissipated, then success can be gauged only performatively, that is, in deed. The impossibility of straightforwardly arguing for the perspective of radical immanence seems in any event to disallow that either Wittgenstein or Spinoza could establish it in the airy form philosophical engagements usually take for granted, another consideration that perhaps more conspicuously shows the performative dimension to be indispensable.

• • •

I have mentioned differences between the ways Wittgenstein and Spinoza engage the perspective of radical immanence, but the exact nature of the impossibility (or possibility) of a position outside the world/thought/language remains to be clarified. To rectify this omission and to explicate the principle behind those differences, I start by distinguishing conceptual possibilities (or impossibilities) from purely logical ones. Borrowing a key element of Wittgenstein's later work, I regard conceptual possibilities as those allowed by the *grammar* subtending our thoughts and uses of language (the "language games"). In other words, I take conceptual possibilities as synonymous with grammatical possibilities in the later Wittgenstein's sense of the term.[5] In contrast, I regard logical possibilities much as the *Tractatus* seems to countenance them: the bare possibilities underpinning all conceptual distinctions, the possibilities "remaining"[6] after all conceptual issues have been clarified by "the one and only complete analysis" (TLP 3.25).

An example from the history of science might help to clarify this distinction. In Newtonian physics, a wave used to be defined as the propagation of a medium's disturbances. This definition makes the conceptual relation between "wave" and "medium" an analytic one; hence, the claim that a wave can propagate in vacuum, without a medium in which disturbances might be propagated, becomes a logical contradiction. But the advent of the special theory of relativity forced us to change our views on the matter. It became established that some waves (electromagnetic ones) do propagate in the absence of any medium. Introducing this novel conceptual possibility with respect to the facts it made us understand was tantamount to widening the previously available grammatical space, a widening capable of conceptually accommodating the novel kind of wave along with the old even as the old concept of a wave was reshaped in terms of the novel grammatical space. What matters here is that after the reconceptualization, the *logical* impossibility implied by the contradiction disappeared. It was retrospectively interpreted as a *grammatical* impossibility, the old grammatical space's inability to accommodate the novel concept, with logical consistency being reinstated in the process.

The lesson should be clear. Conceptual revolutions of this kind ("paradigm shifts" in Kuhn's terminology) widen the grammatical space available in ways that would have been impossible to conceive before the revolution, but after the fact, such widening can be seen to not touch logic as such. Logical possibilities cannot change by conceptual revolutions, but grammatical possibilities can and do. One of the burdens of this book is to show that the *Tractatus* sanctions this distinction between grammatical and logical possibility and thus might help elucidate the nature of a conceptual revolution in science.[7] At the present juncture, this distinction helps us understand how the historical distance separating Wittgenstein from Spinoza can be cashed out philosophically, a point that determines one major constraint on my conjoint reading of the *Tractatus* and the *Ethics*.

Spinoza worked during the irresistible advent of the scientific revolution, which was establishing a radically novel way of conceiving the workings of the world (and hence a novel grammar) that would remain unshakable up to the beginnings of the twentieth century. Independent of any changes in the physical theories at play, the deeper way of conceiving things was taken as final; no radical conceptual change could ever come to challenge foundations. The important point here, however, is that this finality equates grammatical impossibility with logical impossibility: grammar bans the (grammatically) inconceivable, but if the grammar is final, then by definition there can be no novel grammatical space on whose basis one might retrospectively interpret the previously inconceivable as a "mere" grammatical impossibility that has been overcome. Hence the inconceivable becomes synonymous with the logically impossible. It follows that within the framework of thought governing Spinoza's time, and independent of the ways in which different philosophers or theologians were considering the inconceivable in relation to God's powers or attributes (Mason 2000), the inconceivable could not then be split between the logically impossible and the grammatically impossible. Grammatical possibilities and logical possibilities perforce ran together.

Wittgenstein, however, worked in a time of major revolutions in physics as well as revolutionary advances in the conception of logic. The developments in physics were showing, among other things, that concepts taken as unshakable could change radically and—as would become clarified much later—come to be replaced by incommensurable namesakes without upsetting logic in the process.[8] Concurrently, the advances in logic allowed explicating, among other things, how the conceptual could and should be rigorously distinguished from the logical. Thus the means for distinguishing grammatical from logical possibility had become perfectly available, even if the notion of grammar needed more time to gestate and attain philosophical dignity in the later Wittgenstein's hands.

Therefore, once we disregard all other kinds of differences separating Wittgenstein from Spinoza (philosophical vocabularies and agendas, modes of argumentation, broader cultural factors, and so forth), the crux of the matter seems to be simple: the *philosophical* distance separating the two is that conditioned by the *historical* fact that Spinoza lacked a way of distinguishing the grammatical from the logical, whereas Wittgenstein did not. Thus acknowledging that the implacability of their reasoning compelled each to rely on and put to use only the highest standards of rigorous thought his time could afford, we might say that Spinoza could establish the perspective of radical immanence only at the conceptual level, for he was limited to the means then exhibiting those standards—namely, the deductive structures of geometrical order. With those means, the result of his toil could at best take the form of an unshakable philosophical theory.

For Wittgenstein, this was not enough. The revolutionary developments in the physics of his day were showing that establishing a theory at the conceptual level does not secure the theory for good, no matter how accurately aspects of the world might appear to comply with its dictates and no matter how rigorously the reasoning establishing it had been exercised. It is always possible that some part or aspect of the world might prove recalcitrant, leaving the theory open to the concomitant radical change.[9] Therefore all theories are vulnerable to the possibility of radical change, even if this possibility remains abstract and effectively idle until a conceptual revolution materializes it retroactively, making it manifest at the same time. The philosophical message is that any theory whatsoever, postclassical physical theories included, will always be open to the abstract possibility in question. To secure the perspective of radical immanence for good, then, Wittgenstein had to establish it at the level where these abstract possibilities themselves reside.

At the same time, the developments in logic (determined significantly by Wittgenstein's own work) had rendered logic capable of treating precisely such abstract possibilities. These are the bare (i.e., conceptually empty) possibilities that underlie[10] all language (and all thought and all the world) and constitute logic itself. Wittgenstein states this succinctly: "Logic treats of every possibility, and all possibilities are its facts" (TLP 2.0121). It follows that he can permanently secure the perspective of radical immanence by establishing it at the level of these bare possibilities, of logic as such. No theory whatsoever, be it philosophical, scientific, or something else, can reside at this level, for it underpins, and thus allows for, not just any theory and any grammar but also any thought, any piece of language, and anything at all in the world.

Therefore, establishing the perspective of radical immanence at the level of logic and with the means of logic cannot amount to composing a philosophical theory. On the contrary, establishing this perspective at that level reveals the secret of philosophical theories generally. By talking about the world, thought, and language in the standard all-encompassing terms, such theories entitle themselves to an external position that excludes them from the family of bona fide theories, for what they take as their object cannot be the object of any theory. Even if they might offer elucidatory insights, they ultimately amount to nothing. All the characteristics of the strategy I rehearsed previously—the two strategic movements and the self-destructive end—find a clearer expression in the *Tractatus* than in the *Ethics* because, in trying to establish the perspective of radical immanence at the level of logic and by means of logic, Wittgenstein was forced to go "below" theories and deductive reasoning in general to see what makes them possible. By the same token, he was forced to "discover" that the strategy in question was the only strategy he could possibly deploy.

Lest this appear too one-sided, we might argue on Spinoza's behalf as follows. First, the grammatically inconceivable, taken here strictly in relation to science, is not subjective or psychological in the sense that somebody remains blind to a possibility up to the moment when he or she realizes that it is conceptually possible after all. The grammatically inconceivable is *objective* in the sense that no one can overcome it simply by thinking more deeply than others have. Certainly, no grammatical change can arise ex nihilo; only an act of the imagination can bring it about. But the act of imagination implicated in scientific change differs from the kind involved in, say, literature; science is a *normative* enterprise, and hence the act of imagination leading to such a change should prove compelling, at least up to the limits set by scientific practice. Within those limits, the relevant part or aspect of the world should be conceived—by everyone—as the theory coming out of the grammatical change says it should. It is this normative force that allows the novel theory to win the day, simultaneously widening the grammatical space.

A grammatical change in science is radical indeed, and hence the difficulty of bringing it about is correspondingly massive. Since grammar operates from the background to silently determine our concepts and our intuitions, as well as the ways in which concepts interconnect and intuitions marry with concepts—since, if we follow McDowell (1996), we should consider concepts as going all the way down, passively organizing even our barest experiences—such a change demands that we alter at a single blow practically everything determining how we understand things in nature so that our understanding comes to conform to the strictures of the novel grammatical space. What is at stake here is so demanding that, to use Kuhn's turn of phrase, those having undergone the transition to the novel grammatical space live in a different world from the one inhabited by those who have not. For all these reasons, grammatical impossibilities hardly differ from logical impossibilities, while effectively (not abstractly) distinguishing the two can be achieved only ex post facto and only from the vantage point instituted by the novel grammatical space. Such distinction can be had only *post festum*.

Given all this, and because the distinction between grammatical and logical impossibility was literally unthinkable in Spinoza's time, nobody could expect Spinoza to have followed a strategy based on this distinction as clearly as Wittgenstein did. On the contrary, it is a tribute to his implacable reasoning that Spinoza deployed his own strategy in a way that, short of fully realizing the double nature of what he was doing, acknowledging it expressly and drawing all its consequences, matches in all essentials the strategy Wittgenstein deployed.

Yet there is more to add on Spinoza's behalf, from the positive side this time, something at the heart of my undertaking.

Recall that the strategy required for establishing the perspective of radical immanence unfurls along two movements, with the first advancing particular philosophical content. Now if Wittgenstein does manage to establish the perspective of radical immanence at the logical level, as he takes himself to have accomplished, then the outcome of his toil is logically "unassailable and definitive." One might argue from there that the philosophical content he advances along the first movement of his strategy is the only philosophical content that can be advanced for the purpose. Given this, if Spinoza, in seeking the same result, reasons as implacably as Wittgenstein does, then the philosophical content the latter advances should match, in its essentials, Spinoza's philosophical theory on all philosophical subjects that both the *Tractatus* and the *Ethics* address (barring, of course, differences in philosophical vocabulary). To find what Wittgenstein advances in the *Tractatus*, then, we might do worse than to consider what Spinoza advances in the *Ethics*.

Admittedly, this is a strong claim that must be justified in the pages to follow. But if it can be demonstrated, then there are almost no bounds to the admiration one should bestow on Spinoza's philosophical acumen: barring again various inessential differences, Spinoza managed to advance content that parallels Wittgenstein's, but he did so almost three centuries earlier, when everything around him was pushing in different directions and the materials that later allowed Wittgenstein to conceive and conduct his endeavor were literally unthinkable.

In pinning down the specific lines of thought that both Wittgenstein and Spinoza advance, and thus carrying on with my own task, Spinoza's treatise may even prove to be more helpful than Wittgenstein's, as the authors themselves suggest. Thus Wittgenstein openly declares that his work "is not a textbook" (TLP Pr ¶1), thereby admitting that the *Tractatus* could well be assessed—to use a quintessential understatement—as overly concise. In contrast, Spinoza acknowledges equally openly that he has written the *Ethics* "in order to point out the road" to its readers (E V p42s), for "nowhere can each individual display the extent of his skill and genius more than in so educating men that they come at last to live under the sway of their own reason" (E IV Ap §9). Such a description might reasonably enough be taken to suggest that Spinoza sought to compose a helpful textbook. As I will show, some disparities between the *Tractatus* and the *Ethics* might stem from these different modes of presentation.

• • •

The task I have assigned myself might appear tantamount to trying to understand the fundamentals of the *Tractatus* by relying on the *Ethics* while simultaneously trying to understand the fundamentals of the *Ethics* by relying on the *Tractatus*. But this formulation exceeds my ambitions. The fundamentals of the *Tractatus* involve the logical apparatus Wittgenstein introduces, the formalization he proposes, and the conclusions he draws about mathematics and its relation to logic. The following pages barely touch on these matters.[11] The fundamentals of the *Ethics* involve the way Spinoza characterizes the specific emotions in the later parts of his treatise, their role in individuals' lives, and his proposals for overcoming their nefarious effects. I address these matters only to the extent that they help clarify how he cashes out the perspective of radical immanence. More accurately, the texts to be compared are basically limited to an informal *Tractatus* and the first two parts of the *Ethics,* those in which the perspective of radical immanence is introduced and its main ingredients are drawn out.

But even with the investigation limited in this fashion, my account of the two treatises does not aspire to be more than large-scale overview of what they have to say, a constraint imposed by the material at hand. To examine the purported match with even a minimum of conscientiousness, I am obliged to look at many subjects of standard philosophical concern, particularly those on which the two treatises appear to disagree. If I sought to do justice to the material available, however, I would have to write not a monograph but an (idiosyncratic) encyclopedia of philosophy. This is no rhetorical flourish. Both the *Ethics* and the *Tractatus* aim at constituting complete philosophical treatises: each constitutes an attempt to cover all the subjects of major philosophical concern in its time and to solve or dissolve (at least in the essentials) all the corresponding philosophical problems. The scope is even broader in the case of Wittgenstein, for in the *Tractatus* he seems to tackle philosophical subjects that practically none of his contemporaries would have considered important, doing so, moreover, in a way that baffled readers then and continues to baffle them now. To treat all those subjects in detail while taking into account the volumes of insightful interpretations written on those subjects and on either of the treatises would be a Herculean task, one far beyond my capacities.

• • •

In chapter 1 I introduce the two authors to each other, talk about their texts' surface-level similarities and affinities, underscore the intellectual rigor characterizing both, and examine how this rigor relates to philosophical method for either man. The chapter closes by pinning down the core of the two endeavors. If Spinoza undertakes to demonstrate the impossibility of a position overarching the world and thought and language at the ontological level, Wittgenstein undertakes the same at the logical level.

In chapter 2 I examine the frame in which the *Tractatus* and the *Ethics* are set. I discuss the purpose governing the composition of each treatise and the goal each author takes himself to have established. I try to clarify how purpose precedes the activity it governs even though it can be manifested only in the deployment of that activity, whether the author recognizes this or not. I emphasize philosophical activity. The ethical intent and hence the ethical character of both treatises enter at this juncture—obviously intent and purpose are closely connected—while the impossibility of an external position from which to render universal moral injunctions means that both treatises must uphold an ethics of responsibility.

I discuss how these matters relate to the end each author takes his treatise to have attained. Philosophical worries cease at that end, and philosophical silence reigns as those who have scrupulously followed either treatise come to see the world as it really is, *sub specie aeterni* (Wittgenstein) or *sub specie aeternitatis* (Spinoza). Concurrently, I try to clarify what Spinoza means by his "third kind" of knowledge, since it bears an indivisible connection to the end in question.

Furthermore, I examine how Spinoza's thesis on the eternity of the mind might be understood as not too outlandish, for it is interwoven with the way Spinoza understands the human body and its capacities and with what might be called "expert action." This discussion is left open until chapter 6, where I attempt to unravel the connection between the eternity of the mind and the possibility (or rather impossibility) of an afterlife. Wittgenstein's thoughts on those matters appear only briefly here; a more ample examination is reserved for other chapters, although the way he considers life is discussed here as a prerequisite for that.

The following two chapters are devoted to the strategy deployed by the *Tractatus*. Chapter 3 begins by presenting a rough general schema on the relations among history, philosophy, the history of philosophy, and the history of science to argue that radical scientific change takes philosophy by surprise and provokes its more or less drastic reorganization. This schema sets forth a lens for envisioning the *Ethics* and the *Tractatus*.

On the basis of this schema, I try to explicate grammar, how grammar—or rather its germ—is characterized in the *Tractatus* and how it is implicated in radical scientific change. I discuss the prerequisites of the "eureka" leap associated with coming to understand a radically novel scientific theory and seek to identify these prerequisites through a dialogue between a teacher of the novel theory and students eager to learn it while still confined within the old grammatical bounds. This discussion helps illuminate the strategy deployed by the *Tractatus*. I argue that the eureka leap is not only a mental but also a bodily experience, and this allows discussing how Wittgenstein regards the human body. Along the way, I try to pinpoint the differences between coming to understand a radically novel scientific theory and coming to understand a philosophical treatise such as the *Tractatus*.

Chapter 4 confronts the strategy of the *Tractatus* as such. It starts by discussing strategy in general and goes on to a preliminary examination of the strategy Spinoza deploys in the *Ethics*; in addition, I lay out my own strategy for making sense of the *Tractatus*. I present the two movements of Wittgenstein's strategy, examine what "working from within" entails, and discuss how the three terms—*showing, elucidation,* and *nonsense*—key to Wittgenstein's project are put to work. I propose a classification of the *Tractatus*'s propositions in accordance with the strategy these conjointly make up and clarify how the performative dimension enters the picture.

In chapter 5 I discuss the operative plans the *Ethics* and the *Tractatus* set out and their overall structures. I also suggest an explanation for a surprising fact, namely, that both writers begin their treatises abruptly from ontology: Spinoza, by laying out the general coordinates of his conception of God; and Wittgenstein, by laying out the general coordinates of his understanding of the world. I call this surprising because given the position of each author in the philosophical developments of the corresponding period, one would expect Spinoza to start from epistemology, following Descartes, and Wittgenstein to start from logic and its relation to language, following Frege and Russell. I discuss the constraints these plans, structures, and beginnings impose on the match in philosophical content and thus complete the preparation for undertaking the examination of that match in chapters 6 though 8.

Chapter 6 engages the category of substance in three rounds. The first round addresses Spinoza's conception; the second round, Wittgenstein's; and the third examines the match between the two conceptions by bringing together the strands emerging from the first two rounds. The chapter begins by discussing Spinoza's reasons for equating Substance with God and God with Nature. I argue that the latter equation makes Spinoza a kind of naturalist and thus ushers him into the contemporary philosophical landscape, where Wittgenstein resides. I locate some particular characteristics of the language Spinoza employs for establishing his claims on Substance; for example, he indirectly admits that he often uses what he himself calls confusing terms, and he precedes his definitions with a first-person clause. I argue that these features validate the claim that Spinoza's strategy incorporates a second movement, one that—less overtly and barring his theory's self-destruction—matches Wittgenstein's second movement in the essential elements. I also specify some fundamental ingredients of the way Wittgenstein regards the relation of language to thought. These ingredients suggest that Wittgenstein's overall understanding of language and thought might not differ greatly from Spinoza's.

To examine this last claim, I start from Spinoza's overall conception of language as related to his epistemology. I try to clarify how imagination and memory matter here, as well as why Spinoza considers language to be a fundamentally bodily function. Here we can see why he distinguishes three kinds of knowledge: the best, "third kind" of knowledge, which is intuitive knowledge of the essence of things (discussed in chapter 2 as involving "blessedness" in a way to which Wittgenstein would not object); adequate knowledge, "the second kind," which is knowledge by adequate concepts; and inadequate knowledge, "the first kind," which is casual and unreliable knowledge. I argue that Wittgenstein considers thought to be something much like Spinoza's second kind and that although Wittgenstein envisages imagination from the altogether different angle of its relation to logic, he ought not object overmuch to the core of Spinoza's understanding of imagination and its function.

Spinoza (and Wittgenstein) see knowledge and thought as connected to language, so that I make an effort to go deeper into Spinoza's conception of the latter, to underscore that Spinoza understands the fundamental linguistic relation, that of signification, as involving two horns. On the one hand, a bodily affection, such as hearing a sound or seeing the written rendition of a given word, *stands in for* another bodily item, namely, traces of previous bodily affections produced by the same sound or sight. The sum of these traces, which Spinoza calls "habit," constitutes the first horn of the signification relation. The other horn stems from Spinoza's general thesis on body-mind identity, according to which any bodily item necessarily has a mental correlate, which is what Spinoza calls an "idea." Spinoza regards the ideas making up memory to be the mental correlates of the bodily affections making up habit. Hence the second horn of the signifying relation takes the following form: the mental correlates (the ideas) of our bodies' now hearing or seeing written the given word *express* in the mind the mental correlates (the ideas in memory) of the traces previously left on our bodies by the same sound or sight, that is, by the corresponding habit. The two horns of the relation are thus that of "standing in for" (within the body) and that of "expressing" (within the mind).

Although Wittgenstein follows a different route, he comes very close to the same understanding of the signifying relation, at least once we take into account three things: first, that he views language as a part of the human organism and hence a bodily function; second, that a proposition is a fact that stands in for another fact; and third, that a proposition has two aspects, one perceptible by the senses and the other mental in that it expresses a thought. The vindication of the match as such has to wait for chapter 8, where, as I said, I seek to spell out how Spinoza understands the mind-body identity thesis and what Wittgenstein has to say on the matter. It is noteworthy that Spinoza concludes that language is confusing by its very essence, so to speak, while Wittgenstein holds that language disguises thought. I suggest that "confusion" and "disguise" bear significant similarities, basing this on a demonstration that Spinoza understands the linguistic expression of adequate knowledge—by means of what he calls "common notions"—in much the same way as Wittgenstein understands propositions to express thoughts and picture facts univocally.

Chapter 6 continues with a discussion of the way Wittgenstein understands substance. I argue that the objects he says form "the substance of the world" (TLP 2.021) are purely logical, implying no ontological commitments or entailments. Since a logical possibility cannot float freely by itself—a possibility is always a possibility *of* or a possibility *that*—it requires by definition "something" substantive on which to anchor it. Wittgenstein's objects are thus merely the substantive, utterly "thin" bearers of logical possibilities and nothing more. Further on, after discussing how Spinoza characterizes bodies in general and why he believes that atoms cannot exist in nature, I show that Spinoza is obliged to introduce what he calls "simplest bodies" for reasons having more to do with logic than with physics. These reasons are impressively close to Wittgenstein's reasons for introducing objects.

The chapter closes by completing the discussion on the eternity of the mind, explicating why eternity of the mind not only cannot imply afterlife for Spinoza but necessarily excludes it. Here, then, emerges the third round of my discussion of substance, where I present a succinct overview of the previous material.

Chapter 7 is devoted to clarifying Spinoza's overall epistemology, with an emphasis on the proper interpretation of Attributes. I first explicate an interpretation by which Spinoza's notion of ideas comes to match Wittgenstein's notion of thoughts and Wittgenstein's category of facts comes to match Spinoza's extended modes. I further argue that Spinoza's Attributes may be taken as objective perspectives on Substance, much as one might view a major science, such as physics or psychology. Since Spinoza holds that there are infinitely many Attributes, while human reason has access to only two—Extension and Thought—I tentatively explore how Spinoza might have conceived of inaccessible Attributes. To that effect, I ask whether he considered them necessarily inaccessible or only contingently so and examine the extent to which much later endeavors—namely, Freud's approach to human subjectivity and Marx's approach to society and history—might count as Attributes of this sort.

Further on, I argue that Spinoza's failure to distinguish the conceptual (the grammatical) from the purely logical forced him to run together two "parts" of each Attribute, ones that, on Wittgenstein's way of viewing things, should be carefully kept distinct. These two parts are, on the one hand, the logical core of each Attribute, which ensures that it is—exactly like all the others—an objective perspective on Substance, and, on the other hand, the specificity of the Attribute, which can be laid out with concepts proper only to it. Once this distinction is made, the logical core in question might be construed as amounting to a Wittgensteinian manifold of logical possibilities, and the infinite number of Spinozistic Attributes might be regarded as corresponding to the indefinite number of Wittgensteinian objects. That is, the logical core of any specific Spinozistic Attribute can be identified with the manifold of logical possibilities defining, or attached to, a particular Wittgensteinian object, with this particularity subtending that specificity. This allows me to round off my discussion of the manner and extent to which Spinoza's overall epistemology matches the relevant logical requirements set forth by Wittgenstein.

Chapter 8 focuses on possibility, as either Spinoza or Wittgenstein conceive it, as well as on the apparently divergent ways the two portray necessity. I commence by examining how Spinoza understands the naturally possible yet inexistent modes and how this understanding relates to his physics. With respect to this physics, I argue that Spinoza seems to have thought that the natural science of his day should come up with some kind of natural history rather than take the form Newton and his followers gave to it. To make better sense of this last claim, I appeal to more contemporary physical theories or conceptions of physics. It is remarkable that the main instigators of those theories or particular conceptions—namely, Boltzmann and Hertz—numbered among Wittgenstein's heroes in natural science.

Since natural science is a particular way of ordering and connecting things in nature, Spinoza's famous thesis that "the order and connection of ideas is the same as the order and the connection of things" (E II p7) comes to the forefront. At this point I turn to this thesis, which has already cropped up in connection to various issues of epistemology, addressing what order and connection might amount to for Spinoza and what Wittgenstein might be able to say on the matter. I argue that the common order and connection characterizing all Spinozistic Attributes largely parallel the "form" and "structure" Wittgenstein discusses concerning objects and their associated manifolds of logical possibilities. This means that the order and connection in question is what Spinoza would have called simply "logic." This discussion helps clarify Spinoza's epistemology and its connection to the logical realm underpinning it—again, one invisible to him as a distinct realm.

At this point I collect all the previously discussed strands of the body-mind identity thesis to arrive at the surprising conclusion that Wittgenstein can be read as holding a version of it, a link that can be made once we identify Wittgenstein's world with the body of Spinoza's God and the mind of Wittgenstein's "metaphysical subject" (TLP 5.641) with the mind of Spinoza's God. On this basis, solipsism's identity with realism (TLP 5.64) becomes a large-scale version of the mind-body identity thesis. I argue further that, given Wittgenstein's understanding of the human body, the match could be extended, with qualifications, to cover finite human bodies as well.[12] I return to Wittgenstein's physics and his reasons for considering a scientific law (and causality itself) to be not a matter of logic but rather a function of the conceptual network that constitutes a scientific theory. This is, however, a function that no scientific conceptual network can omit: we can think only in terms of concepts, not at the level of logic as such; scientific concepts are connected in conceptual networks by means of the corresponding laws; and only connections of this sort are thinkable (TLP 6.361). Each such network characterizes a scientific law in its own way, and it is up to science to explain why anyone should abide by one law rather than another. Yet what interests Wittgenstein on the matter is not how science comes up with such networks and how it assesses their differences. As always, he focuses on logical possibility, that is, on what makes such networks and the differences among them possible. On this basis, and given that Spinoza could not go below the conceptual, the implacable necessity of natural laws that Spinoza endorses conflicts less than one might expect with Wittgenstein's statement that "there is only *logical* necessity" (TLP 6.37).

The chapter concludes by zeroing in on God and logic and the relation holding between the two for Spinoza and for Wittgenstein. I confirm that the *Ethics* and the *Tractatus* align to an impressive extent and that most of the remaining differences might be explained by the historical distance separating them. Each author believed himself to have written a complete philosophical treatise, though the texts' differing formats create additional disparities between the finished products of the authors' toil.

Having completed my principal task, in the last chapter, "Exodus," I discuss in outline why these treatises continue to be philosophically important and how the work of Spinoza and that of Wittgenstein evolved after each man had reached what he took to be a satisfactory end to the tasks he had assigned himself. The persisting question as to the relation between Wittgenstein's early and later work is broached but discussed only through the lens of the older Wittgenstein's conception of grammar and its relation to logic, which I use to examine radical grammatical change in science. On the basis of this examination, I conclude that Wittgenstein's later work *consummates* the perspective of radical immanence: if logic gives way to grammar, as the older Wittgenstein maintains, then radical grammatical breaks in the history of science cannot be forestalled by means of logic and philosophy even in principle. All possibility of a view *sub specie aeterni*—arguably the remaining vestige of the external vantage point—is thereby completely banned: history and its surprises acquire a kind of predominance over logic and philosophy. This, I suggest, is why Wittgenstein acknowledged the motto governing the whole of his later work to be Goethe's dictum "Im Anfang war die Tat" (in the beginning was the deed).

• • •

At this point a demanding reader might feel fed up with even just this synopsis of the discussion to come. Unclear claims proliferate, and the claims ostensibly justifying them are too vague or convoluted to bear serious scrutiny. The chapters seem to follow little discernible order, the connections linking my claims seem nebulous, and the issues that most people take to be the main contributions of the *Ethics* and the *Tractatus* are either openly excluded from the discussion or lost in the haze. But this should be expected: the *Ethics* and the *Tractatus* are extremely difficult texts, and hence any effort to read them anachronistically in conjunction will only compound the difficulties.

In my defense, I ask only that the reader concede, at least for the sake of argument, my central claim to be tenable, for if the *Ethics* and the *Tractatus* do indeed aim at much the same goal, then relying on one text to understand the other might temper rather than aggravate the all too apparent difficulties of the treatises themselves. Proceeding in this way does involve disrupting the flow of the texts at issue, disregarding the order they follow, mixing up the two, jumping from one philosophical topic to another—in short, engaging in all kinds of philosophically dubious moves. These are unavoidable factors of this kind of project, ones that might well give the demanding reader an overall impression of muddle and disorder, if not downright confusion. Hence, if this reader is to keep an open mind, I should expose at the outset all the unorthodox moves I will allow myself. Apart from asking for the reader's patience and goodwill, I cannot do more.

First of all, I aim at no more than a large-scale overview of these treatises, or rather, of what I take as their core. I do not inspect all the philosophical subjects that the *Ethics* or the *Tractatus* covers, and merely sketch the arguments concerning those I do inspect. These sketches are unevenly distributed and unevenly worked out, with only a few offering a more rounded discussion. The historical developments to which I sometimes must appeal are drafted even more roughly. The issue steering my choices is the extent to which an examination of these subjects might contribute to our understanding of either treatise in relation to the perspective of radical immanence.

Second, to understand what either Spinoza or Wittgenstein advances, we should try reading their texts as closely as possible. Given each text's notorious difficulties, anyone attempting to do this will need help—in the best of possible worlds, that of some intellectually rigorous commentator. As the reader may have inferred from the preceding discussion, my guiding hypothesis is that Wittgenstein provides such an ideal commentator for Spinoza and vice versa. The reason I read the *Ethics* and the *Tractatus* in conjunction emerges from this hypothesis, just as what follows is intended to put it to the test. Hence my principle of reading might be stated thus: under the constraints mentioned, I read the *Ethics* by appealing for the required help to the *Tractatus* while I read the *Tractatus* by appealing for the required help to the *Ethics*.

Nevertheless, this principle of reading requires a vital qualification, for in the form just stated, it implies abilities far greater than those I could deploy. Hence, I appeal to additional help in applying this principle—namely, the vast secondary literature. Yet this literature itself is of gigantic proportions, and the complexity of the material it addresses, enormous. To manage this surfeit of material, I followed a simple principle, though one perhaps unorthodox in its abruptness or even brutality: I considered only those contributions that have shaped or qualified the general coordinates of my interpretation of either text. I take practically no heed of past and present contributions going in different directions, no matter how influential, and I disregard almost all philosophical exchanges concerning them in one way or another. Moreover, I use only what suits my purposes without explaining why I use it and without thoroughly investigating whether the author's overall approach could legitimize the loan. This attitude clearly risks making my text appear overly dogmatic, and I accept the charge insofar as the text itself does not succeed in alleviating it.

This way of proceeding implies no disrespect for the contributions I overlook, even if it might give the impression of mimicking the arrogance both Spinoza and Wittgenstein seem to exhibit in that respect. In my case, this way of proceeding follows from the modest ambitions of the present work, namely, to produce only what one might call a first-level understanding. I do not try to adjudicate any contentious subject of Spinoza or Wittgenstein scholarship because a first-level understanding forms the *prerequisite framework* wherein such contentions can be competently debated. Nonetheless, I have outlined the literature I used and the reasons for selecting it in the preface.

The project might be more gently characterized as equivalent to what the French call "*explication de texte*." This is a close reading of the text itself with a minimum of asides. In using this approach, one takes no notice of issues related to the text's ancestry; one discounts the response it constitutes with respect to its immediate predecessors or contemporary opponents; one disregards the author's later work, as well as any effects the text might have produced; one sets aside controversial interpretations or, if they cannot be ignored, focuses on the passages in the text that underlie them; and finally, one aims simply to clarify the order the text follows, the topics it engages, the arguments it presents, its turns of phrase, and so forth. To do their job of clarification, the comments about such matters might be inspired from sources foreign to the text but familiar to the intended audience of the explication. *Explication de texte* is practiced for the most part in teaching contexts, for its only ambition is to promote the kind of first-level understanding I have in mind.

Yet my *explication de texte* is a more complex affair, for it involves two texts instead of one, while neither is explicated in its own right; instead, the explication of one relies on the explication of the other. This interexplication must follow a comprehensible flow without presupposing too much on the readers' part and without overly taxing their patience and goodwill. These requirements impose constraints leading me to adopt unusual stylistic devices. I quote more than is customary, grafting passages into my text and turning them into integral parts of it (though never without first indicating their origins). On occasion, then, I let my text speak in the voice of Spinoza or of Wittgenstein. To preserve the syntax, however, I often had to alter the grammatical structure of the borrowed passages or even break them up. I sometimes incorporate one part of the passage in one location of my text and the other part in another, though always trying not to distort the author's intended meaning. Finally, although I sought to respect the local contexts in which these passages occur, such respect is not always immediately apparent, so that Spinoza or Wittgenstein might emerge at times as unrecognizable.

Even if the reader makes allowances for these unusual moves, however, the appeal to the technique of *explication de texte* cannot shield my account from all criticism. *Explication de texte* is never as innocent as the definition of it might seem to imply. Anyone crafting an explication inevitably imports something of his or her deeper views. This inevitable bias, then, frequently causes the explication to edge closer to reconstruction than its author would care to admit. And so it is with my own text, too.

I read the *Ethics* and the *Tractatus* through the lens provided by philosophy and the history of science, although neither treatise appears as licensing such a viewpoint. In fact, I explicitly argue that neither the *Ethics* nor the *Tractatus* occupies itself overmuch with science and that each sets forth the deep philosophical reasons for this reticence. Be that as it may, my vantage point—my bias—is rooted in the intricate discussions on scientific change following in Kuhn's wake. My general outlook and my bolder interpretative suggestions as well as most of my examples come from these quarters. To be perfectly frank, I should add that a particular French whiff flavors my account because my involvement with issues of scientific change has been significantly influenced by what one might call the French tradition in philosophy and historiography of science.

Furthermore, the large number of direct quotations from the *Ethics* and the *Tractatus* makes it difficult to neatly separate what Spinoza or Wittgenstein holds from what I am asserting in their names. Thus, rather than let Spinoza or Wittgenstein speak in his own voice, I may appear to be doing the opposite, subsuming their voices to my own. I do strive to effect the required separation: obvious stylistic moves are designed to pin down the places where I depart from the letter of a text to follow a tangent of my own. But this can hardly save the explicatory character of the account; rather than constitute a bona fide explication, it risks appearing as reconstruction with a vengeance, as a reconstruction whose vantage point is expressly censured by its very object, a reconstruction remaining groundless and authorizing itself, so to speak, by the unhindered liberty of a mere fiat. To put Nietzsche's trope to a novel use, instead of proceeding to an explication with the finely tuned instruments that the best philosophical traditions have perfected, I might appear to be philosophizing with a sledgehammer (Kaufmann 1982, 463–68).

I mention Nietzsche here to suggest that the matters I address involve not just Spinoza and Wittgenstein alone but many other philosophers as well. Nonetheless, I read Spinoza and Wittgenstein as engaging in an earnest conversation unfettered by what other interlocutors might add to what the two have to say to each other. In the last analysis, this is why I refrain from taking into account relevant contributions of other authors and from circumspectly examining their views in relation to mine. I do at times mention other philosophers by name or allude to their work by a well-known catchword, yet I do this almost by free association, trying merely to intimate a more general philosophical context. Thus, apart from the few places where I find it necessary to do so, I barely discuss even Descartes or Frege and Russell, the immediate predecessors and direct or indirect influences on the two authors occupying these pages. At the same time, I allow Nietzsche to stand for all interlocutors of Spinoza and Wittgenstein by letting him open each chapter. As I try to explicate at the beginning of chapter 1, there is a way, or so I would like to claim, of allowing him to be considered a kind of mediator between the two.

What is the demanding reader to make of all this? The evocation of explication seems not to have safeguarded my account but rather kicked the bucket empty: I have come to admit that my explication amounts to a tendentious, self-authorized reconstruction with no established grounds on which to stand. To the extent that the main question bothering that reader is the difficulty, if not impossibility, of bringing Spinoza and Wittgenstein anachronistically together, my admission of all these dubious moves will have merely highlighted this difficulty. Thus the reader may very well abandon his or her efforts at this point, considering the case as hopeless. Spinoza and Wittgenstein will be much better off, such a reader might say, if left alone to pursue their courses along their separate paths; most likely, my efforts to tame the work of either by bringing the two together cannot far outclass the efforts expended by the protagonist of an old joke: a rather frail and not too intelligent-looking man boasts at a bar that he has managed to beat the world champion of boxing as well as the world champion of chess. To the repeated questions from his gaping audience he reluctantly comes to reply: "Well, that was not too difficult after all; I have beaten the world champion of boxing at chess and the world champion of chess at boxing."

A reader unmoved by the joke, finding it insufficiently serious for the matters at hand, can evoke a more dignified version of the same indictment. This is poem XI, 305, of the Palatine Anthology (fourth century AD), whose title is a man's proper name alluding to Pallas Athena, the goddess of wisdom. Here is the poem in my own rough translation:

PALLADAS

Son of effrontery, nursling of folly, most ignorant of men,
Tell me, how dare you swagger since you know nothing?
When with grammarians, you act the Platonist.
And when they ask you about Plato's doctrines,
You declare yourself immediately a grammarian.
You flee from one to the other.
A Platonist? Of course not;
And of grammar you know nothing.

Be that as it may, my will—my conatus—forces me to hold on and persevere. What remains at this stage is to respectfully ask my kind readers to withhold final judgment, no matter how harsh, until they hear me out.

CHAPTER ONE

Mutual Introductions

> I am afraid that we are not rid of God because we have still faith in grammar.
>
> —Friedrich Nietzsche, *Twilight of the Idols,* III, 5

NIETZSCHE ACKNOWLEDGED that God might well not be the traditional God of monotheism, a superperson endowed with all the proper anthropomorphic attributes elevated to the superlative. He understood perfectly that God is fundamentally a *transcending court of ultimate appeal* and an *overarching position* put there to collect whatever might satisfy humanity's deep-rooted need for incontestable authority. We today have witnessed how science, progress, the party, the market, or this or that "ism" have been promoted separately or conjointly, concurrently or consecutively, to quasi-transcendent authorities filling God's position and coming to enjoy therein unmitigated respect and all the appropriate honors. Nietzsche's philosophical acumen made him perceive that even modest grammar could be raised to that status.

If Nietzsche was historically positioned to appreciate that Spinoza's philosophy had been committed to destroying the anthropomorphic God of his age, to replacing the divine with the sum of what is worldly and to drawing the implications of that gesture with implacable intellectual rigor,[1] he could not have foreseen how and in what form grammar would finally make it to the throne. Nietzsche could not have foreseen, that is, the spectacular developments awaiting the time-honored discipline of logic at the hands of Frege, Russell, and others, the way these developments would eventually compel many philosophers to apply most of their toil to the philosophical study of language in its relation to logic, and the fact that it would be logic rather than grammar that would finally win the crown.

By the same token, Nietzsche could not have foreseen that one of the protagonists of this turn toward logic and language, a tense young man named Ludwig

Wittgenstein, would be practically the only one who, amid the clamor and the attendant misapprehensions, would keep his mind cool and vehemently counter this enthroning movement in an uncompromising appeal to the new levels of conceptual rigor that this logic itself had attained. Nor could he have foreseen that it would be the same Wittgenstein who would invoke grammar in his later work not as another name of God but as a means for loosening the link between logic and language and concomitantly for further undermining the godlike status that virtually all his contemporaries, not heeding him, had been attributing to logic. If he could have foreseen all this, then Nietzsche would have recognized in Wittgenstein—exactly as he did in Spinoza—another kindred spirit, somebody whose time had not yet come, somebody who, like Nietzsche himself, would have to be born posthumously.

These remarks have no function beyond explicating how Nietzsche's quip might set the stage for the kind of encounter between Spinoza and Wittgenstein I will be examining. While the anthropomorphic God of the Scholastics, and even of Descartes, had in the seventeenth century long occupied the transcending status Spinoza considered himself to have demolished, Wittgenstein's times were witnessing logic's irresistible rise to fundamentally the same overarching position. But despite the volumes dedicated to demonstrating that the *Tractatus* was a major force behind the enthroning movement and that it should thus be read as a celebration of the new deity, I will argue the exact opposite: in the *Tractatus* Wittgenstein aims at—and takes himself as definitively succeeding in—demoting this new god, one of the latter-day successors, if we take Nietzsche at his word, of the personalized God whom Spinoza had set out to overthrow.[2]

Thus if Spinoza's unremitting effort to bring down the old God does share something fundamental with Wittgenstein's equally unremitting effort to oppose the rise of a new one, Nietzsche might be considered, historically and perhaps also theoretically, as the mediator between the two, as the one who lays the common ground allowing them to meet and come to converse with each other. But this is not the place to discuss the particulars of this mediation.[3]

Similarities and Affinities

Wittgenstein wrote at a time wholly different from Spinoza's. Almost three centuries, a very different set of concerns, and a radically changed philosophical landscape separate the two. Spinoza worked at the time when the modern world we still inhabit was being born through almost unprecedented social, political, and ideological upheavals, the time when, concomitantly and to a large extent determinatively, the novel "natural philosophy" was seeing the light of day.[4] These were the times when traditional ways of philosophizing were being wiped away to the profit of novel philosophical programs that responded to those cataclysmic changes by trying to form a consistent whole of the novelties emerging in what we would call today ontology, epistemology, philosophy of science, philosophy of mind, and social and political philosophy. A wholly new philosophical terrain opened up, one whose principal features still dictate most of the pathways our thinking is allowed to follow. I must stress, for reasons I will show in a moment, that geometry—and mathematics generally—was being simultaneously elevated, as Galileo's famous formulation had it, to the status of the unique "language" in which the whole book of nature is written. This is the language unifying the heavenly and earthly realms, the rigorous language that, once deciphered in reference to the fundamental phenomenon of motion, could make us understand the workings of the world above us, around us, and at our feet.

The social, political, and ideological upheavals witnessed in Wittgenstein's times, though different, were no less dramatic, but they did not mark philosophy to an equal extent. Much more determinative for the discipline was the fact that many of the certainties regarding the foundations of mathematics and physics seemed to have almost evaporated.[5] The emergence of non-Euclidean geometries, the efforts to render mathematical analysis formally rigorous and thereby independent of geometrical and temporal intuitions, Cantor's set theory and his awe-inspiring attempt to harness the infinite, and the formulation of the special theory of relativity and the beginnings of quantum mechanics largely formed the circumstances out of which—and in response to which—sprang the new logic[6] of Frege and Russell. This new logic promised to remove all doubts connected to that enormous scientific and epistemological commotion by clearly showing how language functions and how thought is expressed in language. The new logic was thus envisioned as a kind of superscience whose subject matter forms the core of thinking in general and of cognitive enterprises in particular. Eventually, most of its advocates would claim that its application to the analysis of language could demonstrate that most of the so-called perennial philosophical problems arise out of linguistic misuses and hence should be wiped out by the appropriate logico-linguistic proofs. For many this appeared then as the only truly philosophical task.

The new logic of Frege and Russell brought with it extraordinary promises and an electrifying new vision not unlike the promises brought forth by the scientific revolution and the vision inspiring those who were working out its legacy. This similarity in overall philosophical climate presents features of interest here.

In Spinoza's period, the works of Copernicus, Galileo, Kepler, and many others had already started turning natural philosophy into a new discipline relatively independent from the rest of philosophy. The novel endeavor had begun to set its own agenda, to formulate its own problems, and to develop the specific means, theoretical as well as observational or experimental, for addressing them. Philosophy was consequently forced to reorganize drastically in order to accommodate the new points of view on nature and at least the fundamental ingredients of the new results attained. We can locate the beginnings of a twofold historical process here. On the one hand, natural philosophy would develop into the natural sciences—physics, chemistry, biology—which would eventually separate themselves from philosophy; on the other hand, "modern philosophy," as it set out with Descartes and Francis Bacon and culminated in Kant's awakening from his "dogmatic slumber," would sever its ties with its medieval predecessor and start building the general framework by which all philosophical programs more or less still have to abide.

In Wittgenstein's period, the revolutionary developments within mathematics and physics had been convincing many philosophers that, ironically, Kant's "Copernican revolution" had led philosophy to give in to another spell of "dogmatic slumber." To their embarrassment, they were obliged to acknowledge that, precisely since Kant, most had been taking for granted that no big surprises could be forthcoming from the major sciences, for at least the foundations of both mathematics and physics had been secured beyond possible doubt. But the "foundational crisis" in mathematics and the revolutions in physics obliged them to turn resolutely against Kant[7] and try to reorganize their discipline drastically. The new logic of Frege and Russell and the "scientific" philosophy developed in Vienna, in Berlin, and elsewhere were their main responses to that demand.[8]

The predicaments facing Spinoza and Wittgenstein were thus structurally very similar. In both cases, philosophy was provoked to respond to developments that, in a sense, had arisen independent of and outside it. In both cases, these developments were assaulting common assumptions concerning our knowledge of the external world, and hence they could not be passed over by philosophy, for they were targeting one of its main pillars. In both cases, the changes involved were exceptionally deep, forcing philosophy to reform profoundly so as to cope with them. In both cases, the process of reforming had started before the relevant author at issue here had his say. And finally, both Spinoza and Wittgenstein found themselves deeply unsatisfied with the methods deployed and the results attained by precisely those who had opened the respective novel philosophical terrains.

In this last respect, both Spinoza and Wittgenstein reacted in a manner revealing a striking affinity in philosophical temperament. They both royally ignored all issues of priority, philosophical ascendance, source of inspiration, or influence, and each explicitly took to task only those whom he considered as his direct philosophical forebears and, by the same token, his main philosophical opponents: Descartes for Spinoza and Frege and Russell for Wittgenstein. These were the figures who had opened, to their everlasting credit, the novel philosophical terrains, but they had left the work only half done. This lets us place the work of either Spinoza or Wittgenstein in its philosophical context: conservatively, we may consider the work of each as remedying the flaws in the works of his predecessors, dissolving their internal tensions so as to complete the required reorganization of the philosophical terrain at issue. More accurately, for there is no way to take either Spinoza or Wittgenstein as conservatives, we may consider their efforts as aiming at nothing less than the definitive transformation of the whole of philosophy, a transformation that would completely work out the novel philosophical terrain and thus fully exhaust its potential.

Rigor

Spinoza and Wittgenstein envisaged the possibility of thoroughly working out the new philosophy in question because it had opened possibilities of—and the demand for—higher standards of conceptual rigor. Thus both Spinoza and Wittgenstein seem to have judged that although this rigor should organize any endeavor

lity and singularity, not only of the validity of the argument but also
promoting their own goals. Always following the egocentric path,
culminates in later writings where Spinoza elaborates the notion that
t of the supreme good depends on others possessing it as well, since
intrinsic part of the world in relation to which harmony is sought.
rs thus serves the egocentric purpose, while the egocentric purpose is
mocratic.[18]

rns method narrowly conceived, the difference between Spinoza and
qually sweeping. The TEI explicitly holds that method must be based
f our *prior possession* of a true idea (TEI §33), and this entails that
plication of method is tantamount to the mind's being "directed ac-
standard of [some] given true idea" (TEI §38). This seemingly simple
together with the accompanying argument for it, does not constitute
ion of the Cartesian approach in the *Discourse*. The difference is so
e may suppose it to have been among the principal reasons Spinoza
e TEI, leaving it forever unfinished.

subordination of method to a prior fact entails that method cannot
kind of presuppositionless atmosphere that Descartes's conception
ms to imply. Facts of the world or facts about the mind are required
grounding method. Second and conjointly, the subordination of
e fact that we already possess a true idea entails that the essence of
ot as thin as Descartes seemed, at least initially,[19] to conceive it. The
nding of method therefore requires a fuller conception of the human
capacities. Lastly, the subordination of method to the fact of our pos-
idea entails that an adequate exposition of method cannot precede
tent of the true idea in question or the ways we can assess its truth.
t of ideas in general, the ways we can assess their truth as well as the
rrelate and might be connected to the grounding of method, can is-
a fully developed philosophical system. Only such a system could
tent of at least the fundamental true ideas, show how we can assess
d display the ways they relate both with one another and with the
method.

on, only such a system could come up with both an adequate concep-
ind and a description of the facts about the world or the mind that
method. Consequently, method ought to follow and not to precede
content; it is this content that will show ex post facto whether there is
ed for investigating method per se. My speculation therefore acquires
asons such as these may well number among those that led Spinoza
ny independent talk of method and concentrate on the composition
phical system. If this is the case, then Spinoza separates himself by
even from his avant-garde contemporaries who had turned method
stone of effective inquiry.[20]

a forsakes the issue of method after his failed attempt at circumscrib-
enstein seems indifferent to it from the very beginning. The *Tractatus*

engaging the novel philosophical terrain, their respective predecessors had failed to wholly acknowledge the demand, responsibly assume the task, and scrupulously abide by its strictures. In the case of Spinoza, since geometry had proved capable of accounting for natural phenomena, its formal rigor should be mustered for or- ganizing philosophy as well. Hence the new philosophy should be presented ac- cording to its ethos, and the *Ethics* is indeed written *more geometrico*. The case of Wittgenstein is formally identical. During the period when he wrote, the new logic of Frege and Russell represented the novel standard of philosophical strictness. His own writing should therefore unfold following the requisite austerity of *its* corre- sponding ethos, and indeed, we might say that the *Tractatus* is written *more logico*. And as I will show in what follows, the *mos* or ethos involved in either case does carry the implied ethical connotations.[9]

Both Spinoza and Wittgenstein adopted a tone determined by this deep ethi- cal commitment to formal rigor. Both believed in unremitting intellectual honesty that relies on nothing but the sheer power of the intellect, that does not fear its own conclusions or aspire to anything it itself cannot establish. Both are merciless in op- posing authoritative dogma and rejecting vacuous thought, facile ideological con- structs, or emotional underpinnings. Theirs is an intellectual firmness immune to the familiar terror of the authoritative view or lack of consent as well as impervious to the temptation of lowering their standards and accepting less than what the am- bition driving either was dictating. Neither of the two would claim to have achieved anything if his results had fallen short of his self-imposed superlative aims. Hence neither would allow his work to go to the public unless he had made sure of having completed his task. What Hampshire (1987, 23) has to say of Spinoza covers almost verbatim Wittgenstein as well:

> No other philosopher of equal stature has made such exalted claims for phi- losophy, or had such a clear vision of the scope and range of pure philo- sophical thinking. . . . [Spinoza] devoted his whole life to the execution of [his] design and he was confident that he had finally succeeded, at least in great outline. The only instrument which he allowed himself or thought necessary to his purpose was his own power of logical reasoning; at no point does he appeal to authority or revelation or to common consent; nor does he anywhere rely on literary artifice or try to reinforce rational argument by indirect appeal to emotion.

It should be added that both Spinoza and Wittgenstein could afford to be merciless toward their opponents because they were first and foremost merciless toward themselves. Both viewed philosophical thought as directly involving their very selves, and hence what is to be presented to others should apply to them first. Thus Spinoza opens his philosophical quest (in his *Treatise for the Emendation of the Intellect*) in the first person, aiming expressly at his own deeper gratification, while even a cursory reading of the *Ethics* makes clear that his is a "practical" phi- losophy[10] pervading the entire realm of human life, that of its author first of all. In almost exactly the same vein, Wittgenstein characterizes philosophy as not theory but activity (TLP 4.112), and he later acknowledges that "working in philosophy. . . is really more a working on oneself" (1984, 16). For both men, therefore, philosophy is an all-consuming undertaking that should be exercised by drawing on all the available intellectual and bodily resources, with unabated intensity, and without any concession to whatever does not directly promote it.

This way of relating to philosophy fostered striking similarities between the biographies of the two authors.[11] Thus both renounced family wealth to follow an ascetic form of life, and both eschewed their growing fame to devote themselves fully to their work. Characteristically, Spinoza declined a highly prestigious position at the University of Heidelberg in order to continue his labor undistracted, while Wittgenstein decided to abandon philosophy altogether and become a simple schoolteacher after considering his work to have been completed. Both Spinoza and Wittgenstein disregarded the respective authorities of the day, fully assuming the consequences,[12] while both were averse to publishing for publishing's sake, an attitude reflected in the fact that most texts by the two were published posthumously. Finally, both proved capable of gathering a group of devoted students and talented collaborators who could take care of the material each had left behind.

We see, therefore, that the formal rigor with which Spinoza and Wittgenstein tried to carry out their philosophical tasks went hand in hand with the rigor they both adopted in conducting their lives. But this rigor is a value pertaining to the methodological dimension of inquiry. Accordingly, if rigor looms large in the overall philosophical outlook of either man, we should ask whether their attitudes toward philosophical method in general reveal equally telling similarities. This is all the more important because philosophical method constituted a major concern for most of the philosophers participating in the new philosophy during both Spinoza's and Wittgenstein's periods.

Method

With respect to the issues important here, Descartes opened the new era of philosophy by establishing four major points and clarifying the relations among them. First, his work brings to the forefront the unlimited power of individual human reasoning. This is a power capable of rediscovering the medieval proofs for the existence of God but starting this time from scratch, that is, from hyperbolic doubt. Second, this power is distributed equally to everybody. As the first phrases of the *Discourse on Method* make clear, the man on the street is endowed with the capacity of exercising that power to become as good a philosopher as any other. Third, extension—or the concept of the *res extensa*—offers the philosophical ground legitimizing the use of geometry in the then emerging field of natural philosophy. Geometry is by definition the science of extension, so that the unification of the heavenly and earthly realms (Koyré 1957) proves philosophically well founded, while the new science of motion can employ its resources without philosophical qualms. Fourth, individual reason can exercise its capacities in full only if it follows the correct method strictly. This method, supplied by Descartes himself, is capable of unifying all inquiry—thereby dimming if not eradicating differences in subject

matter—and thus leading human rea[...] value. Method becomes thus the touch[...] knowledge. (Significantly, its deep disa[...] temological issues notwithstanding, th[...] or less at the same time across the Ch[...] despite its very different emphasis on c[...] ducted experiments.)[13]

Spinoza was duly impressed with [...] began his own philosophical quest by t[...] noza's *Treatise on the Emendation of th[...]* of the *Discourse on Method*. Nonethe[...] first-person format and address the re[...] that led their authors to the principal [...] largely end there. The very different ph[...] constructed very roughly as follows.[15]

First, if Descartes's work aspires [...] mat of the *Discourse* is appropriate: ce[...] individual. It is therefore no coincide[...] in Descartes's system attaches to a sel[...] follows, so to speak, from the self-refe[...] does not issue from some principle an[...] world. The "I" whose existence is prove[...] isolated, and fully incorporeal mind; i[...] ture of its carrier (the corresponding [...] this "I" represents anyone (for anyone [...] argument for him- or herself), we may [...] with no other properties or qualities. [...] relation between mind and body, the [...] istence of other minds present massiv[...] epistemological level at which it unfol[...] also inherently egocentric in that its fu[...] minimally thin, and incorporeal "I" wl[...]

Spinoza's principal aim, however, [...] good" itself, the *summum bonum* (TE[...] marily for his own purpose, the first-p[...] But the approach here is egocentric in [...] unlike certainty, the "supreme good" [...] knowledge per se (although it can be [...] good referring solely to the mind of th[...] it is a good involving harmony betwee[...] person as well as harmony between th[...] "I" seeking the supreme good is not de[...] and in this sense does not constitute [...] by first convincing themselves on their [...]

own corpore[...] of its *value* i[...] the approach[...] full enjoyme[...] others are ar[...] Teaching oth[...] inherently de[...]

As conce[...] Descartes is e[...] on the fact o[...] the correct a[...] cording to th[...] qualification, [...] a mere variat[...] drastic that w[...] abandoned th[...]

First, the [...] unfold in the [...] of the self see[...] for adequatel[...] method to th[...] the mind is r[...] adequate gro[...] mind and its [...] sessing a true[...] either the cor[...] But the conte[...] ways they int[...] sue only from[...] secure the co[...] their truth, a[...] grounding of [...]

In additi[...] tion of the m[...] could ground[...] philosophica[...] indeed any ne[...] some meat: r[...] to abandon a[...] of his philoso[...] the same toke[...] into the touc[...]

If Spinoz[...] ing it,[21] Wittg[...]

mentions method only once, at 6.53, just two propositions before the book ends. This seems to imply that for Wittgenstein, just as for Spinoza, method[22] might become an issue only after the philosophical system—whatever this might have signified for him—has been developed to completion. To be more specific, whereas Spinoza rules out the possibility of a prior discussion of method because we first must understand what some "given true idea" that we "already naturally possess" amounts to, Wittgenstein sees no use for such prior discussion, for, according to the remark in question, "the right method in philosophy" consists in merely the following: "to say nothing except what can be said, i.e. the propositions of natural science, i.e. something which has nothing to do with philosophy: and then always, when someone else wanted to say something metaphysical, to demonstrate to him that he had given no meaning to certain signs in his propositions" (TLP 6.53). To be capable of applying this method, we need to have already understood not only how the terms "proposition," "meaning," "philosophy," "something metaphysical," and "demonstration" are intended here but also what this rather cryptic remark is meant to convey and to whom it is addressed. In short, we need to have already understood the whole of the *Tractatus*.

Perhaps we might go one step further in this comparison. According to TLP 5.5563, "All propositions of our colloquial language are actually, just as they are, logically completely in order." If this logical order can be somehow correlated to Spinoza's "given true idea," then Wittgenstein could be taken as implying that we "already naturally possess" everything we might require to proceed with philosophical tasks. Hence no prior room is left for method, the issue being restricted to Wittgenstein's previously cited remark. Thus, though the two texts proceed quite differently, the *Tractatus*, like the *Ethics*, displays the correct exercise of philosophical method through its unfolding.

Method had again become the touchstone of effective inquiry in Wittgenstein's times. This concerned not only philosophical approaches to science—where the unique "scientific method" was considered as definitively demarcating the boundaries separating the scientific from the metaphysical (as logical positivism and logical empiricism had it) or the scientific from the pseudoscientific (as Popper's "falsificationism" had it)—but also the rest of philosophy: the new logic of Frege and Russell was taken as the essential tool capable of completely analyzing language and thus for anchoring philosophy to its allegedly only possible ground, namely, experience considered in this or that way. Paralleling the case of Spinoza, then, Wittgenstein's attitude toward method separates him drastically from his contemporaries.

Again, logic is for Wittgenstein what geometry is for Spinoza in the sense that both represent the highest standard of rigor in the writers' respective periods. For this reason, the *Ethics* is structured *ordine geometrico*.[23] For his part, Wittgenstein asserts that the *Tractatus* is structured *ordine logico*: the numbering of its "separate propositions indicate[s] [their] logical importance" (TLP 1n). But geometry and logic play different roles in the two works. Whereas geometry basically orders the subject matter in the *Ethics* yet is discussed in its own right only incidentally, logic not only orders the exposition of the *Tractatus* but also appears to constitute the book's subject matter. Almost everybody who read it at the time of its publication—and for quite some time afterward—took it to be a philosophical study of logic exclusively, in more or less the Fregean or Russellian conception of it. Both Russell, who wrote the introduction, and Ogden, who did the translation, held this view: as their correspondence testifies, both thought that Wittgenstein's work should be titled, "Philosophical Logic" (Wittgenstein 1973, 2–4).

Wittgenstein (1973, 20) disagreed heatedly, however, stating that he could not make any sense of this title. He was irritated, I take it, not only because his text clearly argues that no distinction can be drawn between "philosophical" and any kind of "nonphilosophical" logic but also because the same text plainly states that "the whole philosophy of logic"—what the purported philosophical study of logic would be expected to expound—"is contained in the fact that it is the characteristic mark of logical propositions that one can perceive by the symbol alone that they are true" (TLP 6.113). Clearly, the *Tractatus* does treat subjects not covered by this somewhat tangential and quite haughty remark. It follows that despite appearances, the *Tractatus* can have logic as its subject matter, at least in the accepted sense of "subject matter," only incidentally, even if what Wittgenstein has to say (or to show) regarding it is far from being exhausted thereby. The analogy with the role of geometry in the *Ethics* is thus preserved, at least up to a point. It was noticing affinities such as this that led Moore (Monk 1990) to propose the title *Tractatus Logico-Philosophicus* for Wittgenstein's work, with its author finally accepting the title without much demur.

Wittgenstein and Spinoza thus exhibit analogous attitudes toward method even as their works bear the similarities and affinities just noted. Absent a shared guiding insight, however, these features might be merely coincidental. It is to this issue that I now turn.

Insight

Again, I claim that in endeavoring to annihilate the transcending authority and overarching position of God as generally conceived in the seventeenth century, Spinoza was espousing and working out the philosophical perspective that, to use the largely forgotten terminology of his day, I call the perspective of radical immanence.[24]

This perspective uncompromisingly refuses to countenance God as transcending nature in any way, taking *nature* here in the most general conception of the term. This is Nature with capital N, comprising not only bodies and ideas but absolutely everything, including stones, stars, leaves, trees, worms, dogs, men and women, circles, triangles, laws, proofs, words, images, passions, memories, habits, fictitious stories, political regimes, and social institutions.[25] For Spinoza, Nature is not created by God but is of perfectly equal rank, being strictly identical to Him. As he expressly asserts in the preface to part IV of the *Ethics,* the terms *God* and *Nature* can be used interchangeably: *Deus sive Natura.* It follows that God can be replaced by Nature in all propositions in which He figures and hence that the major proposition E I p15, stating that "whatever is, is in God, and nothing can be or be conceived without God," can be phrased equivalently as "whatever is, is in Nature, and nothing can be or be conceived without Nature," for "there can be no thing external to God [Nature]" (E I p15d). For Spinoza, then, espousing the perspective of radical immanence is equivalent to holding that a position transcending Nature—lying outside and overarching it—is absolutely inadmissible: it can neither be nor be conceived, so that taking anyone or anything, however exalted, as occupying it amounts to total confusion or mere nonsense.

In identifying God with Nature, Spinoza strips God of all qualities, powers, and prerogatives that had formed the main subject matter of Scholastic philosophy and to which Descartes, at least according to Spinoza, remained attached.[26] Thus Spinoza's God neither creates the world or humanity nor intervenes anywhere. He is not the most illustrious person, either the infinitely benevolent Father or some impartially stern sort of Orwellian Big Brother. He does not possess will, "the sanctuary of ignorance" (E I Ap), or moral virtues and does not issue moral injunctions: His doing anything whatsoever "with the good in mind . . . is an absurdity" (E I p33s2). In short, He does not bear any image and likeness to humankind, even elevated to the perfectly eminent. Talk of eminence has no place in Spinoza (Deleuze 1968): he equates degrees on an axis leading to perfection with degrees of reality (E II def6), that is, degrees of the power to act on one's own.[27]

In addition, E I p18 affirms that "God is the immanent, not the transitive, cause of all things." It follows that no final causes act within the world, for these are "figments of the imagination"; that the world's fate is not fixed by divine design, for it is "madness to believe that God delights in harmony"; and that the world cannot be ordered about at will, for miracles are concocted by superstition (E I Ap). Rather, "God [Nature] acts solely from the laws of His [Its] own nature, constrained by none" (E I p17). These are the immutable laws "producing things in the only way and order that is the case" (E I p33), the laws natural philosophy had set out to discover. With the proviso that causal entailment can be identified with deductive entailment, we can see that the perspective of radical immanence amounts to a kind of naturalism. This is a thoroughly contemporary perspective.

Part I of the *Ethics* is devoted to obliterating the personlike God of Spinoza's contemporaries by using the full rigor of the geometrical order to demolish the overarching position He was taken to occupy; the work as a whole systematically draws the implications of that conclusion. Replacing God and His creations by Nature and Its workings is at the same time conducting the rigorous proof that nothing can occupy a position external to the world. As a result, there can be no entity, even if perfect, and no being, even if supreme, that could possibly come to dwell in such a place. Spinoza's work ontologically destroys the position that any godlike entity could ever be taken as occupying. Destroying this position at the ontological level was the philosophical insight guiding Spinoza's work, achieving this destruction was the aim of his philosophical theory, and the *Ethics* carries out this destruction with the implacable rigor of geometry.

Wittgenstein shares exactly the same insight. The only difference is that the new God he set out to demote had a secular appearance and bore a different name. Again, in this era the new logic of Frege and Russell was being elevated to the status of the contemporary deity. The pathways that led to this elevation have been amply rehearsed in the literature, albeit not in this terminology, for, among other reasons, most of Wittgenstein's contemporaries, as well as quite a few of his heirs, had been—and some continue to be—taken in by it. Logic was raised to this exalted status basically because it appeared to be capable of fully analyzing language. Linguistic analysis, as it began to be promoted then, could rely on the unprecedented rigor of the new logic to make "metaphysical" problems understood for what they are, thus coming to demonstrate, allegedly beyond appeal, that most, if not all, could be definitively dissolved.

As buttressed by the new logic, the "linguistic turn" in question[28] thus elevated philosophy of language to the rank of *prima philosophia*. Logic now seemed to occupy a vantage point from which it could survey the whole realm of meaning and sense and thus distinguish sense from nonsense and meaningfulness from meaninglessness once and for all. Furthermore, because the sense and meaningfulness of linguistic expressions were considered to be identical to the semantic rendition of thought (Coffa 1993), the same vantage point appeared to let logic encompass the dominion of thought in its entirety. Valid thought could thus be separated from invalid thought, while a concomitant appeal to experience (spelled out in terms of sense data or something else) could apparently provide logic the means for marking out truth and thereby for making final pronouncements on what is and what is not the case, on what there is and on what there can be. In short, this course sailed logic from a position overarching language through a position overarching thought to a position overarching the world—in fact, all "possible worlds." The end of the voyage was exactly the place God used to occupy when Spinoza undertook his effort to destroy it. Unlike its predecessor, however, the new God had not been bestowed with the power to intervene materially in the world and exercise divine will in perfect freedom.[29]

Wittgenstein's disagreement is radical, for according to his arguments, the movement leading to the enthronement of the new God cannot even get off the ground. For one thing, to be capable of demarcating the boundaries between sense and nonsense and between meaningfulness and meaninglessness in the requisite general terms, logic should occupy a position from which it could look on language as its subject matter. But TLP 6.111 makes clear that logic cannot be equated to a natural science, while TLP 6.13 holds, in even stronger terms, that "logic is not a theory." In not being a theory, logic cannot have a subject matter, which implies that logic cannot be abstracted away from language and be placed in an external position from which it could survey language. Moreover, marking out the logical propositions does not lead far: "The propositions of logic are tautologies" (TLP 6.1), which as such "say nothing" (TLP 6.11). This clarifies the sense in which logic is not a theory: by lacking the capacity to say anything, the propositions of logic are incapable of composing any theory whatsoever.

In addition, logic is in no position to play a distinct self-sufficient normative engaging the novel philosophical terrain, their respective predecessors had failed to wholly acknowledge the demand, responsibly assume the task, and scrupulously abide by its strictures. In the case of Spinoza, since geometry had proved capable of accounting for natural phenomena, its formal rigor should be mustered for organizing philosophy as well. Hence the new philosophy should be presented according to its ethos, and the *Ethics* is indeed written *more geometrico*. The case of Wittgenstein is formally identical. During the period when he wrote, the new logic of Frege and Russell represented the novel standard of philosophical strictness. His own writing should therefore unfold following the requisite austerity of *its* corresponding ethos, and indeed, we might say that the *Tractatus* is written *more logico*. And as I will show in what follows, the *mos* or ethos involved in either case does carry the implied ethical connotations.[9]

Both Spinoza and Wittgenstein adopted a tone determined by this deep ethical commitment to formal rigor. Both believed in unremitting intellectual honesty that relies on nothing but the sheer power of the intellect, that does not fear its own conclusions or aspire to anything it itself cannot establish. Both are merciless in opposing authoritative dogma and rejecting vacuous thought, facile ideological constructs, or emotional underpinnings. Theirs is an intellectual firmness immune to the familiar terror of the authoritative view or lack of consent as well as impervious to the temptation of lowering their standards and accepting less than what the ambition driving either was dictating. Neither of the two would claim to have achieved anything if his results had fallen short of his self-imposed superlative aims. Hence neither would allow his work to go to the public unless he had made sure of having completed his task. What Hampshire (1987, 23) has to say of Spinoza covers almost verbatim Wittgenstein as well:

> No other philosopher of equal stature has made such exalted claims for philosophy, or had such a clear vision of the scope and range of pure philosophical thinking. . . . [Spinoza] devoted his whole life to the execution of [his] design and he was confident that he had finally succeeded, at least in great outline. The only instrument which he allowed himself or thought necessary to his purpose was his own power of logical reasoning; at no point does he appeal to authority or revelation or to common consent; nor does he anywhere rely on literary artifice or try to reinforce rational argument by indirect appeal to emotion.

It should be added that both Spinoza and Wittgenstein could afford to be merciless toward their opponents because they were first and foremost merciless toward themselves. Both viewed philosophical thought as directly involving their very selves, and hence what is to be presented to others should apply to them first. Thus Spinoza opens his philosophical quest (in his *Treatise for the Emendation of the Intellect*) in the first person, aiming expressly at his own deeper gratification, while even a cursory reading of the *Ethics* makes clear that his is a "practical" philosophy[10] pervading the entire realm of human life, that of its author first of all. In almost exactly the same vein, Wittgenstein characterizes philosophy as not theory own corporeality and singularity, not only of the validity of the argument but also of its *value* in promoting their own goals. Always following the egocentric path, the approach culminates in later writings where Spinoza elaborates the notion that full enjoyment of the supreme good depends on others possessing it as well, since others are an intrinsic part of the world in relation to which harmony is sought. Teaching others thus serves the egocentric purpose, while the egocentric purpose is inherently democratic.[18]

As concerns method narrowly conceived, the difference between Spinoza and Descartes is equally sweeping. The TEI explicitly holds that method must be based on the fact of our *prior possession* of a true idea (TEI §33), and this entails that the correct application of method is tantamount to the mind's being "directed according to the standard of [some] given true idea" (TEI §38). This seemingly simple qualification, together with the accompanying argument for it, does not constitute a mere variation of the Cartesian approach in the *Discourse*. The difference is so drastic that we may suppose it to have been among the principal reasons Spinoza abandoned the TEI, leaving it forever unfinished.

First, the subordination of method to a prior fact entails that method cannot unfold in the kind of presuppositionless atmosphere that Descartes's conception of the self seems to imply. Facts of the world or facts about the mind are required for adequately grounding method. Second and conjointly, the subordination of method to the fact that we already possess a true idea entails that the essence of the mind is not as thin as Descartes seemed, at least initially,[19] to conceive it. The adequate grounding of method therefore requires a fuller conception of the human mind and its capacities. Lastly, the subordination of method to the fact of our possessing a true idea entails that an adequate exposition of method cannot precede either the content of the true idea in question or the ways we can assess its truth. But the content of ideas in general, the ways we can assess their truth as well as the ways they interrelate and might be connected to the grounding of method, can issue only from a fully developed philosophical system. Only such a system could secure the content of at least the fundamental true ideas, show how we can assess their truth, and display the ways they relate both with one another and with the grounding of method.

In addition, only such a system could come up with both an adequate conception of the mind and a description of the facts about the world or the mind that could ground method. Consequently, method ought to follow and not to precede philosophical content; it is this content that will show ex post facto whether there is indeed any need for investigating method per se. My speculation therefore acquires some meat: reasons such as these may well number among those that led Spinoza to abandon any independent talk of method and concentrate on the composition of his philosophical system. If this is the case, then Spinoza separates himself by the same token even from his avant-garde contemporaries who had turned method into the touchstone of effective inquiry.[20]

If Spinoza forsakes the issue of method after his failed attempt at circumscribing it,[21] Wittgenstein seems indifferent to it from the very beginning. The *Tractatus* mentions method only once, at 6.53, just two propositions before the book ends. This seems to imply that for Wittgenstein, just as for Spinoza, method[22] might become an issue only after the philosophical system—whatever this might have signified for him—has been developed to completion. To be more specific, whereas Spinoza rules out the possibility of a prior discussion of method because we first must understand what some "given true idea" that we "already naturally possess" amounts to, Wittgenstein sees no use for such prior discussion, for, according to the remark in question, "the right method in philosophy" consists in merely the following: "to say nothing except what can be said, i.e. the propositions of natural science, i.e. something which has nothing to do with philosophy: and then always, when someone else wanted to say something metaphysical, to demonstrate to him that he had given no meaning to certain signs in his propositions" (TLP 6.53). To be capable of applying this method, we need to have already understood not only how the terms "proposition," "meaning," "philosophy," "something metaphysical," and "demonstration" are intended here but also what this rather cryptic remark is meant to convey and to whom it is addressed. In short, we need to have already understood the whole of the *Tractatus.*

Perhaps we might go one step further in this comparison. According to TLP 5.5563, "All propositions of our colloquial language are actually, just as they are, logically completely in order." If this logical order can be somehow correlated to Spinoza's "given true idea," then Wittgenstein could be taken as implying that we "already naturally possess" everything we might require to proceed with philosophical tasks. Hence no prior room is left for method, the issue being restricted to Wittgenstein's previously cited remark. Thus, though the two texts proceed quite differently, the *Tractatus,* like the *Ethics,* displays the correct exercise of philosophical method through its unfolding.

Method had again become the touchstone of effective inquiry in Wittgenstein's times. This concerned not only philosophical approaches to science—where the unique "scientific method" was considered as definitively demarcating the boundaries separating the scientific from the metaphysical (as logical positivism and logical empiricism had it) or the scientific from the pseudoscientific (as Popper's "falsificationism" had it)—but also the rest of philosophy: the new logic of Frege and Russell was taken as the essential tool capable of completely analyzing language and thus for anchoring philosophy to its allegedly only possible ground, namely, experience considered in this or that way. Paralleling the case of Spinoza, then, Wittgenstein's attitude toward method separates him drastically from his contemporaries.

Again, logic is for Wittgenstein what geometry is for Spinoza in the sense that both represent the highest standard of rigor in the writers' respective periods. For this reason, the *Ethics* is structured *ordine geometrico.*[23] For his part, Wittgenstein asserts that the *Tractatus* is structured *ordine logico*: the numbering of its "separate propositions indicate[s] [their] logical importance" (TLP 1n). But geometry and logic play different roles in the two works. Whereas geometry basically orders the subject matter in the *Ethics* yet is discussed in its own right only incidentally, logic not only orders the exposition of the *Tractatus* but also appears to constitute the book's subject matter. Almost everybody who read it at the time of its publication—and for quite some time afterward—took it to be a philosophical study of logic exclusively, in more or less the Fregean or Russellian conception of it. Both Russell, who wrote the introduction, and Ogden, who did the translation, held this view: as their correspondence testifies, both thought that Wittgenstein's work should be titled, "Philosophical Logic" (Wittgenstein 1973, 2–4).

Wittgenstein (1973, 20) disagreed heatedly, however, stating that he could not make any sense of this title. He was irritated, I take it, not only because his text clearly argues that no distinction can be drawn between "philosophical" and any kind of "nonphilosophical" logic but also because the same text plainly states that "the whole philosophy of logic"—what the purported philosophical study of logic would be expected to expound—"is contained in the fact that it is the characteristic mark of logical propositions that one can perceive by the symbol alone that they are true" (TLP 6.113). Clearly, the *Tractatus* does treat subjects not covered by this somewhat tangential and quite haughty remark. It follows that despite appearances, the *Tractatus* can have logic as its subject matter, at least in the accepted sense of "subject matter," only incidentally, even if what Wittgenstein has to say (or to show) regarding it is far from being exhausted thereby. The analogy with the role of geometry in the *Ethics* is thus preserved, at least up to a point. It was noticing affinities such as this that led Moore (Monk 1990) to propose the title *Tractatus Logico-Philosophicus* for Wittgenstein's work, with its author finally accepting the title without much demur.

Wittgenstein and Spinoza thus exhibit analogous attitudes toward method even as their works bear the similarities and affinities just noted. Absent a shared guiding insight, however, these features might be merely coincidental. It is to this issue that I now turn.

Insight

Again, I claim that in endeavoring to annihilate the transcending authority and overarching position of God as generally conceived in the seventeenth century, Spinoza was espousing and working out the philosophical perspective that, to use the largely forgotten terminology of his day, I call the perspective of radical immanence.[24]

This perspective uncompromisingly refuses to countenance God as transcending nature in any way, taking *nature* here in the most general conception of the term. This is Nature with capital N, comprising not only bodies and ideas but absolutely everything, including stones, stars, leaves, trees, worms, dogs, men and women, circles, triangles, laws, proofs, words, images, passions, memories, habits, fictitious stories, political regimes, and social institutions.[25] For Spinoza, Nature is not created by God but is of perfectly equal rank, being strictly identical to Him. As he expressly asserts in the preface to part IV of the *Ethics,* the terms *God* and *Nature* can be used interchangeably: *Deus sive Natura.* It follows that God can be replaced by Nature in all propositions in which He figures and hence that the major proposition E I p15, stating that "whatever is, is in God, and nothing can be or be conceived without God," can be phrased equivalently as "whatever is, is in Nature, and nothing can be or be conceived without Nature," for "there can be no thing external to God [Nature]" (E I p15d). For Spinoza, then, espousing the perspective of radical immanence is equivalent to holding that a position transcending Nature— lying outside and overarching it—is absolutely inadmissible: it can neither be nor be conceived, so that taking anyone or anything, however exalted, as occupying it amounts to total confusion or mere nonsense.

In identifying God with Nature, Spinoza strips God of all qualities, powers, and prerogatives that had formed the main subject matter of Scholastic philosophy and to which Descartes, at least according to Spinoza, remained attached.[26] Thus Spinoza's God neither creates the world or humanity nor intervenes anywhere. He is not the most illustrious person, either the infinitely benevolent Father or some impartially stern sort of Orwellian Big Brother. He does not possess will, "the sanctuary of ignorance" (E I Ap), or moral virtues and does not issue moral injunctions: His doing anything whatsoever "with the good in mind . . . is an absurdity" (E I p33s2). In short, He does not bear any image and likeness to humankind, even elevated to the perfectly eminent. Talk of eminence has no place in Spinoza (Deleuze 1968): he equates degrees on an axis leading to perfection with degrees of reality (E II def6), that is, degrees of the power to act on one's own.[27]

In addition, E I p18 affirms that "God is the immanent, not the transitive, cause of all things." It follows that no final causes act within the world, for these are "figments of the imagination"; that the world's fate is not fixed by divine design, for it is "madness to believe that God delights in harmony"; and that the world cannot be ordered about at will, for miracles are concocted by superstition (E I Ap). Rather, "God [Nature] acts solely from the laws of His [Its] own nature, constrained by none" (E I p17). These are the immutable laws "producing things in the only way and order that is the case" (E I p33), the laws natural philosophy had set out to discover. With the proviso that causal entailment can be identified with deductive entailment, we can see that the perspective of radical immanence amounts to a kind of naturalism. This is a thoroughly contemporary perspective.

Part I of the *Ethics* is devoted to obliterating the personlike God of Spinoza's contemporaries by using the full rigor of the geometrical order to demolish the overarching position He was taken to occupy; the work as a whole systematically draws the implications of that conclusion. Replacing God and His creations by Nature and Its workings is at the same time conducting the rigorous proof that nothing can occupy a position external to the world. As a result, there can be no entity, even if perfect, and no being, even if supreme, that could possibly come to dwell in such a place. Spinoza's work ontologically destroys the position that any godlike entity could ever be taken as occupying. Destroying this position at the ontological level was the philosophical insight guiding Spinoza's work, achieving this destruction was the aim of his philosophical theory, and the *Ethics* carries out this destruction with the implacable rigor of geometry.

Wittgenstein shares exactly the same insight. The only difference is that the new God he set out to demote had a secular appearance and bore a different name. Again, in this era the new logic of Frege and Russell was being elevated to the status of the contemporary deity. The pathways that led to this elevation have been amply rehearsed in the literature, albeit not in this terminology, for, among other reasons, most of Wittgenstein's contemporaries, as well as quite a few of his heirs, had been—and some continue to be—taken in by it. Logic was raised to this exalted status basically because it appeared to be capable of fully analyzing language. Linguistic analysis, as it began to be promoted then, could rely on the unprecedented rigor of the new logic to make "metaphysical" problems understood for what they are, thus coming to demonstrate, allegedly beyond appeal, that most, if not all, could be definitively dissolved.

As buttressed by the new logic, the "linguistic turn" in question[28] thus elevated philosophy of language to the rank of *prima philosophia*. Logic now seemed to occupy a vantage point from which it could survey the whole realm of meaning and sense and thus distinguish sense from nonsense and meaningfulness from meaninglessness once and for all. Furthermore, because the sense and meaningfulness of linguistic expressions were considered to be identical to the semantic rendition of thought (Coffa 1993), the same vantage point appeared to let logic encompass the dominion of thought in its entirety. Valid thought could thus be separated from invalid thought, while a concomitant appeal to experience (spelled out in terms of sense data or something else) could apparently provide logic the means for marking out truth and thereby for making final pronouncements on what is and what is not the case, on what there is and on what there can be. In short, this course sailed logic from a position overarching language through a position overarching thought to a position overarching the world—in fact, all "possible worlds." The end of the voyage was exactly the place God used to occupy when Spinoza undertook his effort to destroy it. Unlike its predecessor, however, the new God had not been bestowed with the power to intervene materially in the world and exercise divine will in perfect freedom.[29]

Wittgenstein's disagreement is radical, for according to his arguments, the movement leading to the enthronement of the new God cannot even get off the ground. For one thing, to be capable of demarcating the boundaries between sense and nonsense and between meaningfulness and meaninglessness in the requisite general terms, logic should occupy a position from which it could look on language as its subject matter. But TLP 6.111 makes clear that logic cannot be equated to a natural science, while TLP 6.13 holds, in even stronger terms, that "logic is not a theory." In not being a theory, logic cannot have a subject matter, which implies that logic cannot be abstracted away from language and be placed in an external position from which it could survey language. Moreover, marking out the logical propositions does not lead far: "The propositions of logic are tautologies" (TLP 6.1), which as such "say nothing" (TLP 6.11). This clarifies the sense in which logic is not a theory: by lacking the capacity to say anything, the propositions of logic are incapable of composing any theory whatsoever.

In addition, logic is in no position to play a distinct self-sufficient normative role; equivalently, it cannot regiment from outside the correct logical order. As I said, Wittgenstein holds that "*all* propositions of our colloquial language are *actually, just as they are, logically completely in order*" (TLP 5.5563, emphasis added). Concomitantly, since "the correct explanation of logical propositions must give them a peculiar position *among* all propositions" (TLP 6.112, emphasis added), logical propositions can do their work only within language. It follows that logic cannot be placed outside language, for, to adopt Spinoza's term, it is immanent in it.

Insofar as language is the medium for expressing thought, the claim that logic is immanent in language implies that "we *cannot* think illogically" (TLP 5.4731).[30] It follows that logic is immanent in thought, too. But logic is not just immanent in language and in thought; it is at the same time "all-embracing" (TLP 5.511) as regards both language and thought. It follows that logic cannot form a subject matter that thought could somehow harness. Apart from what can be covered under the "whole philosophy of logic," there can be no theory *of* logic, for—since the all-embracing character of logic with respect to thought covers all possible thought—there can be no ground on which the premises of that alleged theory could secure a foothold. Yet we should acknowledge that developments subsequent to the *Tractatus* appear to contradict this. Philosophy of logic appears to have become a bona fide field of study that ostensibly aims at a deeper understanding of logic and hence at correcting the possible shortcomings of its conceptualization and the infelicities of the associated formalizations.

Nevertheless, issues of formalism or symbolism aside, the developments in question have no bearing on the immanence of logic in language and hence on the possibility of a general theory of logic. A simple argument can demonstrate this. Consider any "deviant logic" (Haack 1978). It takes only a moment's reflection to realize that the axioms or rules of inference governing it must have been formulated using "standard" logic and that the theorems established do not and cannot obey the "deviant" strictures of that logic. For example, the theorems of many valued logics must themselves be governed by bivalence, while the theorems of paraconsistent logic cannot relate paraconsistently to one another. This is exactly the way Wittgenstein would have it: "The laws of [standard] logic[31] cannot themselves obey [the] further logical laws" (TLP 6.123) under which some finer conceptualization of logic might subsume them.[32]

Thus any effort to turn logic into a proper object of study so as to come up with a proper theory of it (at the level of some sort of "metalogic") must rely on logic, and hence no theory, at least in the accepted sense of the term, could ever come out of this circle. It is impossible to place logic in a position outside thought that could allow thought to carry out the corresponding thorough examination: since logic "embraces" all thought, no such position could exist, while the appeal to any criteria for carrying out the examination could be only logic in action. It follows that there can be no metalogic, while deviant logics and all associated formalizations are at best mere calculi—perhaps useful for this or that circumscribable purpose—that are developed in thought and within colloquial language by means of the "standard" logic immanent in both.[33]

The issue here may perhaps be further clarified. All attempts to state some uncontroversial formulation of the "laws" of logic—no matter how incisive or sophisticated—have failed to silence objections. It is as if the very attempt to formulate laws, ones that should be incontestable by definition, turns them into contestable hypotheses. To give but one elementary example, the fact that material implication disregards relevance has made the "standard" formulation of the laws of logic a matter of ongoing controversies. But if we focus on such controversies, we cannot fail to realize that, *to be understandable at all,* all arguments involved must effectively exercise the very laws whose express formulation is precisely at issue. It seems that the effective exercise of the ostensible laws of logic necessarily outruns all efforts at their express formulation and that what is essentially incontestable in their being exercised will always elude what can be stated about them. At the same time, even as this essential incontestability always eludes its express formulation, it remains at work nevertheless, being *displayed* in and by that work, which is to say, in action. The claim that all propositions within colloquial language are "logically completely in order" implies that the "laws" of logic are *always* in action, making any "proposition of our colloquial language" conform to them, that is, follow the logical order, while these so-called laws themselves are being effectively displayed in action *as* that logical order. As I will demonstrate, what Wittgenstein calls "showing" throughout the *Tractatus* amounts precisely to the kind of displaying involved here.

The elusiveness under discussion can be explained once we seriously consider the notion that logic is immanent in thought. This immanence in thought implies that logic determines the functioning of thought from inside thought even as there is no outside where this functioning could be placed so as to be assessed in its own right and, of course, by its own lights. Consequently, coming to think the laws of logic in their essential incontestability would imply that thought can isolate and take out of itself, as it were, the fundament of its functioning so as to bring this fundament—the laws in question—to the view of thought. To achieve such a feat, thought should be, or should be capable of becoming, perfectly self-transparent. But even if it possesses or could somehow come to possess the corresponding capacity, this self-transparency cannot be definitely delimited *as such*: if logic is all-pervasive, then it determines the functioning of thought *throughout*. Hence there is no way that the fundament of this functioning—the logical laws sought—can be isolated within thought. And of course what cannot be definitely isolated and delimited in thought cannot be expressly stated in language. The laws of logic cannot be expressly formulated in their essential incontestability precisely because of this. But Wittgenstein does not leave things here. What he calls "the complete analysis of the proposition" (TLP 3.25) constitutes the means through which he "shows," or displays, in principle if not in fact, how the essentially incontestable "laws" of logic work in language and thereby, albeit indirectly, how they work in thought as well.

Now, since "thought is the logical picture of the facts" (TLP 3), the immanence of logic in thought makes logic the backbone of that picture. And since logic is all-pervasive, this backbone takes all facts in stride. But since the "totality of facts" makes up "the world" (TLP 1.1), the "logical picture" constituting thought is a full picture of the world, leaving nothing outside. Logic and the world are thus perfectly on a par with each other, the one constituting the backbone of the full picture of the other. Hence Wittgenstein becomes entitled to say that logic constitutes the "reflection of the world" (TLP 6.13). The conclusion is that logic cannot be placed anywhere other than where it is at work, for it can have no outside whatsoever, and it cannot overarch any area of major philosophical concern (language, thought, the world), for it is either immanent in that concern or simply its "reflection."

The title Wittgenstein finally accepted for his work involves not only logic but also philosophy. The hyphen connecting the two seems to imply that, in tune with the then reigning views, he too considered philosophy to be inseparable from the new logic. If this is the case, then logic's immanence in language and thought should have consequences on the place occupied by philosophy: if the two are inseparably connected, consistency dictates that philosophy, too, can find no niche in some external and overarching vantage point. The impossibility logic encounters in this respect should be imparted to philosophy as well.

On the one hand, the *Tractatus* bears out this implication. Wittgenstein agrees that philosophy may not be "one of the natural sciences" (TLP 4.111), commenting that the word "philosophy" might well "mean something that stands above or below the natural sciences, but not beside [them]" (ibid.). But he adds that philosophy's distinct status arises not from its ability to survey language, thought, and the world from the outside but from its being "not a theory but an activity" (ibid.). It follows that it is not philosophy's job to provide the knowledge of *any* domain of inquiry and even less to survey any domain whatsoever; it simply is not that kind of thing.

On the other hand, Wittgenstein does attribute a job to philosophy: it "should limit the thinkable and thereby the unthinkable" (TLP 4.114), or equivalently, it should "draw a limit to thinking" (TLP Pr ¶3). Yet the line demarcating the boundary between the thinkable and the unthinkable cannot be traced from some position that could stand outside thought, for there can be no such position: its existence would enable us "to think both sides of this limit [and hence] to think what cannot be thought" (ibid.). In other words, the line in question can be traced only "from within the thinkable" and only "through the thinkable" (TLP 4.114). Wittgenstein does not miss out on the perspective of radical immanence: there can be no external position overarching thought—as well as, of course, whatever thought can be the thought *of*—for the existence of such position would be strictly identical to the possibility of being able to think both sides of the limit in question and hence with the self-contradicting possibility of thinking what cannot be thought. We may then infer that for Wittgenstein, indulging in the unthinkable amounts to entitling oneself, directly or indirectly, to the existence of this external overarching position. The inference will be detailed in chapter 4.

Wittgenstein's claim that "philosophy is not a theory but an activity" further specifies the way he espouses the philosophical perspective at issue here. If philosophy is an activity, then, like any activity, it should have a purpose. Wittgenstein identifies this purpose with "the logical clarification of thought" and details it as "making clear and delimiting sharply the thoughts which otherwise are, as it were, opaque and blurred" (TLP 4.112).[34] In other words, philosophy is not the kind of activity that can add up to the composition of some theory, for the work of its practitioners, irrespective of what they themselves might believe they are doing, can at best produce a "clarification of thought" from within thought and can "consist solely of elucidations" (ibid.). By consistently pursuing this purpose, philosophy demarcates the unthinkable from within and through the thinkable; equivalently, it is by philosophy's arriving at such elucidations that the unthinkable is effectively ruled out.

That the unthinkable is not automatically ruled out, as one might expect it to be, that a particular complex activity is required for carrying out this apparently trivial task, results from the fact that thought can be accessed only in and through language—and hence the limit in question "can be drawn only in language" (TLP Pr ¶4)—while "language disguises the thought" (TLP 4.002). Wittgenstein points out that this is a "disguise" making philosophers "misunderstand the logic of our language" (TLP Pr ¶2); this misunderstanding blurs the line distinguishing the thinkable from the unthinkable, making philosophers attempt, unaware, to indulge in the latter. But this attempt to think the unthinkable as criticized by Wittgenstein may be regarded as equivalent to the presupposition, direct or indirect, explicit or implicit, of the possibility of an external vantage point overarching thought (and the world and language). Concomitantly, this presupposition is tantamount to elevating logic, in tandem with its associated linguistic philosophy, to that overarching position.

Thus Wittgenstein's goal becomes understandable. It is no other than revealing the core of the "misunderstanding" at issue—entitling oneself, directly or indirectly, to an overarching position—and by the same token displaying with crystalline clarity that such a position is logically impossible. Displaying this logical impossibility amounts to annihilating that position logically, that is, once and for all. To achieve this goal, Wittgenstein traces the line demarcating the boundary between the thinkable and the unthinkable, a line that can be drawn only in language. Nonetheless, the line in question "*can* be drawn in language," because language provides something that can "represent," as it were, the unthinkable at the linguistic level, namely, mere "nonsense" (TLP Pr ¶4).[35]

In other words, Wittgenstein can proceed to the annihilation of the external position, but only indirectly. He can only "show" (TLP Pr ¶2 and ¶8) or "display clearly the speakable" (TLP 4.115) by tracing the line separating what can thus be spoken from what "lies at the other side of the limit," namely, mere "nonsense" (TLP Pr ¶4) or "the unspeakable" (4.115). But since the "significant proposition is the thought" (TLP 4), while "a proposition is the model of reality as we *think* it" (TLP 4.01, emphasis added), it follows that what can be "said clearly" amounts to the linguistic expression of what can be "thought clearly" (TLP 4.116). Hence the line separating what can be said clearly from mere nonsense or the unspeakable is also the line separating what can be thought clearly from what cannot be thought at all. To trace this line, then, is to annihilate the overarching position in question.

Wittgenstein believed that his "book" (TLP Pr ¶2) had managed to achieve this annihilation in an "unassailable and definitive" way (TLP Pr ¶8). In the philosophical context within which he worked, such annihilation could prove unassailable and definitive only by the lights of the new logic and could be brought about only by deploying its full rigor. This annihilation should be absolutely compelling or, equivalently, *logically necessary*. Wittgenstein believed that he had accomplished just that.

Note, however, that Wittgenstein does not limit what he believes himself to have achieved to the linguistic philosophy of his day: his work "deals with *the* problems of *philosophy* [generally] and shows . . . that the method of formulating *these* problems rests on the misunderstanding of the logic of our language" (TLP Pr ¶2, emphasis added). This shows as nonsense not just the philosophical discourse characterizing the "linguistic turn" but any philosophical discourse oblivious to the misunderstanding at issue, any philosophical discourse presuming, in one way or another, the standpoint of God and entitling itself to it, whether directly or indirectly. The "logic of our language" rightly understood *necessarily* disbars this possibility and hence any such entitlement.

Wittgenstein's view that his work had "solved in essentials *the* problems of philosophy" (TLP Pr ¶8) places his endeavor squarely within the philosophical space and the philosophical tradition, the space and tradition wherein Spinoza's endeavor had been long residing. Hence there is no point of principle preventing us from putting the two authors in a conversation with each other so as to determine whether their two works might match in significant respects. Both share the same philosophical insight; both espouse the perspective of radical immanence, even if in different modes: if Spinoza undertakes to destroy God's position ontologically, Wittgenstein undertakes to annihilate it *logically*. The aim of positing such conversation can thus only be that of explaining this perspective—and these different modes of engaging it—in detail.

If Spinoza and Wittgenstein do share the same philosophical perspective and if what frames an activity in general—and thence philosophical activity—is, on the one hand, the purpose setting it off and, on the other, the end by which it ceases, then we have to explore the purpose driving the composition of the *Ethics* and that of the *Tractatus* as well as the end Spinoza and Wittgenstein take their works as having attained.

CHAPTER TWO

Purposes and Ends

The whole nature of reality lies in its acts. . . . For it there is no other sort of being.

—Friedrich Nietzsche, *Philosophy in the Tragic Age of the Greeks,* §5

WHETHER IT IS SPELLED OUT ontologically, logically, or otherwise, the perspective of radical immanence leaves no room for any higher power to which one can appeal in case of need or any higher authority one can obey in case of doubt. To go on with one's life, one can work only with what the world at large can provide for the purpose, or more specifically, with what the world at large makes present at hand and, from the world at large, can be made ready to hand.[1]

Since the hands in question are the hands of the self, we might say that, in some sense, what lies closest at hand is one's own self. Hence to go on with one's life, to conduct one's life adequately, one ought to care for the self, to work on oneself, and develop one's own resources. The life that can cope with the world at large, the life in which one proceeds by elucidating what is at issue and by clearing one's vision for thought and action, is the philosophically examined life, the traditional *epimeleia heautou,* or care of self.[2] Both Wittgenstein and Spinoza were fully aware of this. Recall that Wittgenstein views working in philosophy as "working on oneself," while Spinoza starts his quest in the first person, seeking the "supreme good" for his own sake. Concomitantly, although neither the *Tractatus* nor the *Ethics* is framed in the first person, it comes to prominent view at some strategic places of the first work and in almost all the scholia of the second, where the geometrical order leaves room for it. For future reference, note that Spinoza presents all his definitions through a first-person clause.

Even if the perspective of radical immanence is inextricably involved with the self, working it out is not a self-absorbing or self-absorbed undertaking. To the world at large belong other selves, and some of them are always present at hand.

Care of the self involves care for others. Spinoza holds that "nothing is more advantageous to man than man" (E IV p37s1), for "the good which every man who pursues virtue aims at for himself he will also desire for the rest of mankind" (E IV p37),[3] while Wittgenstein says that the goal of the *Tractatus* will have been "attained" if his work has "afforded *pleasure* to one person who [has] read it with understanding" (TLP Pr ¶1, emphasis added). We can say that care of the self goes hand in hand with care for others: working on oneself while coping with the world, finding one's place in it and assuming all this responsibly, goes together with helping to make the world better for all.

But others are not compelled to help us conduct our own lives. Barring extreme social or political conditions, the corresponding burden, the burden assumed in carrying out the activities structuring one's life, is exclusively one's own. Others cannot be answerable for anything regarding one's way of adult living, and hence the attendant responsibility is, again, entirely one's own. This is a responsibility with no preset bounds: in being a responsibility toward oneself, it is at the same time a responsibility toward everything that is present at hand and can be made ready to hand, but since what can be made ready to hand might be anything, it is, in the last analysis, a responsibility toward the world at large. As I will show, both Spinoza and Wittgenstein fully assume this responsibility.

According to chapter 1, Spinoza and Wittgenstein espoused the perspective of radical immanence essentially by seeking to annihilate God's overarching position. On this account, both men undertook the task to clear the way for conducting their lives on their own, without need of appealing to any higher authority, assuming in both word and deed the attendant responsibility. In short, both engaged in philosophical activity for exactly the same purpose. But if an activity necessarily has a purpose, it must have an intended end, and there should be ways of evaluating whether and to what extent the activity has attained that end. Here the two authors appear to part company: although both evaluate their philosophical activities as having attained their intended ends, the ends in question look very different.

Again, Wittgenstein considered philosophy to be an elucidatory activity that cannot come up with theories. For him, the "result" of philosophical activity is "not a number of 'philosophical propositions' but to make propositions clear" (TLP 4.112). Hence the intended end of his work could not have been knowledge of any kind; rather, it was only clarifying, at least "in essentials" (TLP Pr ¶8), whatever required clarification. The "unassailable and definitive" (ibid.) success of his work therefore could not have been any kind of philosophical doctrine but was instead "merely" his attaining philosophical "silence" (TLP 7) after the completion of such clarification. For Spinoza, however, the end of the philosophical activity was the "adequate knowledge of the essence of things" (E II p40s2), which is the kind of knowledge he spelled out as "knowledge of the third kind," or "intuition" (E II p40s and V *passim*). And exactly as Wittgenstein would in regard to his own project, Spinoza took himself to have succeeded: "I have now completed all I intended to demonstrate" (E V p42s). In fact, the tone of voice Spinoza adopts with respect to his success is comparable to Wittgenstein's. The latter considers the outcome of his toil "unassailable and definitive"; the former considers the outcome of his to be like "all things excellent . . . as difficult as they are rare" (ibid.). Yet this similarity in tone is not the only issue here: despite the difference just noted, Wittgenstein's philosophical silence is almost identical to Spinoza's "intuition" or "knowledge of the third kind."

To substantiate this claim, we need to take a closer look, determining what philosophical activity is, how its purpose relates to its end, and what values the two authors use to assess purpose and success. This examination will involve investigating how the ethical relates to the emotions for Spinoza and to the will for Wittgenstein and how responsibility enters the picture. The quasi identity of the ends to which they aspire and see themselves as having attained will emerge in this journey.

Activity and Purpose

First of all, irrespective of whether philosophy can be equated to an activity *simpliciter,* as Wittgenstein maintains (TLP 4.112), working in philosophy amounts to engaging in some kind of activity: philosophical labor must be a thinking activity of sorts. For Spinoza, this is fundamental: "By idea I understand a conception of the Mind" (E II d3), where conception should be distinguished from perception because while the latter is "passive," the former constitutes an "*activity* of the Mind" (E II d3ex, emphasis added). Philosophical labor is thus an activity involving ideas.

If working in philosophy is an activity, then philosophical activity should necessarily have a purpose, as all human activities do,[4] and doing philosophy well, that is, in a way that fulfills its purpose, should be distinguishable from doing philosophy badly, that is, in a way not fulfilling its purpose. An activity that fulfills its purpose is an activity that works, one that discharges its tasks. An activity that does not fulfill its purpose is an activity that fails, one that does not discharge the corresponding tasks, even if its practitioners might not recognize the failure. For Wittgenstein, the purpose of philosophical activity is "the logical clarification of thoughts" (TLP 4.112), while for Spinoza, it is to attain "adequate knowledge of the essence of things" (E II p40s2). Hence for Wittgenstein, to do philosophy well is to perform an activity that arrives at the logical clarification of thoughts, and to do philosophy badly is to perform an activity that does not clarify thoughts logically— or to describe it in stronger terms, an activity that confuses thought.

But though Spinoza defines doing philosophy well as performing an activity that produces adequate knowledge of the essence of things, doing philosophy badly does not amount simply to failing that purpose. The way he conceives both adequate knowledge of the essence of things and the "road" (E V p42s) leading to it makes him understand doing philosophy badly as, fundamentally, formulating confusions of the kind his contemporaries, Descartes included, were putting forth.[5] Most scholia in the *Ethics* point out and try to eradicate such confusions by laying bare their sources. It follows that one cannot attain adequate knowledge of the essence of things unless one gets rid of such confusions by clarifying thought and that doing philosophy well is inseparable from such clarification. We may say, therefore, that both Wittgenstein and Spinoza hold that performing philosophical activity well clarifies thought, while performing it badly confuses thought, and this holds irrespective of their difference with respect to the activity's intended end.

Now if we consider our lives as structured by the extremely variegated activities we perform, with this or that duration and this or that intensity, ranging from pure passivity to wholly undivided devotion, then we might say that purpose—or intention, which I take here as more or less synonymous—is the way human agency (and perhaps, with appropriate qualifications, the agency of all living beings) *hooks onto the world.*[6] Purpose is what makes us, as human agents (or more generally, us as living beings), engage in any of the activities structuring our lives (including the activity of thinking) and carry them out—or not—to the end. Purpose answers why we have undertaken (willfully or not, fully aware of this "why" or not) to perform an activity at all or to cease performing it. Therefore, as a matter of conceptual analysis, the concept of activity presupposes the concept of purpose, irrespective of whether, in engaging in an activity, we are conscious of our purpose, how and in what conditions the activity becomes carried out, or the effectiveness or ineffectiveness of its performance. It is purpose that sets off the corresponding activity and, in that precise sense, drives or governs it.

Although he does not use the terms "purpose" or "intention," Spinoza makes essentially the same point. He notes how everything, without exception, strives for all it is worth to keep hooking onto the world, "to persist in its own being" (E III p6), calling this "conatus." And he equates conatus with a thing's very essence (E III p7), implying that conatus logically precedes any activity in which the thing engages and is also expressed therein—as far as living beings are concerned—as the corresponding purpose of that activity. We might thus say that purpose precedes the activity it governs through this connection to conatus.

Since it forms the essence of every particular thing, inanimate as well as animate, conatus is ubiquitous, constituting in that quality a main pillar of Spinoza's whole philosophical theory. But as this theory aims at "pointing out the road" for attaining adequate knowledge of the essence of things, Spinoza has to specify the doctrine of the conatus. And this is exactly what he does. Conatus, as it refers specifically to humankind, he calls appetite: "Appetite is . . . nothing else but man's essence" (E III p9s). Appetite itself is defined as "the end for the sake of which we do something" (E IV d7), which is exactly purpose as I have been discussing it.

Here Spinoza makes two distinctions whose importance will soon appear. First, given his conception of mind-body identity, he carefully distinguishes appetite, which is "related to mind and body [taken] together," from will, which is appetite (or conatus peculiar to humans) as "related to the mind alone" (E III p9s). Second, he defines desire as "appetite accompanied by the consciousness thereof" (ibid.). This last qualification allows him to take in the possibility that we are not always conscious of our purposes, for one might not be aware of the goal toward which one's conatus is leading. The phenomenon is quite common. Appetites (or purposes) "are not infrequently so opposed to one another that a man may be drawn in different directions and know not where to turn" (E III p59d1s). We may say, therefore, that the fact that an activity might hide its real direction for the one having undertaken it makes purpose *inaccessible in itself.* I will demonstrate the significance of this in a moment.

Intimately connected with one's conatus are the emotions. In fact, desire, which requires consciousness of the corresponding conatus, constitutes the foundation of them all, for it forms "the very essence of man" (E III p59d1).[7] Of the remaining emotions, Spinoza singles out two, pain and pleasure, which, together with desire, serve to define all the others: "I acknowledge no other primary emotion than these three (i.e.: pain, pleasure and desire)" (E III p11s). Always taking body and mind together, he goes on to define pleasure as the "transition of man from a state of less perfection to that of greater perfection" (E III p59d2) and pain as the opposite (E III p59d3), where the transition at issue brings pain and pleasure down to the conatus again, for it is simply that whereby "a man's power of activity is diminished" (E III p59d3s) in the case of pain or enhanced in the case of pleasure. It follows that, in relation to performing some activity undertaken "with consciousnesses" and tending to increase "our perfection," pleasure is the emotion accompanying the steps toward fulfilling the associated purpose, while pain is the emotion accompanying the steps falling short of that purpose.

To round off the discussion of Spinoza at this stage, note that he takes the conceptual distance separating the emotional from the ethical to be minimal. Thus, he says that he uses the terms "virtue and power [to] mean the same thing," which is to say that virtue amounts to the ethical expression of "man's very essence" (E IV d8), that is, to the ethical expression of the conatus as such. Spinoza clearly specifies human conatus to be the foundation of the ethical in E IV p22: "No virtue can be conceived as prior to this one, namely the conatus to preserve oneself." He retains the ethical valuations of "good" and "bad" (E IV Pr) after criticizing their then current vulgar understanding. He then goes on to maintain that pleasure "is in itself good" while pain "is in itself bad" (E IVp41), and he makes the move from the emotional to the ethical even clearer by holding that the "necessarily good" is whatever "is in agreement with our nature" (with our essence or conatus) (E IVp31), with anything that disagrees with our nature being "necessarily evil" (E IV p30). In summary, then, Spinoza takes good and bad (or evil) to be the only two ethical values that are exclusively and directly referred to one's conatus and hence to the purpose necessarily attached to and preceding one's carrying out any activity.

In this sense, the values typically assessing the performance of an activity, namely "effective" and "ineffective," become fundamentally identical to the moral values of "good" and "bad." Thus, by countenancing no foundation of the ethical other than the "conatus to preserve oneself" and by thus reducing the moral sphere to what is effectively or ineffectively conducive to one's well-being, Spinoza most explicitly and implacably bans God from any moral interference in the conduct of one's life. No moral imperative, no moral injunction, no moral order can issue from God's standpoint or from any position external to one's activities in life. With respect to the ethical, Spinoza most intransigently holds firm to the perspective of radical immanence.

Wittgenstein addresses these issues in a strikingly similar manner. To begin with, he relegates all the emotions to the domain of empirical psychology, however "superficial" it might be (TLP 5.5421). The reasons are not hard to find. For one, since the purpose of philosophical activity is the "logical clarification of thought" (TLP 4.112), where thought is the dispassionate "logical picture of the facts" (TLP 3), emotion has nothing to do with the workings of philosophical activity taken in itself. Concurrently, as the very title of the *Tractatus* has it, Wittgenstein is concerned only with logical(-philosophical) analysis. This entitles him to ignore the obvious effects of emotion on any actual activity. However, he does take account, as he should, of the logical aspect of emotion. This aspect appears to comprise only the logico-linguistic properties of the "propositional attitudes" (hoping that, fearing that, etc.),[8] which from this point of view amount to the so-called opaque contexts. Those he treats in the family of propositions at TLP 5.541—5.5421 on the basis of the model "A believes that p" or "A thinks p" (TLP 5.542). I will look at that family in later chapters.

Second, Wittgenstein does not explicitly mention either purpose or intention. What comes closest is the "will,"[9] considered unambiguously as "what bears or supports [the German uses the noun *Träger*] the ethical" (TLP 6.423). Given the obvious similarity to Spinoza's approach, according to which the ethical is borne exclusively by what I have been calling purpose, we may consider Wittgenstein's "will"—which he appears to assume is present to consciousness—to correspond either to Spinoza's concept of will[10] (i.e., the conatus as referred solely to the mind) or, if we take into consideration the body, to Spinoza's concept of desire. Although Wittgenstein does not appear to be interested in the body, he actually does take it into account in a remarkable way. Hence we may take Wittgenstein to mean by "will" what Spinoza means by "desire," an identification I will attempt to justify in the following pages.

Furthermore, there is for Wittgenstein only a twin ethical valuation of the will, namely "good willing" and "bad willing" (TLP 6.43).[11] The two correspond exactly to Spinoza's twin values of good and bad, which appraise things solely with respect to whether they enhance or diminish one's power to act according to one's nature. In the first case the corresponding emotion is pleasure; in the second, pain. Although Wittgenstein does not mention anything resembling a drive for self-preservation, once his notion of will is taken to correspond to Spinoza's concept of desire, the structural similarity between the two approaches not only is preserved but becomes arresting: the highest expression of Spinoza's pleasure is "blessedness" (E V *passim*) or, synonymously, unadulterated happiness. As I will show in a moment, this might be taken as identical to the emotion Wittgenstein's "happy man" (TLP 6.43) feels.[12] Concomitantly, Spinoza's pain, once elevated to the corresponding degree, might be identified with the emotion Wittgenstein's "unhappy man" feels. In short, for both Spinoza and Wittgenstein, the realm of ethical values includes only good and bad borne exclusively and directly by what I have been calling purpose.

Given this identity, I maintain that the pleasure Wittgenstein invokes in the preface of the *Tractatus* is not gratuitous; it is a fundamental aspect of the way he understands the ethical.

But again, purpose in itself is inaccessible. It is inaccessible for the reason already mentioned—that the corresponding activity might hide its real direction from the person having undertaken it—but also for a second reason that explicates the first at a deeper level: in a sense, it *does not inhabit the world*. For Wittgenstein this is straightforward: obviously purpose cannot be an extramental fact of the world, but since thought is "the logical picture of the facts" (TLP 3), it cannot be *a* thought either. In the case of Spinoza the issue is more involved. To the extent that purpose (i.e., desire) is nothing but "man's essence" (E III p59d1), and since essences cut across all Attributes in that they are expressed concurrently in each, purpose *as such* cannot be a mode and therefore cannot inhabit what Wittgenstein calls "the world,"[13] even if Spinoza takes that essence as a potential object of thought.[14] It follows that purpose is for both merely whatever precedes any of our activities, making us undertake them and carry them out as far as we do.

These ideas bear a straightforward explication: the purpose driving any of our activities cannot inhabit the world and cannot be a thought proper simply because there is no way we can split ourselves in two, one part distancing itself appropriately from the other to take it as a sufficiently delimited object of scrutiny. There is in this respect no way that could guarantee even a modicum of stability and consistency and hence the reliability of assessment. The authority by which our purposes are supposedly assessed and the supposedly assessed purposes themselves are inextricably intertwined "within" an "inner world" in such a way that nothing can secure that the purpose we might consider as circumscribed is *indeed* the purpose at issue. And of course, others cannot give us a helping hand here either.[15]

However, although inaccessible in itself, purpose does manifest itself in the world. What hooks us onto the world, what makes us persist in our living by engaging our various activities, is *displayed in the actual performance* of those activities. Purpose *can* be gauged or assessed, but only thus, that is, *only in action*.

Purpose and Responsibility

The purpose behind the performance of an activity can be assessed through the effectiveness (or ineffectiveness) of that performance. Both Spinoza and Wittgenstein take effectiveness and ineffectiveness as fundamentally identical to the ethical values of good and bad. But these values cannot inhabit the world insofar as the purpose they assess does not. Wittgenstein makes this explicit: "*In* [the world] there is no value" (TLP 6.41). But on the perspective of radical immanence, neither can values be issued from some vantage point overarching the world. Hence, to assess purpose and thus to negotiate the norms of our activities, we are left completely on our own, with no foolproof guidance coming from anywhere at all. Moreover, if purpose can be assessed only in action, what counts for assessing purpose is only what we *do*: we can be held accountable only to and for our activities. But since nobody is compelled to help us in carrying out our activities or can be answerable for them, we ourselves must be accountable for our activities, for and as sustaining them. In Wittgenstein's words, "whatever ethical reward or punishment" exists lies in one's performing "the action itself" (TLP 6.422). Exactly the same holds for Spinoza: "Those who, failing to understand what virtue is, expect God to bestow on them the highest rewards [are] far astray" (E II Ap §1), for "he who loves God cannot endeavor that God should love him in return" (E V p19). It follows that there is no external agency to determine and administer retribution; virtue (or effectiveness) is its own reward. On the perspective of radical immanence and for both authors, purpose can be assessed—and by the same token authorized—only by the one who undertakes the corresponding activity. Purpose is self-authorized and self-assessed.

If the values assessing purpose neither inhabit the world nor constitute thoughts properly speaking, the ethical amounts to a nothing, the nothing accompanying the mere what of purpose noted previously. With respect to this nothing, Wittgenstein is comprehensive: he does not allow ethical values and the ethical in general even to be expressible in language: "All propositions are of equal value" (TLP 6.4), or more explicitly, "there can be no ethical propositions" (TLP 6.42). At this point, Spinoza seems to go his own way. That is, he seems to countenance the linguistic expressibility of values, for throughout the *Ethics* he appeals to values in assessing his contemporaries, for showing them how they go wrong and for "pointing out the road" to "blessedness."

But this need not amount to much more than a difference regarding the intended reception of their works, the difference in an aspect of purpose already underscored. The *Tractatus* aims at a logico-philosophical analysis regardless of the way purposes are assessed in our everyday dealings, and it is addressed to "those who had already had the thoughts that are expressed in it—or similar thoughts. It is not therefore a textbook" (TLP Pr ¶1). In contrast, the *Ethics* aims to "point out the road" to "blessedness," and hence it is directly concerned with the purposes driving such dealings. It is addressed to those who have *not* already had thoughts identical or similar to those expressed in it. Therefore it *is* a textbook: "Nowhere can each individual display the extent of his skill and genius more than in so *educating* men that they come at last to live under the sway of their own reason" (E IV Ap §9, emphasis added). The issue is not exhausted, however, for we have not yet examined what language in general can or cannot express for either Spinoza or Wittgenstein. Chapter 6 will be partly devoted to this.

Be this as it may, the nothing to which the ethical amounts is a strange kind of nothing. It may not inhabit the world, it may not be a proper thought, but nevertheless it relates to everything in the sense that human agency as such is at stake here. If only our activities structure our lives and hook us onto the world, and if the nothing of the ethical grounds all assessments of the purposes driving these activities as good or bad and hence the activities themselves as driven well or badly, then all our dealings with the world hang on the balance of this nothing. The nothing of the ethical is in this sense the "everything," for it is the locus where our entire lives are at risk, where all that is essential in them is to be won or lost.

If purpose is indeed what hooks us onto the world and thus drives our lives, then it stays with us all along. We cannot avoid adjusting and readjusting it, assessing and reassessing it, irrespective both of our awareness of doing this and of the extent to which this might be helpful. Purpose is inseparable from the values assessing it and thus is always laden with ethical intent. But since values neither inhabit the world nor can be advanced from outside, ethical intent must be, first, the nothing of our merely assuming responsibility over our activities and, second and concomitantly, the "everything" of assuming responsibility over all that we do, over how and why we do it. Since no one else can be answerable for our activities, since responsibility can be neither delegated nor assumed by others, only we ourselves can assume it, whether or not we realize and acknowledge this responsibility. And of course, symmetrically, we assume it for our own sakes and nobody else's, for the sake of persisting with our own lives the best we can. In a world with no outside, we are exclusively responsible for our activities and exclusively responsible to ourselves for those activities.[16]

On the perspective of radical immanence, therefore, ethical intent precedes any activity and presides over the effective performance of it, while the moral sphere in its entirety reduces to the nothing of one's assuming responsibility on one's own and for one's own sake. In this sense, all activities[17] carry ethical intent and thus necessarily partake of the ethical,[18] while the ethics of radical immanence is purely an *ethics of responsibility,* however this might be spelled out in its details.[19]

If every activity possess an ethical aspect, if ethical intent precedes and presides over any activity, then ethical intent precedes and presides over philosophical activity as well. Each philosophical work is thus a work of ethics. Both Spinoza and Wittgenstein were fully conscious of this implication. Spinoza admits it by the very title he gives his work, while Wittgenstein, consistent with his position that no ethical (let alone "metaethical") proposition can be expressed in language, leaves his readers to work it out on their own.[20] In addition, since consistency compels philosophical activity to strive at encompassing everything, leaving nothing unexamined, the corresponding responsibility applies to everything, it is a responsibility throughout.

... and the World

Even if the ethical amounts to nothing, the ethical intent of our activities, the twin values by which purpose is addressed, has a bearing on the point of our activities in general, namely, coping with the world. The upshot is important, for from the standpoint of the self engaged in an activity, good or bad willing can change no less than "the limits of the world," making "the world of the happy ... quite another from the world of the unhappy" (TLP 6.43). For Spinoza, the world of the happy, the world of blessedness, is the world fully understood as it really is: if knowledge of the third kind, or intuition, is adequate knowledge of things in their essence, then "the more we understand particular things, the more we understand God [Nature]" (E V p24), and this understanding goes hand in hand with the "highest possible contentment of mind" (E V p27). Although the world in itself does not change, it becomes quite different from the world perceived by those who remain unaware of the way the world works and of their places in it, unconscious of where their desires are driving them, attributing to imaginary causes whatever befalls them and thereby living with minds in turmoil and full of worries.

The twin ethical values of good and bad therefore *color* the world, "painting" it either white, and thus allowing all its variegated colors to reach the eye (the body and the mind) of the happy perceiver, or black, and thus picturing it in the form a miser would have it, a form keeping back all the world's colors, allowing none to reach the eye (the body and the mind) of the unhappy perceiver. In Wittgenstein's words, "The world as a whole must wax or wane" (TLP 6.43) as a result of our painting it with the ethical color of our purpose, the color of our ethical intent.[21]

For Wittgenstein, "ethics and aesthetics are one" (TLP 6.421). If, given the preceding discussion, we take this to signify that aesthetic valuation is limited to the binary opposition of the beautiful and the ugly (or their cognates), that these values are borne by one's will and that they coincide with the ethical values of good and bad, respectively, then the reason Wittgenstein identifies ethics with aesthetics becomes clear: coloring the world with the ethical brush of our good or bad willing amounts to employing at the same time an aesthetic brush. That is, "good willing" is painting the world so as to make it appear beautiful, and "bad willing" is painting the world so as make it appear ugly. This can be transposed readily to Spinoza given his view that the world wherein one finds "the highest possible contentment of the mind" can be perceived only as a beautiful, fully harmonious world where every single thing follows the course dictated by its own nature.

Thus the world of the happy person, as Wittgenstein has it, seem to be identical to the world wherein one finds the highest contentment of the mind, as Spinoza has it. The *Tractatus* and the *Ethics* thus appear to have reached the same end. The evidence marshaled, however, cannot suffice. This identity may well be formal, coincidental, merely apparent, or even illusory. To make good my claim, I must, working on the principle that only the wholly accomplished activity can fully display the purpose having driven it, take both Spinoza and Wittgenstein at their words when they claim that they have completed their work to their satisfaction and look at how each substantiates the endpoint of his toil. Since purpose and end frame[22] all activities, doing this will complete my discussion of the frame of the philosophical activity carried out in the *Ethics* and in the *Tractatus*.

End

The endpoint of Wittgenstein's work is philosophical "silence" (TLP 7). Wittgenstein arrives there after having clarified whatever required clarification and having thus brought to its end the complete philosophical treatise that is the *Tractatus*. The *Tractatus* itself was the "ladder" allowing him and those who followed suit to climb to that end, the ladder that is no longer needed and should be "thrown away" (TLP 6.54). The vantage point reached in this way is that of Wittgenstein's happy man and Spinoza's blessedness, the vantage point allowing Wittgenstein to "see the world rightly" (ibid.) and Spinoza to see it as it really is.[23] Seeing the world rightly, seeing it as it really is, is necessarily indebted to the *ethical* intent of good willing as accompanied by the blessedness felt by the happy person who has reached this end.

But for Wittgenstein, at least, this is the *whole* world. On the one hand, as he himself expresses it (in the first person), his success in arriving at this end has led him to occupy the position of the "metaphysical subject" (TLP 5.641). Because of the "truth of solipsism" (TLP 5.62), this is a position whereby his vista encompasses everything. On the other hand, since "solipsism, if strictly carried out, coincides with pure realism" (TLP 5.64), the "everything" in question can be only the world *as it really is*. Wittgenstein's arrival at the end of his work makes him—as metaphysical subject, not as "a man, not as the human body or the human soul of which psychology treats" (TLP 5.641)—perfectly coextensive with the world as it is and, in this sense, identifies him fully with it. As I will discuss in a moment, Spinoza, too, reaches a similar kind of full identification.

Now since the vista of the metaphysical subject encompasses the *entire* world, the subject in question must take in the world all at once in the form of a something, "a limited whole" (TLP 6.45). For Wittgenstein, this form of relating to the world is "contemplation" (ibid.). To contemplate the world in this way is neither to think of it nor to speak about it. Contemplation is neither thinking nor speaking; it is silently taking in feelingly—specifically, with the feeling of the "happy man"— the object of contemplation. Wittgenstein arrives at this position of philosophical silence only after having arduously climbed up the ladder of the *Tractatus*, which means that his ending his work is tantamount to his getting objectively through the activity of thinking philosophical matters and expressing the outcome in language, that he has finally elucidated all that requires elucidation, that he has finally lain to rest the whole of philosophical activity after having gone through it as thoroughly as was necessary.

To arrive at a place where the whole world as it really is comes into view, and doing so after having objectively exhausted all possibilities of thinking on such matters, is to possess the world's whole truth; it is to posses the "complete truth itself" (TLP 5.5563). This truth is exactly what Wittgenstein believes he "ought to give" (ibid.) in bringing his work to completion, enabling himself thereby to leave all philosophical activity behind him. Because this truth neither adds anything to the world as it is nor subtracts anything from it, it is a "simple thing," despite or rather because of its completeness. The claim that the truth of the *Tractatus* as a whole is "definitive" (TLP Pr ¶8), that nothing can be added to it or subtracted from it, refers precisely to this completeness.

Differences in terminology apart, almost exactly the same holds for Spinoza. Starting from "the fact that we have common notions and adequate ideas of the properties of things," we can exercise our "reason" to attain "knowledge of the second kind" (E II p40s)—knowledge laid out in concepts—as far as this can go. "From this knowledge we can deduce a great many things so as to know them adequately" and thus form the best kind of knowledge, namely, "the third kind of knowledge" (E II p47s), for this third kind "arises from the second" (E V p28). Of course, to attain such knowledge, we must go as thoroughly as possible through all that knowledge of the second kind can offer. Thus, if Wittgenstein reaches the end of his work by marching unerringly, with the full rigor of the logical order, on the road composing the *Tractatus*, Spinoza does the same, *mutatis mutandis,* by marching equally unerringly, with the full rigor of the geometrical order, on the road composing the *Ethics.* In reaching this end, Spinoza, together with those he has educated, attains "knowledge of God [Nature] *alone*" (E II p49s§1, emphasis added), which means that he has arrived at a vantage point allowing him to see "Nature," or equivalently, Wittgenstein's "world," as it really is, for again, knowledge of the third kind is knowledge of the essence of things.

Spinoza explicitly states that knowledge of the third kind is not laid out in concepts. It is knowledge one attains after having completely worked through concepts at the second level. That is, we attain the third kind of knowledge after having exhausted our reasoning capacities and having thus been led to a vantage point allowing us to "understand particular things" exactly as they are in their "essences" (E V p24, II p40s2). This is the vantage point from which we can intuit things as they really are, directly, effortlessly, and spontaneously, with ideas that are sharply distinct and crystal clear.[24] Hence knowledge of the third kind is not knowledge properly so called, which is why Spinoza calls it "intuition." Attaining this vantage point involves feeling the "greatest happiness or blessedness" (E II p49s§1), which is exactly what Wittgenstein's "happy man" feels. Spinoza's intuition thus nicely matches what Wittgenstein calls "contemplation"; as a result, we might conclude that Wittgenstein's "complete truth" is equivalent to Spinoza's "intellectual love of God [Nature]" (E V p32c). I will return to this in a moment.

Clearly, neither Wittgenstein nor Spinoza believe that doing away with philosophical thinking will affect science. Wittgenstein delegates to the province of science the kind of thinking—and the expression of thoughts—that properly belongs to it: "The propositions of natural science have nothing to do with philosophy" (TLP 6.53), the place of which can be only "above or below, not beside, [that of the] natural sciences" (TLP 4.111). Spinoza, in a sense, does the same. He leaves others to occupy themselves with the particulars of the natural science of his day while appropriating from their results only what he considers to have been established beyond doubt. As I will show, he proceeds in this way both because he considers these results to be indispensable in a comprehensive philosophical treatise and because this treatise would thus remain anchored to the science of his day. The digression on physics, or "the nature of bodies" (E II p13), inserted in the middle of part II of the *Ethics,* just after Spinoza defines the human body—together with the fact that Spinoza explicitly considers it a digression—is meant to play exactly this role.[25]

Eternity

After Wittgenstein has relegated the particular workings of the world—the "how" of the world (TLP 6.44)—to the sciences, what remains is the "what" of the world (TLP 5.552), the mere fact that the world *is* (TLP 6.44). For Wittgenstein, doing this is equivalent to "contemplating the world as a limited whole" (TLP 6.45). Spinoza sees things in a quasi-identical manner. Thus, after he has exhausted everything knowledge of the second kind can provide and has concomitantly relegated to science what properly belongs to it, what remains is, notwithstanding any appearances to the contrary, not some substantive philosophical theory but the mere "what" of God (Nature), the mere fact that God (Nature) *is*. (In chapter 6 I show that Spinoza takes "intuiting" God/Nature and all the particular things therein to entail this.) We may infer that contemplating the world as a limited whole and intuiting God/Nature and all the particular things therein should be taken as virtually indistinguishable.

For both Wittgenstein and Spinoza, to contemplate or intuit the world in this way is to contemplate or intuit it "under the form of eternity": "*sub specie aeterni*" (TLP 6.45) or "*sub specie aeternitatis*" (E V *passim*). Wittgenstein explains that by eternity he "understand[s] not endless temporal duration but timelessness" (TLP 6.4311), while Spinoza says that "eternity . . . cannot be explicated through duration of time, even if duration be conceived as without beginning or end" (E I def8ex). They therefore understand eternity in identical terms. Obviously, since death is always an event *in* time, both men deny that eternity, as timelessness, can ever be touched by death.

Now, since death cannot be "lived through," it cannot be an "event *of* life" (TLP 6.4311, emphasis added). It follows that, for Wittgenstein, it is possible that one may live under the form of eternity: "he lives eternally who lives in the present" (TLP 6.4311). On his part, Spinoza maintains that we can "feel and experience that we are eternal" (E V p23s). It thus follows that we live or feel this way precisely when we come to contemplate or intuit *the world* (Nature or God) under the form of eternity. This occurs when we come to "see the world as a limited whole" (TLP 6.45), allowing its "complete truth" to touch us, in the case of Wittgenstein, or equivalently, when we attain "intellectual love [of] God [Nature]" (E V p32c), in the case of Spinoza. Again, Wittgenstein states that to live in the present is to live "eternally," which entails (for, as I will show, Spinoza too seems to endorse this) that we somehow experience ourselves as eternal at the times when we live fully in the present moment, taking in the world as a whole, being totally aware of all its variegated colors, devoid of all worries, with a feeling of utmost happiness or blessedness overwhelming our being. And moments such as this do indeed occur.

Wittgenstein calls the feeling overwhelming us in such moments "mystical" (TLP 6.45), while analogous characterizations have been attached to Spinoza's assertion that we can feel ourselves to be eternal. But the feeling in question, however one would cash out the term *mystical,* need not be incomprehensible, unfathomable, occult, or esoteric, regardless of its being "inexpressible," merely allowing itself to be "shown" (TLP 6.522). If the previous discussions are correct, a philosopher can attain this oceanic feeling by (and only by) effectively exhausting all objective possibilities of thinking matters philosophical. It is not an option somebody could choose by ascetically renouncing the world or thought or language; on the contrary, the feeling is the point to which thinking leads if carried out implacably to its ultimate end. Moreover, it does not lie beyond anyone's powers, even if we do not go through the requisite philosophical toil: as I just claimed and will explicate, *anyone* can experience it at some privileged, if disparate, moments of his or her life.

Spinoza relates the feeling in question to a whole doctrine about the eternity of the mind. In a nutshell, this doctrine might be presented as follows.[26] First, Spinoza holds that "the object of the idea constituting the human mind *is* its [corresponding] body . . . and *nothing else*" (E II p13, emphasis added), which is to say that our minds have only our bodies as their objects. It follows that the idea constituting the human mind—the idea of its body—is and must be only the sum of the ideas (and thereby the knowledge) of what that body does or undergoes, whatever the kind or level of that knowledge might be. Specifying this with respect to our cognitive relations to things external to our bodies, Spinoza holds that the mind can "perceive any external body only through the ideas of the affections [that this external body] causes on [its] body" (E II p26). Hence any knowledge we obtain regarding external things can involve only ideas capturing the affections that external bodies cause in our own. Such knowledge is, as a rule, inadequate, for "the idea of any affection of the human body does not involve adequate knowledge of the human body" (E II p27).

Now for Spinoza, the idea or knowledge constituting one's mind "is in God" (E II p20), though there it is fully adequate. God knows perfectly all the causes responsible for whatever our bodies have experienced, all the causes inciting our bodies to perform their actions, all the effects of all those causes, and all the reasons we have been forming mostly inadequate ideas of these factors. In other words, God knows perfectly all the happenings undergone or initiated by us in their true causal interconnectedness, while we typically lack adequate knowledge of such interconnections. In this sense, the idea of one's body—which is one's "mind and nothing else"—constitutes an eternal truth in the mind of God: "There is necessarily in God an idea which expresses the essence of this or that human body under the form of eternity" (E V p22). More prosaically put, it is an eternal truth that individual X has lived and that, while living, his or her body experienced such and such and caused such and such; all these things are causally interconnected, and the eternal truth in question constitutes the true account of those causal connections.[27]

At the same time, Spinoza holds that the human "mind is eternal insofar as it involves the essence of the body under the form of eternity" (E V p23s); one's mind involves the essence of one's body under this form whenever it conceives things "by eternal necessity though God's essence" (E V p23d), that is, through the immutable "order and connection" of things (E II p7) in God or Nature. In other words, the ideas one's mind possesses in accordance with this immutable order and connection are eternal truths, part of the eternal idea (the eternal truth) of one's body in the mind of God. Hence the part of the human mind consisting in these ideas is eternal as well, while the ideas in question, as eternal truths, are fully adequate. It follows that "we feel and we experience that we are eternal" (E V p23s) when and to the extent that we "feel and experience" precisely this adequacy—that is, when and to the extent that our minds are dominated by eternal truths, by ideas capturing the essence of our bodies as they merge with the immutable order and connection of things in Nature. At these moments and to that extent, we feel and we experience our own bodies as being at one with the world, as occupying the particular place their essences determine, the place reserved for them in the immutable order and connection linking everything.

Blessedness, then, is precisely this feeling of adequacy, this being at one with the immutable order and connection of things in Nature. In other words, we feel and we experience that we are eternal when and to the extent that we feel being in full harmony with the world as it really is, when and to the extent that we are overwhelmed by the "intellectual love of God," a love that is eternal as well (E V p33). The sense of this adequacy constitutes an emotion or feeling—*love* of God— because it itself is not cashed out in concepts; rather, one can attain it, the third kind of knowledge or intuition, only after having gone through the second kind of knowledge as far as this can go. In addition, this love is deemed intellectual both because the "intellect . . . is the eternal part of our mind . . . whereas the part which we have shown to perish (V p21 [Spinoza's reference]) is the imagination" (E V p40c) and because we can arrive at it only "intellectually," that is, only after having worked through the second kind of knowledge.

Bringing in the intellect at this juncture is no doubt also meant to underscore that love of God is neither some gratuitous feeling entertained arbitrarily or at will nor a state of mind imposed by the practices of organized religion. On the contrary, feeling "intellectual love of God" is *justifiably* feeling at one with the world, with the world as it really is, the grounds of justification being as secure as they can possibly be: again, we attain this feeling only after exhausting all that the second kind of knowledge can provide. But attaining this feeling is not an affair of the mind alone; the body is implicated in the process most decisively.

With respect to the body, Spinoza maintains that "nobody as yet has determined the limits of the body's capabilities" (E III p2s), for "the body, solely from the laws of its own nature, can do many things at which its mind is amazed" (E III p2s). Once we come to understand how our bodies can do such amazingly unexpected things, we tame this amazement by our intellects and thereby augment the part of our minds that is eternal, for again, it is the intellect that forms the eternal part of the human mind. At the same time, this activity of the body, as it is being simultaneously understood for what it is "through the intellect alone" (E V p40c),[28] is what forms the "highest possible contentment of [the human] mind" (E V p27). It follows that the moments when we feel at one with the world as it is, when we feel and experience that we are eternal, are precisely the moments during which we are deploying the capabilities of our bodies to the full, and this with an adequate understanding of our actions. Since nobody has yet determined the limits of the body's abilities, nobody can tell how vast the eternal part of the human mind can become with respect to perishable imagination.

The claim that a part of one's mind is eternal does not imply that individual human beings can enjoy eternal life after death; no afterlife, no paradise and no hell, is implied by Spinoza's doctrine of the eternity of the mind. I will, however, postpone the explanation for this until chapter 6, after clarifying the sense in which Spinoza takes individual bodies to be perishable and his conception of individuals in general.

Spinoza's Body

Spinoza's doctrine of the eternity of mind underscores that knowledge of the third kind is not knowledge laid out in concepts. It is a kind of intuition capturing the "essences" of particular things within the relevant causal nexus. That is, it captures each particular thing within the causal context in which this thing finds itself, a context that, since it "surrounds" this particular thing, is singular as well. If we persist in calling this intuition "knowledge," we should identify it with the sort some have termed *expert knowledge* (Dreyfus and Dreyfus 1988; Dreyfus 1992).

Like intuition, expert knowledge does not involve explicit reasoning and is not organized by concepts and their interrelations. It cannot be put in propositional form, and hence, alluding to Wittgenstein, we can call it "silent knowledge." Expert knowledge is knowledge of *how to act* in a given context,[29] while its adequacy is assessed by the corresponding effectiveness. Expert knowledge is thus knowledge that is fully effective in any singular context within the field of the relevant expertise, while its adequacy is displayed through the spontaneous deployment of the expert's capability to respond to the particular circumstances: the expert is capable of turning to his or her advantage whatever these circumstances may provide, no matter how negative this might appear from outside the action. Expert knowledge is thus *powerful* knowledge, knowledge allowing the expert to dominate the circumstances but also, ipso facto, freeing the expert from those circumstances. Such power and such freedom are precisely the marks of Spinoza's third kind of knowledge.

Moreover, as Spinoza too would have it, expert knowledge is *fully embodied* knowledge, knowledge carried by or infusing the body. In acting on the basis of expert knowledge, our minds do not work on their own by processing the associated calculations of givens and implications, of pros and cons, of chances and probabilities for success—that is, by reasoning independently of the body. In expert action, body and mind neither work separately nor are felt as separate; neither assumes precedence over the other, and neither falls short of the other. They work together spontaneously and indistinguishably, in perfect unison, exhibiting thereby the undivided nature of the individual performing the action. Thus expert action manifests the merger of mind and body and displays how this merging works: a body-mind, that is, a person *as* body-mind, knows on his or her own, by his or her inseparable body and mind, what the body should do and what the mind should do and how to act with both as inseparable.

Since expert action is spontaneous effective action, success implies that the expert has spontaneously and accurately taken in all factors determining the context of the action as these have arisen and become relevant to success. To take in all such factors accurately amounts to forming an *accurate dynamic map* of the context in question "on" (or "within") the expert's body-mind. Success shows that this map is indeed accurate, that the expert has taken in the context as it really is. Since the context of action is always singular, while expertise refers to a whole field, it follows that the mark of the expert is the capability of his or her body-mind to form such an accurate map for an entire field of singular contexts, the field of the corresponding expertise.

Successfully performing an activity with wholly undivided purpose[30] on the basis of expert knowledge—devoting oneself fully to this activity and executing all the actions making it up both expertly and fully—is invariably accompanied by a feeling of being at one both with oneself and with the world at large. On the one hand, the expert's body and mind have worked in perfect unison, manifesting thereby that person's whole and undivided nature both to the actor him- or herself and to everybody concerned; on the other hand, success demonstrates that the expert has taken in the world as it really is and hence that he or she has been in full harmony with it. For the expert, the accompanying feeling is one of deep contentment overwhelming the body and mind simultaneously, the feeling that he or she has fully lived the moment of success as a *present* moment, the feeling, precisely, of having experienced eternity.

Examples exhibiting these characteristics are not lacking. Picasso's one-stroke sketches; Mozart's effortless compositions; Michael Jordan's domination of the basketball court in all three dimensions as his teammates and opponents rapidly change position; or even the deep, fully to the point empathy we might experience with respect to a fellow human being—these are cases where all the characteristics just mentioned seem to apply. In each case, expert knowledge refers to a particular field of expertise. Since, however, the limits of the body's capabilities remain undetermined (E III p2s), perhaps nothing prevents any of us, at least in principle, from becoming experts in many or even all of those fields. One who lacks inborn talent, whatever this may mean, will have to compensate by the corresponding toil.[31] In any case, if the considerations at issue hold water, then the doctrine of the eternity of the mind may not be too implausible or arcane.

The *Tractatus* lacks any propositions matching Spinoza's doctrine of the eternity of the mind. Given that such eternity is indivisibly linked to the body and its capabilities, it appears once again that the body does not focus Wittgenstein's interest. When Wittgenstein talks about "*seeing* the world rightly" (TLP 6.54, emphasis added), however, he is, at least according to the previous analysis, referring to a seeing that has left behind the activity of philosophical thinking. This is spontaneous and effortless seeing involving the eye as part of the body, a seeing that goes together with the feeling of contemplation overwhelming Wittgenstein's "happy man." As that which remains after exhausting the activity of thinking about philosophical issues, the feeling in question must be, to use Spinoza's terminology, a *bodily* affection.

In addition to involving the body in this way, Wittgenstein also maintains that "the sign is the part of the symbol perceptible *by the senses*" (TLP 3.32), and in the few places where he mentions experience explicitly, it clearly relates to our sense organs and hence our bodies. Much more important, Wittgenstein considers language to be "part of the human organism" (TLP 4.002). (I return to related issues in chapters 3 and 8.)

Admittedly, neither Spinoza's "intuition" nor Wittgenstein's "contemplation" appear to marry to action very well, which seems to jeopardize much of what I have been saying. To alleviate the worry, however, we have only to note that, despite the exorbitant labor devoted in bringing their works to completion, neither Spinoza nor Wittgenstein supposed that finishing his work would allow him to sit down, as it were, and take in the world passively, intuiting or contemplating it to his heart's content. The standard connotations of the words *contemplation* or *intuition* should be resisted here. As they have been used, these terms neither refer to a state of passivity of body or mind nor imply an *attitude* toward the world or toward one's life. Because they refer exclusively to the objective exhaustion of thinking through philosophical matters, they are meant to name only the feeling of deep contentment associated with the realization that one has rid oneself of confusions and elucidated all that is at issue. And this feeling accompanies the realization that one has done this not in order to take a comfortable seat and stare passively at the world but in order to open the way for clear thought and effective action with respect to the issues in one's life, issues that can henceforth be formulated in real and not imaginary terms. Contemplation or intuition is the feeling of being at one with the world in this sense and in this sense only.

In support of this argument, recall that neither Wittgenstein nor Spinoza embarked on philosophical activity for its own sake. Each believed that going through it would merely eliminate confusion and thus allow him conduct his life in clear terms.[32] That both took their efforts as successful signifies, therefore, that they believed themselves to have exploited to the full everything that philosophy could offer in this respect and in this respect only, putting simultaneously on view *all* it could offer. They had no further need of philosophy because their efforts had brought it to completion, or, equivalently, their efforts had *silenced* it for good.[33] Having thus reached the end of philosophical activity, they could get hold of their gains, however meager or significant, and go on with their lives.

Life

But what might going on with life mean for either man? What might it entail if we took the discussion one step further? What is life, what could life be, given all the previously stated considerations? Trying to answer this question will make Wittgenstein join the discussion more fully.

Again, what constitutes our lives and makes us "persist in our being" is our performing the various activities structuring life. But who or what is this "we"? Who or what is this "I" who carries out activities in living and in order to live?

In discussing the inaccessibility of purpose in itself and in trying to explicate the sense in which purpose neither inhabits the world nor constitutes a thought properly so called, I appealed to the impossibility of reliably splitting ourselves—the so-called inner world—into two parts, one wherein our purposes would reside and another that could circumscribe them, allowing them to be assessed for what they are. Yet as we witness daily, we actually do something much like this, for example, when we speak *of* our bodies or *of* our minds, silently implying that they lie at some distance from we who speak in this way, or when we take ourselves as *having* bodies and their various parts or as *having* minds partitioned in intellect, emotion, and will. That is, the "I" who does this places itself at a position overarching both "its" body and "its" mind, entitling itself thereby—and unproblematically, as a matter of course—to the corresponding illusory possession and privileged access, if not mastery and command. Indeed, our language pushes us in this direction almost inexorably: saying that I have a mind or that I have a body sounds perfectly natural, while saying that I *am* my body, that I *am* my mind, or that the single "I" that I am is both my body and my mind inseparably together and nothing else sounds like a solecism.[34]

In expert action, however, this kind of split is seamed over, as it were. Again, expert action manifests, both to the one performing it and to those the action concerns, *that* mind and body merge into one while simultaneously displaying *how* this merging works: the expert is an inseparable body-mind and acts as an indivisible body-mind. In this way the "I" of illusory possession and privileged access, if not mastery and command, fades out and tends to disappear. In addition, the expert in action is not only at one with his or her own self but also, as the expert's success demonstrates, at one with the world as it really is in the respects relevant to the action. That is why I maintained that the expert's body-mind could be seen as an accurate dynamic map of the world in these respects.

Recall that Spinoza maintains that "nobody as yet has determined the limits of the body's capabilities" (E III p2s) and therefore that in principle nothing prevents us from becoming experts in all fields of expertise. Suppose someone to have attained such universal expertise, reaching a state of body-mind where all the person's actions without exception are expert actions and hence all the activities structuring his or her life are being performed expertly: in such a case, that person's body-mind would be an accurate dynamic map of the world at large. But since the person's life is made up of these activities and only of those activities, it is his or her *life* that becomes the accurate dynamic map of the world at large.[35] As Wittgenstein has it, "The world and life are one" (TLP 5.621), and it should come as no surprise that he holds this. The body-mind situation of expert action makes up the "action analogue" of the static-sounding position in which the metaphysical subject finds itself. Recall that this is the position whereby the "truth of solipsism" (TLP 5.62) allows the vista of this subject to encompass the whole world, and to encompass it as it really is, because "solipsism, if strictly carried out, coincides with pure realism" (TLP 5.64).

Going a step further, we may say that world and life are one not in general but only *for me,* for the me who performs my activities expertly. That is, *my* life and *my* world are one. But my life is precisely what I am, for I am what I am only by and only through the activities, my activities, making up my life. Hence I cannot be anything but my life; I am my life. But since my life and my world are one for me, "I am my world" (TLP 5.63). Finally, since my life when lived expertly constitutes an accurate dynamic map of the world at large, my life, what and who I am, is an accurate map of the world at large on my comparatively microscopic scale—or more accurately, on my *microcosmic* scale, for in being a map of the world, I am in myself a world, that is, a cosmos. Hence I am a microcosm. And since in being accurate maps of the one world, all microcosms must be identical in this sense, I am *"the microcosm"* (ibid., emphasis added).

Recall that a feeling of deep contentment accompanies all cases of completely absorbing expert action and its success. This is the feeling of being at one with oneself and the world as it is, the feeling of living the present moment fully, devoid of all worry, the feeling of experiencing eternity. This, then, is the feeling that results when the overarching "I" retreats and seems to disappear, the feeling whereby the split that marked this "I" and made it rise to its overarching position is seamed over. And since worry, in the general sense, can be generated only by this "I," its disappearance in expert action takes away all such worry. This is then what "solves" the *"problem* of life" (TLP 6.521, emphasis added): living the present moment fully, feeling life purely, feeling life as such, eliminates all worry about this and similar problems, thus making "the problem of life *vanish"* as a problem (ibid.).

I might add, even more tentatively, that here we may have something like a definition of life. For pure life, life without an overarching "I" and the worries it comprises or generates, life as such, is animal life. All the activities that structure the life of an animal are expert activities at its singular scale, while this life accurately maps from its particular vantage point its corresponding world in all that this world affords—or lacks—for sustaining the life.[36] Of course, this world is in a deep sense different from ours: it is not arrived at in the ways I have been describing but is unproblematically always already there; it is not a cosmos in the sense previously employed.[37] But some kind of "blessedness" seems to apply even here. For we actually see and feel the deep contentment in a cat's purring, in the lazy movements of a lion that has satiated its hunger, in the big brown eyes of a cow peacefully grazing in the field. The image of the manger haunts our childhoods, I take it, precisely because of this.

To sum up, life and world are one, seeing the world rightly and acting in the world expertly are one, living fully the present moment and feeling eternity are one. Reaching this end results in unadulterated happiness, and this is the end *for* and *of* Spinoza's and Wittgenstein's philosophical toils.

Nothing

The philosophical vernacular offers a ready characterization of this happy marriage of life and world, an end that Spinoza and Wittgenstein each takes to be the goal of philosophical activity, and one, moreover, that each takes himself to have achieved: at the ethical level, one is to follow the dictates of one's own nature responsibly and do what pleases one most deeply; at the epistemic level, one is to leave to science the particular workings of the world and simply see it as it really is; at the practical level, one is to become an expert. Managing to do that leads to unadulterated happiness.

But at the close of the day, attaining this end seems to mean very little. Doesn't this amount to a quintessential triviality? Isn't this what we all wish to do and try to do, what we all at least believe we are trying to do, one way or another, all the time? Is this more than merely nothing?

Obviously Spinoza does not think that. In his day, undertaking to oust God from His overarching position by ruthlessly destroying His vantage point at the ontological level was an endeavor too extreme to be considered dispassionately even when thought was laying claim to astonishingly novel ground. At the time and with the means at hand, only a philosophical theory solidly grounded on the rising natural science and fully armored against the all too predictable deep misunderstandings could open the novel path and secure it the prospect of being heard. And this is exactly what Spinoza thought he had achieved: the *Ethics* constitutes for him the true philosophical theory whose truth is unassailable and definitive to the same extent that the truth of Euclidean geometry is.

Such an endeavor could not then be a safe exercise in pure philosophical thinking. In Spinoza's day, a most radical enterprise such as his was necessarily also the most perilous, fraught with the gravest of dangers for the very life of the person who undertook it—not to speak of his afterlife—as well as for the fate of his work. His assumption of all these risks, theoretical as well as harshly material, in full awareness and acceptance of responsibility and with unwavering consistency, compelled Spinoza to include the success of his toil among "all things excellent." The tone he adopts for assessing this success is understandably triumphant.

In Wittgenstein's day, before the clouds leading to World War II had gathered, endeavors to defy authorities merely by thinking and expressing thoughts were shielded from most kinds of outside interference. At the same time, most had abandoned the idea of philosophy as an all-consuming undertaking that demands the undivided commitment of its practitioners, viewing it instead as a mundane profession roughly like any other. Anyone working in this capacity risked nothing weightier than a failure to meet established professional standards, a failure that at worst incurred ridicule. To the extent that Wittgenstein was compelled, like most of us, to consider the endpoint of his toil in accord with the stakes of his assuming it, this was the gravest outside risk to which he was open. As was discussed in chapter 1, Wittgenstein readily assumed this risk; as will be confirmed, he assumed it with a vengeance.

Within the linguistic philosophy inexorably ascendant during Wittgenstein's time, the *summum malum* was talking nonsense. Even if anachronistically, one has only to recall Carnap's (1978) attack on Heidegger. The indictment does not refer to the very real political and ideological stakes Carnap faced in taking arms against his fellow philosopher, for it all boils down to the allegation that Heidegger indulged in nonsense. If nonsense was the chief enemy, however, then Wittgenstein, in openly admitting at the end of his work (TLP 6.54) that he himself has been talking nonsense all the time, not only assumes the risk with a vengeance but rises to sheer provocation. And by the same token, the charge Wittgenstein takes on himself becomes correspondingly massive. The task he undertakes is no less than to show that all philosophy self-complacently taking itself as combating nonsense bathes in it fully and, moreover, completely unawares: such philosophy elevates logic to the stature of the new God that can safeguard it from its chief enemy for all eternity, but as this gesture is the source of all philosophical nonsense, it is only nonsense squared.

Completing one's philosophical work, therefore, cannot amount to anything substantial. In Wittgenstein's words, precious "little has been done when [the] problems [of philosophy] have been solved" (TLP Pr ¶8), even if they have been solved in a way that is unassailable and definitive. For solving these particular problems does not describe how things are, does not prescribe how to go on with one's life, and leaves everything exactly as it is. In this sense and for Wittgenstein, at least, his work amounts to little more than nothing. Accordingly, the tone he adopts for assessing his success cannot be triumphant; it is only gloomily victorious.

Silence

The silence reached at the end of Wittgenstein's work is merely the silence attending this nothing. This is a silence unrelated to the "something" that allegedly cannot be said in words, though many commentators seem to think that this unsayable something could somehow be indicated by such silence. Much more prosaically, after the work is done, after the whole of philosophy has been completed, there is nothing but silence because there is *literally nothing* to be said by philosophy that would make the slightest difference. On this analysis, at the end of the philosophical day everything remains exactly as it was before the philosophical turmoil started. To responsibly follow what one's nature dictates, to see the world rightly and to act expertly—exhortations to do these things cannot amount to serious advice of any sort, for this is exactly what we all are trying to do and have always been trying to do in conducting our lives. To come to repeat such empty advice by going through the convoluted ways of philosophy can therefore be only a pure waste of time and mental energy.

To be fair, we should acknowledge an exception. To those who have become entangled in futile philosophical worries confusing their thoughts and bungling up their activities, to the readers who have "already thought" thoughts "similar" (TLP Pr ¶1) to Wittgenstein's but remain ensnared in philosophical worries, the *Tractatus* can graciously offer its help. It can display with implacable rigor the futility of philosophical activity and thus lead such readers out of it while simultaneously "affording" them the accompanying "pleasure" (ibid.)—a significant "pleasure," however, for it partakes of Spinoza's "blessedness" itself.

If, as I said, the "simple" truth that Wittgenstein thought his work "ought to give" is the "complete truth itself" (TLP 5.5563)—the complete truth, unassailable and definitive, as referring outwardly to the world—then this truth is also the truth of Wittgenstein's treatise as referred inwardly to itself. That is, to arrive at the complete truth itself with respect to the world is to arrive at the truth of the *Tractatus* as a (logico-)philosophical treatise, or since philosophy is an activity (TLP 4.112), at the victorious completion of Wittgenstein's own philosophical activity. In other words, the "*truth* of the thoughts communicated" (TLP Pr ¶8) by the *Tractatus* can be unassailable and definitive only insofar as Wittgenstein's laborious activity has been fully successful *in deed*. If this is the case, then it follows that there is now and always was literally nothing for philosophy to say. Despite philosophy's pretensions to the contrary, despite its understandable resistance to such total annulment, philosophy has been shown for what it is beyond appeal and thereby has been silenced for good in its entirety.

Wittgenstein does not take himself as having discursively *proved* that he has managed to silence philosophy altogether. He takes himself as having only *displayed* its futility—but having displayed it beyond possible doubt, in a way that is logically unassailable and definitive—simply by *carrying out philosophical activity to its ultimate end*. The three dicta summarizing the outcome of his work and the truth of his treatise, namely, to responsibly follow the dictates of one's nature, to see the world as it really is, and to act expertly, match exactly Kürnberger's "three words" in the phrase Wittgenstein borrows for the epigraph condensing the sum of his toil: "And whatever a man knows, whatever is not mere rumbling and roaring that he has heard, can be said in three words." To keep the balance, we can bring in Spinoza's dictum that "the human condition would indeed be far happier if it were equally in the power of men to keep silent as to talk. But experience teaches us with abundant examples that nothing is less within men's power than to hold their tongues" (E III p2s).

CHAPTER THREE

Grammar

> What is familiar is what we are used to; and what we are used to
> is most difficult to "know"—that is, to see as a problem; that is,
> to see as strange and distant, as "outside us."
>
> —Friedrich Nietzsche, *Gay Science,* §355

ALMOST 300 YEARS of historical distance separates Spinoza from Wittgenstein. Hence the question inevitably surfaces: why are their works so strikingly similar? Were the associated historical changes not important enough, or is the relevant philosophical activity capable of ignoring—or eradicating—historical change? How does it take account, if it does at all, of historical context?

I have noted some differences distinguishing the *Ethics* from the *Tractatus,* attributing these to disparities in historical context. But it is not enough merely to mark such differences while leaving the striking similarity unquestioned, and this can be understood only on the basis of an adequate answer to the questions just mentioned, questions boiling down to one big question with various horns: how is philosophical activity related to the overall historical process wherein it unfolds? Specifically, how does this overall process affect philosophical activity specifically, and how may it be affected in turn? What consitutes the identity of philosophical activity, and how might this identity evolve historically? What is philosophy properly so called, and what relations might it bear to the history of philosophy and to history in general?

Not My Purpose . . .

These interrelated questions raise broad and important issues—to begin with, about the adequacy of their formulation—which I cannot broach here in the seriousness they deserve. The questions inevitably crop up, however, which obliges me to draw a coarse outline of an answer bearing on what I am trying to do—and not trying to do—in the present work.

To sketch this outline, I must start by declaring that I will not attempt to determine whether and to what extent either the *Ethics* or the *Tractatus* can stand up to contemporary criticism. I undertake simply to read each work through the other so as to help release part of the critical potential they both harbor and, by thus helping them, to help us all to profit from such release. My ambition is restricted to encouraging both these works to participate in the ongoing philosophical discussion more actively than they actually do and, by the same token, to allow them to shed their own light on this discussion.

My reasons for embarking on this project are not different from those making anyone engage in philosophical activity. I believe that a vigorous critical power lies latent in the two works under discussion, which to an extent have long been left unexploited, with only a small exegetical push necessary to activate that power. If the final prepositional phrase marks the burden of the present work, the assertion in the relative clause is readily ascertainable. The *Ethics* seems to have long ago exhausted the critical energy with which it might have initially been endowed; it has lain ever since almost inert, consigned to the province where historians of philosophy continue to do their valuable work unobtrusively.[1] Closer to us, the *Tractatus* seems to have consumed its own critical potential almost in one single spark within a brief period after it was published. In impressive contrast to Wittgenstein's later work, it has been left standing almost on its own, an object of study for Wittgenstein scholars alone. In this sense, this work too tends to become dispatched, if it has not been already dispatched, to the exclusive care of historians of philosophy.

That the history of philosophy thus appears as the dustbin of philosophical activity is no fault of its practitioners. A particular philosophical tradition, still dominant in many respects, has been imperiously disdaining most major issues regarding how philosophy proceeds historically and how a philosophically adequate history of philosophy should be conceived,[2] turning "history of philosophy" into a quasi-autonomous discipline taken to be exercised by somewhat tedious children of a lesser god. For this tradition, work from the past (and on this view, the *Tractatus* belongs to the past) is interesting only to the extent that it can supply us with some particular, narrowly isolated, and almost never sufficient means for tackling our present problems, which are taken to be better situated, better delineated, better formulated, and better solved or dissolved. That some philosophical work of bygone days, be it one hundred or even four hundred years old, might prove not just fully up to our standards but also deeper and more robust than some of the most admirable contemporary works can come then only as a big surprise, if not an embarrassment.

This tradition's continued dominance makes my task harder. Beyond their all too obvious difficulties, which are troubling enough, these texts are also shrouded in a philosophical atmosphere that hides what they are saying or trying to convey and that encourages many philosophers to look condescendingly on all efforts to come to terms with them. Our days are more prosaic than even those of Wittgenstein, and philosophical stakes have become exclusively professional—at least in what we consider to be democratic countries, for in others expressing a simple philosophical point of view can incur the gravest of dangers—with philosophical thinking not having become thereby necessarily deeper. Even a modest ambition such as my own cannot be brought to fruition without perseverance in swimming against a current still going strong.

This current is formed mainly by attitudes, not people. To adopt these attitudes is to forgo questioning the source of the wisdom on which the attitudes are supposed to rest, to tend to superciliously and thoughtlessly dismiss the *Ethics,* as well as other works from the past, for, among other things, its "naïve" claims, its argumentative "flaws," its high tone, and its dated vocabulary. Similarly, it is to tend to dismiss the *Tractatus*—if not superciliously, for it is still difficult to adopt such a stance with respect to Wittgenstein, nevertheless thoughtlessly—by invoking new technical developments in logic, by refusing to countenance its allegedly "mystical" remarks, or simply by ignoring it altogether to the profit of the supposedly more "comprehensible," or more "serviceable," later work of the same author.

Countering such attitudes directly does not form part of my task, for this would be at least premature. Because both the *Ethics* and the *Tractatus* are covered under the haze just described, we must first come to terms with the texts on their own; pushing away the clouds surrounding them even slightly, and thus obtaining a glimpse of what the texts themselves are saying, would already be enough at this stage. Part of this effort involves trying to understand what makes their similarity possible and what their differences may mean.

We can go on, then, by having a cursory look at the relations between the history of philosophy and history in general, on the one hand, and between the history of philosophy and philosophy in general, on the other. Discussing these relations inevitably introduces the history of science in a particularly prominent position.

History and the History of Philosophy

Human history has been witnessing and no doubt will continue to witness major sociopolitical ruptures whose depth and effects nobody could have foreseen or been prepared to manage. These ruptures take in stride most of the ways in which we used to conceive things in the more general sense of the term. Such breaks in human history—including the rise of Christianity, the Renaissance, the scientific revolution, the English Civil War, the French and American revolutions, or the Russian Revolution of 1917—have brought with them major ideological and conceptual upheavals that profoundly affected how we understand the world, our societies, and our places in either; how we understand ourselves and our thinking on such matters; and how we go about, or should go about, conducting our lives.

These conceptual upheavals affected philosophy at its deepest level, compelling it to respond by advancing novel philosophical approaches. If, to use Sellars's felicitous formulation, to do philosophy is to attempt to understand "how things in the broadest possible sense of the term hang together in the broadest possible sense of the term" (Sellars 1991, 1), then the new "things" brought forth by such major ruptures in human history inevitably solicit from philosophy their proper understanding. Moreover, and insofar as such ruptures are deep enough, the old established "things" require a novel philosophical understanding as well: they too change, for they tend to be perceived under the light of the new "things." It follows that major ruptures of this kind do not affect philosophy only superficially; they induce novel philosophical approaches that make it reorganize itself drastically.

Novel things in Sellars's sense do not arise only through major sociopolitical ruptures, however. I launched the discussion in chapter 1 by noting that the different historical contexts within which Spinoza and Wittgenstein undertook their works had not been determined only by the kind of conceptual commotions just mentioned. The same contexts were simultaneously marked by the conceptual disruptions that major *scientific* advances were generating in both periods, advances philosophy could not possibly ignore. It was then, as it had always been,[3] compelled to take such advances into account and reorganize itself drastically. To say the same in received terminology, philosophy is always compelled to respond to the radical paradigm changes occurring in science. It follows that there are two major sources provoking deep changes in the history of philosophy: conceptual upheavals induced by major sociopolitical ruptures and conceptual disruptions generated by radical scientific paradigm change.

The philosophical approaches induced by paradigm change in science do not develop in isolation from those induced by the conceptual upheavals attending deep sociopolitical ruptures. On the contrary, Sellars's formulation implies that the philosophical understanding of things, with the term construed broadly, must be *unified* understanding.[4] It follows that philosophical approaches coming from either direction necessarily inform one another and tend to merge with one another, usually in the form of a philosophical system. Althusser (1990) goes as far as to reduce the whole of philosophy to the role of a mere go-between for the scientific and the sociopolitical spheres, the very "essence" of which is rooted in historical change.

This, then, is the double *external* motor of historical change in philosophy.[5] Novel philosophical approaches are incessantly solicited by new things, brought into being by major historical events of some kind, new things that, along with the old, require unified philosophical understanding. Of course, philosophical understanding is inseparable from the continuous critical reassessment of the old philosophical approaches, which always takes place in conjunction with the conceptual dissection of the novel approaches. This is what makes up the properly *internal* aspect of philosophical activity as it has been going on since its inception. And such critical assessment and conceptual dissection have been going on and must continue to go on, because this is the only way to draw the line between elucidation and confusion, between genuine understanding and the semblance of understanding with respect to both the new things that historical change brings forth and the old things, which tend to change in the process.

The History of Philosophy and Philosophy

Reassessing old philosophical approaches in the light of the new and conceptually dissecting the novel approaches—the everyday internal workings of philosophical activity—require incessantly burrowing into the philosophical field in its entirety. For one thing, to assert what it is and what it claims, every novel philosophical approach must be carefully delineated so as to distinguish it from the approaches that already occupy the philosophical terrain. But effectively tracing these lines makes the novel approach enter deeply into the detail of the work against which it is set. This dialogue, however, allows these other approaches to make headway into the new and to sully the pristine purity with which it first appeared on the scene. Criticism thus develops reciprocally, and philosophical activity finds all the fuel it needs to go on indefinitely.

Even if we are allowed to talk about fundamental philosophical camps—those named after the various philosophical "isms"—the previously stated considerations mean that these camps are incessantly regenerated by the novel approaches flowing into the philosophical scene. This regeneration inevitably trespasses many of the lines that appeared to have marked out those camps once and for all, while the possibility of such a trespass indicates that these demarcations do not lie outside philosophical activity, passively waiting philosophers' gruelling efforts to clear them up by asserting and reasserting them. These lines exist only by and within the efforts that trace them; that is, the battle over each such line is fought anew each time, with all the weapons forged for the purpose and with an outcome that cannot be guaranteed otherwise than by these efforts themselves. To say it by paraphrasing Kant, philosophy forms a battlefield where no everlasting peace seems to be attainable. In this sense, philosophy cannot properly end.

But if philosophy has no proper end, it cannot have a proper beginning either, for the beginning of philosophical activity can only be the first efforts, however hesitant, aiming at *logon didonai* (giving reasons). These are the first systematic and hence relatively autonomous steps whereby, in trying to explain or justify a statement or an act in our everyday dealings, we *give reasons* by appealing to something that lies beyond what is present at hand. Philosophy is then just the activity of giving this kind of reason as it has been ordered, structured, and developed over the course of the centuries. And since giving reasons of this kind may concern anything at all, the "external" motor driving philosophical activity is not limited to paradigm change and to the conceptual upheavals induced by radical sociopolitical ruptures. Proper philosophical understanding, understanding how things hang together, in principle involves all "things" without exception; in this sense, philosophy is all-pervasive.

If philosophy is all-pervasive in this manner, then the understanding proper to it should be fully comprehensive and fully self-consistent: only in this way may one properly understand how each particular thing hangs together with all the other particular things. In addition, to constitute proper understanding, philosophical understanding should be wholly compelling. All philosophical approaches aim at an understanding featuring all those traits, and attaining it constitutes the aim of philosophical activity as such.

If the understanding in question is to encompass all things without exception, then these "things" cannot be limited to that which history has already brought forth in one way or another; these things should include all possible things that history *might* bring forth. Proper philosophical understanding involves all things in this unlimited sense, even if such understanding can be achieved only at some level of abstraction. Since attaining such understanding implies that there can be no more essential things to be understood, once philosophical activity has reached this end, it can have no further reason for existence. Effectively attaining proper philosophical understanding exhausts all that philosophy can offer and completes philosophical activity as such. Aspiring to attain proper philosophical understanding is simultaneously aspiring to end the very activity that aims at attaining such understanding. All philosophical approaches, not just Spinoza's or Wittgenstein's, aspire to put an end to philosophy in this sense, to silence first all other approaches by defeating them in open philosophical battle and, in achieving this, to silence *themselves*.

To the extent that this aim is attained, philosophy will have managed to defeat history, at least "in essentials." That is, if philosophical activity must continue after this victory, it will have to occupy itself only with the inessential. Nothing history could ever produce would ever be in the position to take philosophy by surprise. Among other philosophers, both Spinoza and the young Wittgenstein viewed themselves as having exhausted what philosophy could offer, at least in the essentials, as having thus completed philosophical activity as such and silenced all philosophy for good. They were convinced that the outcome of their toil could never allow history to take *it* by surprise.

Philosophy and the History of Science

Again, radical paradigm change is a major, totally unforeseeable historical event that does take philosophy by surprise. Such events shake philosophy to its very core, for they present it with with particular "things," always in Sellars's sense, that philosophy had not merely overlooked; rather, it could not possibly have envisioned them. A radical paradigm change involves the introduction of something that was *in principle* inaccessible to philosophy up to that moment. And as I will show, this thing was inaccessible because it had hitherto been *literally inconceivable* in any way whatsoever.[6] Both the *Ethics* and the *Tractatus* constitute aggressive responses to such a thorough shake-up; their aim was not merely to philosophically tame the corresponding particular challenge but, much more ambitiously, to reestablish the precedence of philosophy over history by making the former invulnerable to similar challenges that might threaten it in the future. Understanding how the works of Spinoza and Wittgenstein relate to radical paradigm changes in the history of science will help us understand both what the two approaches share and the principle of their remaining differences.

Historians of science have been meticulous in underlining that no two paradigm changes can proceed in an identical manner. It follows that, to assess any radical paradigm change's effects on philosophy, we must take into account the specific content of the change in most of its intricate details. This kind of work, however, falls outside my present concerns, which here are limited to a single difference between the historical contexts within which Spinoza and Wittgenstein worked. Giving the term *paradigm* the requisite latitude, we can say that Wittgenstein could benefit from lessons drawn from two radical paradigm changes in the history of science, while Spinoza benefited from only one—a change, moreover, that had not yet been brought to its relative conclusion during his time.

A radical paradigm change shakes philosophy to its core, for it demonstrates retroactively that the grammatical space subtending all uses of language up to that moment—and hence the language of philosophy as well—has been too narrow. In other words, paradigm change *widens the grammatical space available.*[7] For example, counter to all our intuitions and to all theories based on such intuitions, the scientific revolution opened the grammatical space in a way that allowed us to conceive of rest and uniform motion as being essentially identical, of planets as turning in elliptical orbits around the Sun (or other planets) while *defined* as revolving in circles around the Earth, or of instantaneous action at a distance with nothing to mediate the interaction, all literally inconceivable notions on the basis of the grammar subtending colloquial understanding. As if such conceptual disruptions were not enough, the scientific developments witnessed in Wittgenstein's times widened still further the available grammatical space by rendering it capable of hosting "impossible" non-Euclidean geometries, infinite sets of infinitely increasing cardinality, waves propagating in the absence of a carrier whose disturbance they nevertheless are, or trajectories whose defining characteristics (position and velocity) cannot be determined simultaneously, even in principle—all weird things, again literally inconceivable before the relevant paradigm change.

The difference between these two contexts of paradigm change is telling not merely because the grammatical space available in Wittgenstein's time grew wider—while drastically changing in the process—than did that available in Spinoza's time. With respect to the matters at issue here, the more important point is that from the seventeenth to the turn of the twentieth century, the revolutionary scientific advances brought forth in Spinoza's time had been consistently thought to provide the ultimate ground for solving, at least in rough outline, all fundamental scientific problems. The indomitable optimism characterizing the Enlightenment and the associated idea of boundless progress find their deepest roots here: human history had found a novel absolute commencement and could travel only forward on a road thus irreversibly opened up. Therefore, to the extent that Spinoza had taken adequate account of the corresponding unprecedented achievements, and given the formal rigor of his reasoning, he was fully justified in asserting that he had established the "road to salvation" or, equivalently, that he had definitively destroyed God's vantage point at the ontological level: using what was available to him, he had come to secure philosophy's precedence over history for good and thus to silence philosophy for good.

In contrast, Wittgenstein's times were characterized by the unwelcome surprise that the fundamental scientific problems had *not* been solved even in their essentials. In chapter 1 I discussed how this wholly unanticipated shock spurred the demand for an approach that could safeguard philosophy from history's ever again surprising it in such ways and how this demand promoted logic, in Frege's and Russell's sense of the subject, to the exalted status of the new deity: if logic, together with its associated linguistic philosophy, can analyze language down to its logical ground, thus forestalling the grammatical widening that radical paradigm change might bring forth[8] and thereby philosophically defusing the attendant conceptual changes, then it could justifiably claim that it had solved, or rather dissolved, all *philosophical* problems for good, at least in their essentials. No further surprise coming from history could ever again shake it up. This is the exact point where Wittgenstein's approach converges with that of his contemporaries but also the point where it separates most resolutely from them. By exhibiting logic's connections to language, the *Tractatus* does fulfill the demand of his contemporaries; at the same time, however, it annihilates the godlike status attributed to logic by the same means. The success of the endeavor not only secures philosophy from future surprises but also silences it for good.

This is, then, Wittgenstein's relative advantage over Spinoza, and this is what measures the historical distance separating the two men: the fundamental *philosophical* difference between the *Ethics* and the *Tractatus* boils down to that between seeing or failing to see that the all-decisive philosophical stake at issue here is the status of grammar with respect to logic. Logic in Wittgenstein's sense was not available to Spinoza, while Wittgenstein took logic as underlying all possible conceptual issues and thus as untouchable by any grammatical widening. Hence if there is nothing beyond logic and if logic is the core of philosophy, then to the extent that the *Tractatus* succeeds in its task, philosophy has indeed defeated history definitively.

Thus hindsight allows us to recognize that the *Ethics* could not win a final victory over history. Spinoza was in no position to foresee—because it was then absolutely inconceivable—the grammatical changes that were to come and hence the need to distinguish the logical from the conceptual or the grammatical. This was only Wittgenstein's prerogative, for again, once the *Tractatus* succeeds in its task, philosophy defeats history all in all: its logical armor secures it from any future grammatical, and *a fortiori* conceptual, surprise. As he himself states succinctly, "there can *never* be surprises in logic" (TLP 6.125). The discussion of the preceding chapter shows why this ultimate victory adds up to little more than nothing.

Nonetheless, the similarity between Spinoza's and Wittgenstein's approaches arises from the fact that they belong to the *same philosophical camp,* the one defined by the perspective of radical immanence, whose main enemy is any philosophical outlook that invokes, directly or indirectly, the possibility of a position overarching the world, thought, and language. The implacable formal rigor to which both works submit themselves places them at the heart of this camp: defining their philosophical enemy that sharply and holding to their purpose as tenaciously as they do makes the two works not just similar but almost identical.

When Wittgenstein demonstrates that the existence of God's vantage point is logically impossible, he is also demonstrating that philosophical views relying on or implying the possibility of this vantage point are untenable because it is logically impossible to hold them. But then the perspective of radical immanence ceases to be a philosophical perspective to begin with: espousing it becomes *logically necessary.* And this, I contend, was exactly what Wittgenstein intended to do with the *Tractatus,* the finished product of his toil constituting for him nothing short of the effective—the "unassailable and definitive"—demonstration of precisely this claim.

This demonstration is far from trivial, and tracing its general contours will occupy both this chapter and the next; its difficulty, moroever, requires presenting it gradually, one step at a time. At this stage I can proceed with only a small first step. Provisionally ignoring the point just made, I will talk of the perspective of radical immanence as if it were a bona fide philosophical perspective countering other philosophical perspectives. Similarly, I will be taking Wittgenstein as piling up all his tangible opponents—any philosophical approach, present or past, consistent with the existence of an external vantage point—in the quasi-logical and in this sense ahistorical form of a *generic* opponent.

That is, to explain why there can be no logical room for the philosophical perspectives countering the perspective of radical immanence and hence none for that perspective itself, I must first assume that philosophical camps can meaningfully exist, but *only two* of them. Thus Spinoza fully retains his place in the conversation, for he would not at all object to piling up his own opponents in an analogous fashion. It is perhaps worth adding that in discussing Lenin's position in philosophy, Althusser (1972) draws a similar line between the two possible camps, though in the apparently very different terms of "materialism" and "idealism."

The History of Science, Language, and Philosophy Again

Wittgenstein's advantage over Spinoza further allows the later writer to clearly view why philosophical approaches belonging to the two "bottom" philosophical camps just specified can never come out in a pure logical form but always appear in different, historically determined guises. According to TLP 4.002, "colloquial language disguises the thought; so that from the external form of the clothes one cannot infer the form of the thought they clothe." It follows that the guises in question are made up from the "clothes" that any philosophical approach must wear, and it must wear them because no philosophical approach can be an exercise in or of pure logic, capable of presenting itself in the corresponding formally immaculate nudity; concomitantly, each has to assert itself by tracing the line dividing it from the *tangible* and hence logically impure opponents it finds as already occupying the philosophical terrain. Colloquial language is the only terrain where philosophical battles can be fought, just as the understanding proper to philosophy, like all understanding, can be expressed and negotiated only in colloquial language. Thus the guises in question are nothing but the various ways of engaging in philosophical activity as it appears in history and is determined by history.

Colloquial language, moreover, as working outside science, does *not* reform grammatically after a radical paradigm change; among other indicators, pedagogical research attests[9] that a novel counterintuitive scientific theory has to overcome all the attendant resistances. But a philosophical approach is compelled to take such grammatical widening into account even while unfolding within colloquial language. It follows that the language formulating any philosophical approach must reach a compromise between these two demands pushing in opposite directions. Effectively achieving such compromise makes philosophical language often sound abstruse.

The language of the *Tractatus,* however, can hardly be characterized as abstruse; the text's all too apparent difficulties seem to lie elsewhere. But as commentators have noted and as biographical accounts confirm, Wittgenstein was exceptionally sensitive to language, worrying particularly about the aptness of the linguistic expression of his thoughts. Thus, as regards the *Tractatus,* he does not hesitate to admit that he "has fallen far short of the possible" in this respect simply because his "powers were insufficient to cope with the task" (TLP Pr ¶7). He consequently urges "others to come and do it better" (ibid.). Generations of commentators attracted by the *Tractatus* have apparently heeded the urge and have tried their best if not to do it better at least to understand what the author of that work sought to establish and how he went about trying to achieve his aims. But perhaps even those closest—or taking themselves as closest—to Wittgenstein's intentions have fallen short of this modest rhetorical goal.

Even independent of such entanglements, a simple question cannot fail to arise at this point: what might be the upper limit of the possible that Wittgenstein fears he has missed? Is there such a limit in the first place? Can there be a single *best* way of expressing what Wittgenstein intended to convey? If one can exist, it will have to be couched in colloquial language, no matter the cost in difficulty or abstruseness. But can colloquial language be made to withdraw enough to allow this best way to emerge? Is it possible for what Wittgenstein intended to convey to appear in colloquial language in the unadulterated form of a pure per se?

By maintaining that he has finally solved the problems of philosophy, albeit only in the essentials, Wittgenstein appears to be entertaining the idea that this might be possible after all. Nonetheless, his call for others to try to do better seems to leave open a narrow path for what I have been maintaining, namely, that no bottom line can ever be traced definitively in philosophy, or, to use Althusser's arresting formulation, that the "lonely hour" of such a "last instance" never comes (Althusser 1996a). It might be fair to say therefore that the issue remains open for the young Wittgenstein, for it is uncertain how he would have responded if pressed to take a firm stand on it. But if this is uncertain for the Wittgenstein of the *Tractatus,* the older Wittgenstein most certainly gives an unequivocal negative answer: such finality is simply out of the question. I myself have already taken the side of the older Wittgenstein: the very term *grammar,* which I have been using freely, is a direct loan from his later work.[10]

Be that as it may, the young Wittgenstein's position on the matter does not have a direct bearing on what follows, and the time has come to invest this terminological loan by explicating what I take grammar to be and how it relates to my project here. Doing this involves three closely interconnected steps: first, explicating what the grammar of paradigm change amounts to independent of either the *Tractatus* or the *Ethics*; second, determining whether the *Tractatus* engages paradigm change, grammar, and the relation of grammar to logic; and third, assessing how deploying this key element of Wittgenstein's later work might contribute to understanding his earlier work. The focus on Wittgenstein at the expense of Spinoza is a direct consequence of the central point made earlier, namely, that the main difference between the two involves the relation of grammar to logic, which is a nonissue for Spinoza, and lies chiefly in the fact that Wittgenstein enjoyed the advantage of two major grammatical shifts in the history of science, whereas Spinoza had only one.

I will start by discussing a particular aspect of one major conceptual revolution in twentieth-century physics, namely, the passage from classical mechanics to the special theory of relativity (STR). To simplify matters, I will consider this passage as consisting in the establishment of just one novel concept, that of the electromagnetic field. To explain the passage without going too far adrift, I will take full advantage of all that the *Tractatus* has to offer with respect to this explanation.

Grammar and Paradigm Change

The physics involved in the passage from classical mechanics to the STR began with Maxwell's electromagnetic theory (1864), which, among other things, unified electricity, magnetism, and optics.[11] A salient result of that achievement was the prediction of electromagnetic waves, the existence of which was later confirmed experimentally by Hertz. What matters here is that, for the wisdom of the period, these waves, precisely by being waves—the propagation of a medium's disturbances, according to the classical definition—necessarily required the existence of a corresponding material carrier. Honoring a long tradition, this carrier was called the ether. For no obvious reason, however, all efforts to pin down the ether's properties failed, making its character more and more elusive and mysterious. A crisis in Kuhn's (1962) sense of the term thus settled in with no obvious way out. This was the context in which Einstein published his paper "On the Electrodynamics of Moving Bodies" (1923, originally published in 1905), in which he introduced the STR. Stating it as simply as possible, this theory obviated any need for the ether, for it posited that electromagnetic waves manifested the hitherto totally unsuspected existence of an additional kind of physical entity, altogether different from particles and media, the electromagnetic *field*. The most prominent characteristic of the electromagnetic field is precisely that it is a kind of wave that *can* move by itself, without the support of any medium whatsoever, that is, in vacuum. The ether thus proves "superfluous." The existence of this entity had been hitherto totally unsuspected simply because it was literally inconceivable in the terms of classical mechanics. Accordingly, understanding the STR hinges on understanding, among other things, how such a weird entity can be conceived.

Why is the electromagnetic field inconceivable in the terms of classical mechanics? Simply because the electromagnetic waves manifesting the field's existence are characterized as waves, and a wave is classically *defined* as the propagation of a medium's disturbances. This is no ad hoc definition giving content to a technical term. This is a definition that is intended to capture—and does indeed capture—all our everyday experience with waves (ocean waves, sound waves, and so forth) in terms of a physical theory well established on all counts. In other words, the term *wave* merely names, with all the necessary mathematical niceties, a particular form of movement that we variously experience, the familiar movement constituted by the more or less periodic disturbances of a material medium. But all this is to say that the physical concept of a wave is *analytically* related to the physical concept of a medium. Hence the alleged existence of waves that can propagate in vacuum, in the absence of the material medium the propagation of whose disturbances they nevertheless are, involves a self-contradiction. It is like maintaining that some bachelors may be married males although we have defined the term *bachelor* as a never married male.

Yet the subsequent successes of Einstein's ideas have obliged physicists to come to terms with this contradiction and further articulate the corresponding novel paradigm in various directions. Elsewhere (Baltas 2004) I analyze how such contradictions arise in the history of science and what accommodating them entails. In particular, I maintain that the process begins with the introduction of a novel, highly imaginative conceptual distiction. This distinction is highly imaginative in that it reaches out of the contradiction, so to speak, going beyond the available grammatical space. Thus, this conceptual distinction inevitably appears to be impossibly nonsensical when it is introduced. Eventually, however, this leap into the ungrammatical proves to have the capacity to create a conceptual context around it, to widen the grammatical space available to the inquiry and thus allow the erection of a novel scientific theory carrying all the requisite normative force. Within the grammatical space thus widened, the contradiction disappears: the breakthrough establishes a logically consistent and grammatically coherent conceptual system centered on the imaginative distinction initially introduced. This is a novel conceptual system, radically different from the old.

Any scientific conceptual system of the sort at issue here (such as that of classical mechanics) harbors background "assumptions" that can be pictured as the grammatical "hinges" (Wittgenstein 1969, §§341–43) silently fastening our understanding of it. A crisis situation occurs when the process of inquiry is tripped up by some such "assumption." This is a crisis situation because nobody can comprehend why the current theory is not working: the work of these "assumptions" is to secure the understanding of the conceptual system implicated by operating from the background to guarantee its overall grammatical coherence. In that capacity, they and their role are necessarily taken fully for granted; researchers unthinkingly entertain them without the inkling of a suspicion that they could possibly be questioned. Even if somebody comes to state any of them explicitly, nothing dramatic happens, for such a statement merely reiterates the obvious. All this is to say that the grammatical possibility of questioning such "assumptions" cannot arise in normal[12] circumstances, while *resistance* to its arising is concomitantly at work.[13] For all these reasons, these beliefs cannot amount to proper assumptions.

As Wittgenstein makes clear in *On Certainty,* the existence and silent functioning of such "assumptions" is inescapable. In that work, however, emphasis lies on their positive role as the indispensable grammatical hinges that must stay put for the door of understanding to move. The additional point I am trying to make is that even though these background beliefs, these grammatical "hinges," are necessary for understanding, they occasionally constitute fundamental *obstacles* to understanding as well. Hence, to keep the same metaphor, the conceptual breakthrough of paradigm change and the concomitant grammatical break amounts to the realization that some of these hinges have rusted. This realization accompanies the introduction of the novel conceptual distinction, a distinction inconceivable within the grammatical space determined by the old hinges, the distinction around which the grammatical space is widened and the novel conceptual system is established. At the grammatical level, advancing this distinction and widening the grammatical space is tantamount to removing old hinges and replacing them with others, those securing the overall grammatical coherence of the novel conceptual system and thereby fastening our understanding of the conceptual breakthrough and its outcome. Grammatical "assumptions" in general are thus here to stay; insofar as we are linguistic animals, we are grammatical animals as well.

A few further points bear mentioning. First, the widened grammatical space can host a *re*interpretation of the old conceptual system and sometimes a translation of parts of the old conceptual system in terms of the new (the STR exemplifies this phenomenon). Hence, within the bounds set by the novel conceptual system, these parts—precisely as reinterpreted/translated—can remain useful to scientific practice. However, the reverse interpretation/translation is blocked: the old grammatical space cannot accommodate what the new allows. This makes the two paradigms *asymmetrical,* and the incommensurability between the two, in Kuhn's (1962) sense of the term, is based on this asymmetry. The phenomena of communication breakdown discussed in the literature are due precisely to this asymmetry: those who have not gone through the grammatical break inevitably take for granted the background beliefs fastening the understanding of the old paradigm.

Second, after the conceptual tumult has settled down and the new conceptual system has become established, the widened grammatical space subtending it is considered to have been instituted *for good*: on the one hand, all further inquiry is based on it; on the other hand, returning to the old grammatical space becomes grammatically impossible. For it is obviously grammatically impossible to push back into the background and reacquire blindness to the novel grammatical possibility that has been opened, the possibility that allowed us not only to come to terms with the crisis situation but also, and more significant, to promote successful research along the associated novel directions.

It is important to note, however, that this widening of the grammatical space does not greatly affect what happens outside the process of scientific inquiry. I noted that colloquial language does not grammatically reform itself in such cases, and colloquial understanding continues to go on more or less as before: the novel paradigm remains highly counterintuitive, requiring initiation as well as instruction for one to understand it properly. To achieve such understanding, prospective students must somehow reenact the experience scientists have undergone. This situation is of course not limited to the introduction of the STR or even to contemporary science; pedagogical research amply demonstrates that even physics graduates spontaneously conceive the world in pre–classical mechanics terms.[14] Therefore, to teach science effectively, one must take careful account of such unavoidable resistance; at the same time, popular presentations of science tend to flood the market precisely in order to appease it.

Moreover, after this widened grammatical space has proved its capacity to host novel, previously inconceivable scientific results and after the attendant grammatical peace has been instituted, practicing scientists tend to look condescendingly on the replaced paradigm: the old contradiction becomes retrospectively interpreted as a kind of oversight whose gripping power manifested only our failure to notice a possibility that was there all along. This is, on the one hand, our alleged failure to come up with the conceptual distinction at issue well before we actually did and, on the other, the shortcoming of our scientific past with respect to the glorious scientific present. The success of the breakthrough is taken to demonstrate that the possibility must have always been lying "out there," waiting to be dis-covered by the cleverest among us. The almost irresistible tendency of practicing scientists to espouse what has been dubbed the "Whig conception of history" finds its roots here, as well as some apparent reasons for its justification (Baltas 1994).

It should be clear that this appraisal mischaracterizes the relation between the scientific present and the scientific past and that no cleverness can be involved here. Coming to see a novel grammatical possibility in the way that, for example, Einstein did always involves allowing imagination free rein, but within scientific activity, imagination can neither be appealed to nor be relied on unless it it is backed up by the appropriative normative force—that is, by particularly solid scientific success, the kind of incontrovertible success required by the stringent criteria of this sober and highly disciplined activity. Imaginative proposals of all kinds may well be put forth in a crisis situation, but few of them will be justified by rock-hard results that force a widening of the grammatical space. The others remain mere flights of the imagination that, though perhaps admirable in themselves, are soon forgotten if they fail to provide a coherent and effective explanation for experimental results.

Note that to effectively widen the grammatical space and hence regain grammatical coherence as described is simultaneously to reinstate *logical* consistency: to come up with grammatical room for the novel conceptual distinction is concurrently to eliminate the logical contradiction. Putnam (2000) discusses a simple Wittgensteinian example that, though it concerns colloquial language in its everyday uses, may be suggestive of what befalls logic in paradigm change. Thus the logical standoff forced by the contradictory order "Come to the ball wearing and not wearing clothes!" can be overpowered by the addressee's imagining the possibility of wearing a fishnet. The grammatical space embracing the concept of wearing clothes is thus enlarged to accommodate a novel distinction, inconceivable except by an act of imagination and the new thought this act brings with it: before this act/thought enlarges the set of grammatical possibilities, the background-level hinges governing the uses of the concept at hand may lead us to assume that wearing or not wearing clothes has nothing to do with fishnets. By the same token, the logical contradiction disappears as such after it: once we arrive at the answer, "the *riddle* does not exist" (TLP 6.5).

What distinguishes scientific practice from attempts to solves such "riddles" is the requirement that the conceptual and experimental implications of the thought produced by the imaginative act be borne out not by colloquial language usage (which does *not* bear out novel scientific concepts in many cases) but by the workings of the world. What I have been calling the normative force of the imaginative act has its source here.

From Grammar to Logic and Back

The fact that even the most radical grammatical change cannot touch logic as such brings the *Tractatus* back into the picture.[15] For one thing, a successful challenge to an apparently self-evident claim cannot censure logic, for self-evidence is not the mark of logical propositions (TLP 5.4731, 6.1271). Logic "takes care of itself" (TLP 4.73), and thus it always finds a way to pass unscathed through the corresponding act of imagination. But more important, logic is not impugned even by an act of imagination that successfully defies—by exhibiting the capacity to muster and implement the requisite normative force—what, before the act, was logically secure. Again, a grammatical break that manages to overcome a bona fide contradiction institutes a widened grammatical space that is logically consistent. The new grammatical space has to bow to logic in exactly the same way that the old grammatical space did, which means that logic is reinstated. In short, no act of imagination whatsoever can ignore or disparage logic; "logic fills the world" (TLP 5.61), and therefore even the wildest imaginative fancy must remain worldly and hence logical: "It is clear that however different from the real world an imagined world might be, it must have something in common—a form—with the real world" (TLP 2.022).

It follows that the novel grammatical possibilities borne by widened grammatical spaces cannot be seen as implying novel *logical* possibilities. As "there can *never* be surprises in logic" (TLP 6.125), neither can there be inventions or discoveries in it. Logical possibilities cannot be created, constructed, or abstracted away, nor can they rest dormant, passively in the offing, waiting to be disclosed. Since "logic treats of every possibility, and all possibilities are its facts" (TLP 2.0121), considering logical possibilities as prone to any of these situations would be placing logic outside language, thought, and the world and putting it out of action. The conclusion is that, to the extent that the *Tractatus* takes up grammar at all, the grammatical "level" must remain distinct from the logical "level."

So does the *Tractatus* tackle grammar and its associated issues? In fact, it does, despite what a first reading might seem to suggest.

To begin with, the *Tractatus* contains an exact analogue of the previously discussed grammatical hinges: "the enormously complicated silent adjustments [allowing us] to understand colloquial language" (TLP 4.002).[16] These silent adjustments are presumably intended to capture the workings of that which Wittgenstein would later discuss as the grammatical "riverbed" of language (Wittengenstein 1969, §§96–99). Thus, if the hinges of that work are intended to capture what remains *grammatically fixed* in a given context of linguistic usage (a "language game"), then the adjustments of the *Tractatus* are intended, symmetrically, to capture the *grammatical leeway* permitting us to move grammatically from one linguistic context to another. In both cases Wittgenstein seems to mean what is commonly shared in the background by everybody participating in a linguistic community, even if his hinges tend to focus at a given linguistic (and not merely linguistic) context, while his adjustments tend to encompass the grammatical bedrock of colloquial language throughout.

Restricting the terminology to that of the *Tractatus*, then, we may say that to understand propositions in general, we are indebted "silently" (i.e., without being aware of their workings) precisely to those adjustments, which provide the necessary leeway for understanding hitherto unencountered propositions made by our linguistic peers as they advance these propositions in various and continuously changing circumstances or contexts. This leeway lets us recognize what is grammatical and what is not in any given circumstance; it also determines how we distinguish sense from nonsense and meaningfulness from meaninglessness. Consequently, to say that some grammatical hinges are removed in the passage from some paradigm to its successor is to say that some *re*adjustments have taken place in the part of the "riverbed" silently determining the sense and the meaning of the concepts involved.

Given this, Wittgenstein's claim that "man possesses the capacity of constructing languages in which any sense can be expressed" (TLP 4.002) can be taken as pointing at, among other things, the phenomenon of paradigm change, for "every sense" includes nonclassical senses as well. This is no mere speculation. The intellectual upheavals brought about by the developments in physics—at least after the publication of Einstein's relativity paper in 1905—clearly touched him, as the passage at TLP 6.341–342 demonstrates. In these uncharacteristically lengthy remarks, Wittgenstein uses the idea of (conceptual) "nets" or "networks" of different shapes (square, triangular, hexagonal, and so forth) covering a surface to discuss how different possible theories of mechanics—such as, presumably, classical mechanics and the STR—relate to the world, and he comments on the relation such networks bear to logic. The remarks are nested within the family of propositions from TLP 6.3 to TLP 6.4, where Wittgenstein lays out his understanding of natural science.

These remarks indicate that conceptual distinctions, conceptual upheavals, and the workings of the grammatical level are not Wittgenstein's prime concern in the *Tractatus*. Such matters engage only the "form of the clothes" (TLP 4.002) constituting colloquial language and involve only the ways and means for arriving at a "finer," "simpler," or "more accurate" (TLP 6.342) description of the world. But exploring and elaborating such issues is the business of science and not that of Wittgenstein's logico-*philosophical* treatise. For the *Tractatus* as a *logico*-philosophical treatise, the issue is that "the world . . . *can* be described in the particular way in which as a matter of fact it is described" (TLP 6.342, emphasis added) and no more.

What we may discover or invent as a matter of fact is irrelevant to that treatise, for we "must have to deal with what makes [such discovery or invention] possible" in the first place (TLP 5.555).[17]

Accordingly, to bring matters down to the level of purely logical rather than grammatical possibilities, Wittgenstein seeks to provide, at least in principle, "the one and only complete analysis" (TLP 3.25). At this level, each "elementary proposition" (TLP 4.221) consists simply of a "concatenation of names" (TLP 4.22), each of which stands for a "simple object" (TLP 2.02). At this level, the only level of purely logical possibilities, all the "enormously complicated silent adjustments" *have already been addressed.* Developments at the level of concepts (from "electromagnetic field" to "wearing clothes"), as well as the ways scientists handle the corresponding conceptual issues, are a different matter altogether, one concerning the "form of the clothes" constituting colloquial language.

The possibility of reinterpreting the *content* of a given proposition *p* (e.g., "wearing clothes") by an act of imagination so as to make a logical contradiction of the standard form "*p* and not-*p*" disappear seems to suggest that those silent adjustments are implicated in determining even whether ostensibly aboveboard contradictions such as the one at hand are real or can be only apparent. Wittgenstein accepts that "colloquial language [generally] disguises the thought" (TLP 4.002) and admits that "Russell's merit" was "to have shown that the apparent logical form of the proposition need not be its real form" (TLP 4.0031). However, the *Tractatus* does not make perfectly clear whether Wittgenstein thought that such a disguise or concealment of "real logical form" might implicate even bona fide contradictions or, equivalently, that the silent adjustments could extend so far that they determine whether contradictions—or analytic relations such as that defining the concept of a wave in classical mechanics—might not really be what they appear to be.

There is, moreover, no textual evidence indicating that the young Wittgenstein would have subscribed to what I have been forwarding in the last two sections, namely, that radical paradigm change could involve a *grammatical* dimension, that the attendant conceptual change could come to challenge an *analytic* relation holding between concepts of the old paradigm, and that the reinterpretation of those concepts in the context of the novel paradigm could come to reinstate logical consistency and grammatical coherence.

Be that as it may, the young Wittgenstein does seem to offer the requisite latitude to the silent adjustments in question. Among the tasks he assigned to "the one and only complete analysis" was that of doing away with all the "disguise" with which these adjustments adorn propositions and, concomitantly, that of eradicating *all possible ambiguity* regarding propositional content. That is, any contended proposition could be finally brought down to the level of "elementary propositions" that consist in mere "concatenations of names" (TLP 4.22). As I will show in chapter 6, elementary propositions are logically independent from one another (TLP 5.134), and as mere concatenations of names, they do not involve logical constants. Hence no elementary proposition can contradict or become implicated in an analytic relation with another. This is to repeat that at the level of the one and only complete analysis, all silent adjustments have been addressed and hence no issue of real versus apparent contradiction (or analytic relation) can be raised. By the same token, since the analysis in question is the only one possible, no radical conceptual change can possibly contest its outcome.

Wittgenstein explicitly considered this "complete analysis" to be possible, at least in principle, and asserted that it could deal with all *logical* issues associated with the *scientific* change of his times (I discuss this in chapter 8), because carrying out this analysis comes to display all logical possibilities regarding language—all possible ways by which "*any* sense can be expressed" (TLP 4.002, emphasis added)—and hence all logical possibilities concerning conceptual issues generally. The possiblitiy of the analysis in question is tantamount to the possibility that all facts of the world can be described at the conceptual level—the level of scientific theories—according to some "single plan" (TLP 6.343). This implies that the attainment of an all-inclusive scientific theory is logically possible, but it is logically possible, too, that more such all-inclusive theories ("single plans") may come about: the world "*can* be described more simply by one system of mechanics than by another" (TLP 6.342).

Wittgenstein's acknowledgment that more such "single plans" are logically possible marks the historical advantage he held over Spinoza. Spinoza was historically in no position to disengage the logical from the conceptual, which made envisaging things at the level of logical possibilities impossible for him; consequently, the factors of scientific change important for Wittgenstein (and *a fortiori* for Kuhn) were nonissues for him.

As Medina (2002) notes, however, at least one reason leading Wittgenstein to change his views in a way that tends to downplay logic to the profit of grammar is related, first, to his difficulty in formulating an operational principle that would allow the analysis leading to elementary propositions and, second, to the question whether it makes effective sense to countenance elementary propositions at all. This is to say, among other things, that the relation between logic and grammar is not as clear in the *Tractatus* as I am making it appear to be. I reserve a more detailed discussion of this relation for the final chapter, "Exodus."

Teacher and Student

Leaving scientists safely alone to do what they must do in a crisis situation, I now turn to the related situation of a physics teacher facing students eager to understand a radically novel theory, such as the STR. The idea is that Wittgenstein's position with respect to his readers might be profitably considered as analogous to that of the physics teacher facing these students. The similarity rests on the assumption that Wittgenstein intends to convey something radically different from the content a typical reader of the *Tractatus* might expect to find in a philosophical treatise and that such novelty can be expressed only in colloquial, even if "philosophical," language, so that grammar is inevitably implicated in one way or another. Nevertheless, a significant difference separates the two situations: in the first, the issue is coming to understand a particular something, namely, a well-established scientific theory; in the second, it is performing philosophical activity in a certain way.

We can see this analogy by asking certain questions: How can physicists start teaching the novel paradigm of the STR to students who are still entrenched in the old? How can we make them conceive the electromagnetic field, the disturbances of nothing that propagate by themselves within nothing? What kind of helpful picture can we draw to depict such an impossible state of affairs?

First of all, it would be impossible to make such students understand us simply by expounding the STR and the evidence for it. Since their system of concepts (that of classical mechanics) is not only self-consistent but also incommensurable with ours, we cannot straightforwardly prove our case as we could if we were sharing the same grammatical background with only the "adjustments" it allows. Our entry into the novel conceptual system has "readjusted" how we understand all the relevant concepts of both the old and the new appraoches, while the students have not yet made the passage; they have not yet understood the STR and hence not yet realized what talking from the vantage point of that theory might mean—such talk appears nonsensical to them—as well as what this might imply or entail. In such cases, proof, evidence, and all that are insufficient; *in themselves,* they are *powerless* to do the job, and this irrespective of the students' reasoning capacities. And as our everyday discursive practices demonstrate, when proof and evidence are thus powerless, we have to resort to elucidations in Frege's sense of the term.[18]

Second, to get these students to understand us is to make them rationally forgo their conceptual system and rationally join ours. No argument from authority and no kind of psychological coercion should come into play as means of persuasion. As Spinoza would have it too, only the intellect is to be involved. To reach the students intellectually, then, the elucidations we forward should be such as to build a kind of bridge or ladder between us and them, one that can make the necessary concessions to their failure of understanding us, that enables us to meet them halfway.[19] To be a means for connecting us, this ladder must comprise elements from both systems and thus employ locutions and pictures somehow common to both. Since each system is self-consistent and the two are incommensurable, the ladder connecting the two will necessarily constitute nonsense from the point of view of either.[20] Elucidation and nonsense thus seem to go hand in hand.

Third, the aim of meeting halfway on the ladder presupposes a certain kind of "good willing" (TLP 6.43) on the sudents' part. Nobody can teach anything to anyone who wills badly in the relevant respect, that is, somebody who is unwilling to learn. Obviously we cannot require the students to have already entertained the ideas we are trying to teach them, as Wittgenstein demands of his readers (TLP Pr ¶1), for if they had, they would have no need of us, but we are nevertheless compelled to require them to have some inkling of similar thoughts, even if only in the form assenting to the project of learning what we are trying to teach. And this requirement arises not merely because it is the only way in which the students can respond to the "good will" *we* lay out in our effort to teach them but mainly because it is a sine qua non of all teaching. Ethical intent presides over the activity of teaching as well.

Given all this, consider a case that may provide a perfect example for the kind of answers I am seeking. Addressing himself to laypersons—that is, to prospective students of the sort at hand—Stephen Weinberg, a Nobel laureate in physics and grand master of exposition, defines the electromagnetic field as "a taut membrane without the membrane" (Weinberg 1977, 17).

This obviously self-destroying "definition" provides exactly the kind of ladder needed in these situations: its first part refers to classical mechanics (taut membranes are legitimate objects of that theory), while the second sits squarely within the STR. The fact that the ostensible definition destroys itself points at the incommensurability of the two conceptual systems, while the fact that it creates some kind of "picture" helps elucidate the concept at issue, even if this picture is impossible: as is suggested by the notion of the vanished Chesire Cat's enduring smile (a smile, that is, without a cat), we may well have something in mind when trying to picture an elusive taut membrane that does not exist. Obviously, when the characterization is considered in itself, in no relation to the purpose for formulating it in the context at hand, its status as a self-contradiction makes it nonsensical: it *precludes* by what it is saying what it *presupposes* in order to convey what it says, and hence somebody altogether foreign to physics can make no heads or tails of it.

Weinberg's definition thus amounts to constructing one fundamental rung of the necessary ladder. This is an elucidatory rung despite its being nonsensical, or rather, it is elucidatory because it is nonsensical. Its construction amounts to a judicious use of nonsense, with Weinberg employing it both in conformity to his purpose in the context at hand and with the efficiency required for decisively striking what he is targeting. Weinberg's nonsensical definition indeed manages to hit the nail on the head, something Wittgenstein hopes his own project to accomplish (TLP Pr ¶7).

This can perhaps be said more clearly. In being self-contradictory, Weinberg's definition lays bare the impossibility of understanding the STR in the terms of classical mechanics; indeed, it self-destructs precisely by referring to them. It is this self-admitted failure, however, that makes the characterization effective in the given context: only by openly exhibiting, or "*showing*," that it is impossible to come to understand the concept of the electromagnetic field in terms of classical concepts can the definition pave the way for properly understanding the new concept. Hence, to make its point, the definition *should* destroy itself, and should do so *thoroughly,* to the point of becoming nonsensical. In this way, and only in this way, its failure becomes the condition for its success. To say the same thing differently, by its very enunciation in the given context, Weinberg's "definition" in and by itself performs the particular kind of not understanding to which unenlightened students should submit if they are to understand the new concept. Such acknowledged submission to not understanding (the new concepts in terms of the old) forms the first step toward properly understanding the new concept and thus the workings of the novel paradigm in its entirety. The students must force down the resistance inevitably arising when they hear the nonsense we are advancing. In thus intentionally submitting to such nonsense, in working against their own understanding, the students manifest the appropriate "good will" with respect to our efforts to teach them.

The picture proposed by Weinberg's definition aims at building and conveying for students a kind of picture, or rather, instilling some kind of intuition in them. Once the definition has gotten through, students will be able to go on building intuitions *internal* to the conceptual (and experimental) workings of the STR, intuitions of the sort physicists use and discuss all the time,[21] "intuitions [that do not arise ex nihilo but] are supplied by language itself" (TLP 6.233).[22] The taut membrane without the membrane will then remain the fleeting picture/no picture that bridged the students' way into the STR and helped them understand it. Hence the ladder in question not only can but also should be thrown away, as Wittgenstein recommends of his own work (TLP 6.54), after students have entered into the new paradigm and understood the novel theory. They can and should throw it away when they no longer need to visualize or intuit the electromagnetic field in inappropriate terms, that is, in terms *external* to the physical theory in question.

To use a more martial metaphor, the ladder I have in mind could alternatively be thought of as a kind of siege engine designed to exert pressure on the background "adjustments" that have left no grammatical room for understanding the STR. The aim is to make students proceed to the appropriate readjustment, and the means is our judicious employment of nonsense. Pedagogical success amounts to penetrating the students' conceptual systems, opening them up and taking them from the inside, as it were, by flooding them with ours. Such success then makes the point we want to make emerge and stand sharply out of the background, simultaneously inducing a eureka experience in the students. This is exhibited by an exhilarating cry of "Yes, now I *see!*" by which students mean that they see both the part of the world involved and the concepts they have been entertaining under a new light provided by our own conceptual system, that of the STR. At this point, we have fully flooded their system with ours. The sought after readjustment has taken place, we have *shown* the students what they now *see,* and their understandings can move freely within the grammatical space thus opened. Such understanding amounts to seeing "the world rightly" by the lights of the STR.

Coming to feel the effect of Weinberg's definition fully, moreover, is *sufficient* for becoming clear *on all aspects* of the issue at hand.[23] Once the picture of the taut membrane without the membrane gets through, students see several things in a flash and all at once (i.e., in a eureka moment): the concept we were asking them to understand, the "obstacles"[24] that had been preventing such understanding before—namely, the background adjustments whose silent operation let them understand classical mechanics but left no room for understanding the novel concept—and, thereby, the reason we were compelled to inflict nonsense on them so as to help them open the grammatical space capable of hosting the novel concept. In other words, though they were not talking nonsense when resisting us, for they were simply holding fast to the far from nonsensical classical mechanics, they needed only feel the full force of *our* nonsensical definition to understand that this was *the only way* to get them to understand the contended concept.

Wittgenstein's Body

Everybody who has undergone a eureka experience knows it to be a joyful liberating experience that one feels through one's whole body. Various kinds of jubilatory bodily movements often demonstrate the pleasure felt—to the extent that context and social conventions allow the corresponding free rein—while the characteristic cry manifesting this pleasure ("Yes, I now *see!*") points at a feat of the particular sense organ that seems to connect to understanding most spontaneously. Wittgenstein himself makes exactly the same appeal at the end of his work: understanding the *Tractatus* amounts to *seeing* the world rightly. But the eye is not the only sense organ highlighted in connection to understanding. I have just talked of "silent adjustments," thereby pointing at both the mouth and the ear, with the latter especially salient when difficulties in communication are at issue: we are accused of being *deaf* to what the other is trying to tell us, and incommensurability in cases of scientific exchange is often described as "scientists talking past one another." It seems that for Wittgenstein, colloquial expressions such as these cannot be merely metaphorical, for overarching them is his strong claim that "colloquial language is *part of the human organism* and is no less complicated than it" (TLP 4.002, emphasis added).

If colloquial language is indeed part of the human organism, then understanding, insofar as it involves language, must involve the part of the organism in question, that is, some aspect or capacity of the human body, and the eureka experience makes such involvement of the body fully manifest: on the one hand, the jubilatory bodily movements induced by deep understanding directly display the pleasure felt by the body; on the other hand, appealing to our sense organs to verbally express the experience of such understanding (the "I now *see!*" of the eureka experience) points metonymically at our body's involvement therein. More generally, all linguistic expressions that appeal to our sense organs in connection to understanding can be taken as employing metonyms of the body and hence as expressing what our body feels—joy or frustration, pleasure or pain, in Spinoza's terminology—when we understand or when we fail to understand. It follows that on Wittgenstein's all-encompassing claim, reference to our sense organs in connection to understanding is not at all gratuitous.

The fact that the experience of deep understanding occurs when some grammatical readjustment has taken place in the background implies that this background, the grammatical "riverbed" on which our understanding and our expert usage of language rests, is intimately related to the part of the human organism constituting colloquial language. The grammatical adjustments and readjustments I have been discussing ought to amount, then, to some kind of bodily function.[25] If this is the case, we can understand why the feeling induced by opening a new grammatical possibility is akin to the feeling of acquiring, as it were, a new limb, a new capacity for movement, as well as why we tend to retrospectively perceive the previous state as one of having been in the grips of some kind of cramp. The pleasure felt in such occasions is of a kind with the deep contentment we feel when we are aware of being actively at one with ourselves, with our bodies and our minds indivisibly linked and working in perfect unison.

This could perhaps be taken one step further. If colloquial language and the grammar supporting it make up a part of the human organism, then the human body is somehow inhabited by language. One's body is determined in its functions and in its actions, if not by complete propositions, then at least by the "bits" of meaning that Wittgenstein calls "expressions" or "symbols" (TLP 3.31). Psychoanalysis may help show how such a determination might work, even though I can present here only a rough compendium of the principles guiding this discipline as they might be interwoven with Wittgenstein's ideas.

According to Lacan's conception of psychoanalysis, at least, if the "sign is the part of the symbol perceptible by the senses" (TLP 3.32)—more or less what Lacan, following Saussure, calls the "signifier"[26]—then it somehow acts on our bodies when it is heard (or uttered) and affects them in the corresponding respect. Signs do not float freely by themselves but are put in action to impart their corresponding "symbols" (TLP 3.31)—the sense attached to them (Lacan's and Saussure's "signified")—while in some cases such sign-symbol dyads are laden with telling emotional content. Now in cases where this emotional content cannot be handled adequately, when we suffer trauma, the symbol becomes "repressed" and the sign, detached from its symbol, attaches somehow to a bodily organ or a bodily function affecting its normal operation. Repression hides the initiating cause of such dysfunctions from conscious view; psychoanalysis calls such dysfunctions "symptoms."

On this basis, the aim of clinical psychoanalysis is to overcome repression and reconnect sign and symbol (or symbols: a sign may lie at the intersection of various chains of signification, being subject to the corresponding "*over*determination")[27] at the level of full conscious awareness because only then can the symptom disappear for good. When this happens, one attains a feeling of liberation akin to that of a eureka experience. And this is as it should be: according to psychoanalysis, body and mind form an undivided whole, and hence merely knowing psychoanalytic theory and grasping dispassionately, at the purely intellectual level, what the causes of one's symptom might be will not suffice to rid oneself of it. Grasping at the purely intellectual level should be distinguished from the kind of deep understanding felt in a eureka experience—as expressed by Lacan's "*parole pleine*" (Lacan 2006b)—which involves the whole body.

More generally put, Lacanian psychoanalysis is particularly telling in the present context, for it emphasizes the *constitutive role* language plays in turning an animate physical body, merely endowed with the requisite physiological capacities, into a human subject, that is, in making it put these capacities into action and become properly a talking he or she by simultaneously entering the social order. In other words, language is the agent organizing the merely physical human body into a *socialized* (and hence gendered) human body (Althusser 1976). Though this inference requires a separate treatment in its own right, we might say that the human body conceived in this way appears as identical to the human body for which McDowell's (1996) "second nature" calls.

Wittgenstein's attitude toward psychoanalysis seems to have been ambiguous (Bouveresse 1995; Lear 1999; Heaton 2000), and nowhere does he discuss the human body as such. Nonetheless, life acquires a prominent position in the *Tractatus,* being equated to nothing less than the world: "The world and life are one" (TLP 5.621). Now since only human bodies can bear life in this sense, we might say that the human body with its different organs and its multiform capacities is continuously hovering in Wittgenstein's mind even if it does not become a subject of investigation in its own right. Evidence to that effect is provided by Wittgenstein's later work, where bodily gestures acquire a pride of place in the "language games" Wittgenstein discusses, while the riverbed of language is provided by "forms of *life.*" This resonates with his view that colloquial language makes up part of the human organism, thus specifying a thread linking the *Tractatus* to Wittgenstein's later work. If the idea of the human body, or rather of *his own* body,[28] is (to use standard psychoanlytic terminology) "repressed" in Wittgenstein's thought, his writings manifest the "return of the repressed" at their perhaps most fundamental junctures.

Telling Nonsense

Note that Weinberg's elucidatory "definition" is not a bona fide proposition. Its nonsensical character renders it incapable not only of being true or false but also of being supported by arguments, explained or analyzed further in its own right, or even explicated otherwise than through mere paraphrase. In the given teaching context, it either does the job for which it was designed or it doesn't. That is, its enunciation must be seen as a kind of *action* performed within the activity of teaching, as a move in the strategy we deploy to make our students understand the novel concept of the electromagnetic field. Its value can be assessed only in this respect. That is, the only value it can bear is *effectiveness*: to the extent that it fulfills the purpose for which it was designed, it has worked as what we might call *telling* nonsense.

Of course, not all nonsense can be telling in this sense, and no hard and fast rules exist to specify which particular piece of nonsense can be telling in any given and always singular context. Imagination, therefore, should necessarily be called on in the ways I have tried to describe. The need for such an appeal to the singularity of imagination is strengthened by the fact that no two students understand classical mechanics in exactly the same manner, and usually both students and teachers suffer from blind spots in the understanding, though located at different places for different people.

Even so, nonsense is rarely if ever totally pointless or altogether devoid of purpose. From the nonsense of Lewis Carroll to that of Baron Münchhausen, from the theater of the absurd to the "philosophical" stories of Borges, from the nonsensical languages invented by children during their games to the fundamental stipulation of psychoanalytic treatment, according to which one has to say whatever comes into mind, no matter how inconsequential, silly, or absurd, nonsense has *a point* to make, *does* something to us, *affects* us in various ways. From this point of view, our dealings in colloquial language are largely dealings with nonsense. If we were not straitjacketed into the reigning "conception" of rationality, we could perhaps even risk an oxymoron: in many circumstances, our dealings with nonsense saturate it with meaningfulness. The way nonsense affects us, however, becoming thus performatively meaningful, always depends heavily on our relations to various contexts, linguistic and otherwise. There is no general category "nonsense" that can be divided in the abstract, independent of context and purpose, between telling nonsense and pointless or idle nonsense.

As the overall content of "The Search for Unity: Notes for a History of Quantum Field Theory" attests, Weinberg's elucidatory and thereby nonsensical definition was not an isolated flash of inspiration; it formed part of his pedagogical strategy, whose various steps were probably well laid out in his thought beforehand, and strategy will provide the next focus of discussion. I will try to sketch the particular line of attack that Spinoza and Wittgenstein pursue in order to fulfill the purpose driving their work and to attain the end they intend. I will, moreover, persist in focusing on the *Tractatus* at the expense of the *Ethics,* though without losing view of the similarities between the strategies the two pursue.

CHAPTER FOUR

Strategies

> It is my ambition to say in ten sentences what everyone else says in a whole book—what everyone else does *not* say in a whole book.
>
> —Friedrich Nietzsche, *Twilight of the Idols*, "Skirmishes of an Untimely Man," §51

THE DIFFERENCE IN THE philosophical spaces available to Spinoza and Wittgenstein gave Spinoza's strategy a narrower range of possibilities for unfolding; the wider range Wittgenstein enjoyed let him pursue a more intricate strategy. For this reason, understanding Wittgenstein's strategy may help in understanding the perspective of radical immanence overall and thus provides a useful starting point. And even if Spinoza's strategy appears straightforward enough, a consideration of Wittgenstein might allow a second, more incisive look at that strategy, one that lies below the surface of the *Ethics* and thus provides additional means for situating the constraints that historical context imposed on its composition. Obviously, this second look must come not only after the first but also after the examination of Wittgenstein's strategy.

Spinoza's Strategy: First Round

Spinoza's strategy was aimed at the composition of a true philosophical theory complete in all relevant respects, at least in the essentials thereof. To promote his strategy and compose his theory, Spinoza had to rely on his intellectual forces alone. Descartes had already shown the unlimited power of individual human reason and established that in principle everybody shares this power. Nonetheless, Cartesian philosophy had not prevented Spinoza's contemporaries, Descartes included, from continuing to bathe in confusion, despite the extraordinary feats of the period. Hence, apart from borrowing from them whatever he thought necessary for bringing his strategy to fruition, Spinoza could not generally rely on the work of his contemporaries—and the same applies, a fortiori, to the work of his philosophical ancestors, for irrespective of their philosophical achievements, they were in no position to take into account the radically novel ways of understanding the world marking Spinoza's times.

If Spinoza was to conclusively attain his goal, the strategy leading to it would have to be deployed with the highest standard of intellectual rigor, which at the time was that of the geometrical order. This order could both safeguard the outcome of Spinoza's toil from the predictable objections and misunderstandings and allow it to be taught systematically. The *Ethics,* the achieved goal of Spinoza's strategy, displays all this openly, for everybody to see. Thus Spinoza provides definitions, axioms, postulates, lemmas, proofs, corollaries—all the works of pure deductive reasoning—as strictly interconnected by the geometrical order. Concomitantly, he disperses prefaces, appendices, scholia, and explications throughout the *Ethics,* clarifying in more colloquial terms what could well be hidden from view by the abstract character of deductive reasoning, as well as circumscribing the dominant confusions and dissipating them by bringing them down to their sources. On the face of it, Spinoza's strategy appears to have resulted in the straightforward com-position of a philosophical theory of the purest, most rigorous, and most complete kind. At least on a first round of examination, everything concerning this theory and the strategy leading to it comes out as it should, with no surprises lurking in its shadows.

Wittgenstein's Position

Ironically in the present context, if there is anyone who could consistently dis-miss Spinoza's theory out of hand, it is Wittgenstein: given his view that philosophy aims merely at the clarification of thought, so that philosophical theories, which arise from the "misunderstanding of the logic of our language" (TLP Pr ¶2), boil down to nonsense, the *Ethics's* candid self-presentation as a fully fledged philo-sophical theory appears sufficient to justify him in dismissing it.

Here as elsewhere, however, things are less simple than they initially appear to be. Wittgenstein cannot proffer this dismissive charge, at least out of hand, for he openly pleads guilty to uttering nonsense in his own work;[1] in fact, he thoroughly accepts the charge by admitting that he has been talking nonsense throughout the *Tractatus.* Here is the penultimate remark of that work:

> My propositions are elucidatory in this way: he who understands me finally recognizes them as nonsensical, when he has climbed out through them, on them, over them. (He must, so to speak, throw away the ladder after he has climbed up on it.)
>
> He must surmount those propositions; then he sees the world rightly. (TLP 6.54)

This remark summarizes the strategy Wittgenstein adopts to achieve his goal, that is, to reach the point where he and all his conscientious readers can "see the world rightly." What is most noteworthy here is not the metaphor of the ladder. This could apply equally well to Spinoza's strategy: composing any bona fide theory may well be viewed as building a ladder of propositions, and coming to understand that theory may be seen as climbing up this ladder (De Dijn 2004), as I have already discussed with respect to the STR. But Wittgenstein's strategy differentiates itself sharply in its characterization of the ladder's rungs. In accordance with TLP 4.112, where he says that *any* philosophical work consists "essentially of elucidations," Wittgenstein deems his own philosophical propositions as "elucidatory." But then he goes on to maintain that these propositions, precisely as elucidatory, will ultimately be recognized as nonsensical. This is surprising enough outside the kinds of teaching activity discussed in the previous chapter, but there is more: the propositions in question will be finally recognized as nonsensical not by anybody who happens to read the *Tractatus* (who, by the same token, is *not* expected to recognize them as nonsensical) but only by those who understand the author of those propositions, that is, *Wittgenstein himself*.[2]

Understanding Wittgenstein himself in this context means understanding the purpose that drove him in writing the *Tractatus,* the purpose displayed by the activity of composing that work. Understanding Wittgenstein's strategy, therefore, amounts to understanding how the purpose at issue is displayed in his work and, concomitantly, what *elucidation* and *nonsense,* the key nouns Wittgenstein uses to characterize his strategy, mean and can contribute in this respect. As the ongoing discussions on the issue attest, this is an arduous undertaking.

Thinking of Strategy

Recall the assessment of Weinberg's "definition" of the electromagnetic field as part of his pedagogical strategy, one whose consecutive steps were probably well laid out in his preparatory thought. Using this as a springboard, we might first ask what strategy in general encompasses and how it can be cashed out, as well as whether a strategy can be laid out in thought before it is effectively undertaken. This last question, whether a strategy can be thought *of* in the first place, arises from strategy's intimate connection with purpose: as was discussed in chapter 2, purpose cannot inhabit the world and cannot be a thought in the proper sense of the term, and hence it seems natural to contemplate strategy in such respects.

As regards things in the world—in the Attribute of Extension, according to Spinoza—a strategy is presumed to be an effective way for attaining an intended goal in terms of a succession of tactical steps, or actions, designed to secure intermediate goals. If each such action is executed successfully, then, other things being equal, their cumulative effects are expected to lead to the intended final goal. Therefore, to lay out a strategy in thought prior to its implementation is to think what the intended effect of each tactical step should be, to think each such step as a causal connection between the action to be undertaken and its intended end, and to think all these actions as occurring consecutively in the progression of the strategy.

Now as Wittgenstein has it—and Spinoza would agree here and largely in what follows—"we make to ourselves pictures of facts" (TLP 2.1), pictures that, to the extent that they are not "disguised" by the "clothes" of colloquial language (TLP 4.002) and are thus "clear and delimited sharply" (TLP 4.112), are "logical pictures" of those facts and hence "thoughts" (TLP 3). A fact is "thinkable" if we can make a "picture of it" (TLP 3.001), and hence we *can* think the intended ends of the actions in a strategy by picturing the facts that should result from these actions. If an end is thinkable, it is possible (TLP 3.02); hence, it can result causally from our action.

Nevertheless, since only "what can be described"—or pictured in the above sense—"*can* happen," while what "is excluded by the law of causality cannot be described" (TLP 6.362), we are allowed to conceive of the tactical step as causally bringing about its intended end and compelled to think the relation between this step and that end in causal terms. To think the whole causal sequence of the tactical steps that make up the strategy is to *order* the corresponding actions according to the work we intend them to do. It is this order that makes up the strategy in the narrow sense.

Nothing guarantees that any of the tactical steps will be effective in the way we intend, so that in laying out a strategy in thought, we cannot know whether the intended goal will eventuate: "We cannot [even] *know* whether the sun will rise up tomorrow" (TLP 6.36311). Such lack of knowledge is related to our freedom to choose which particular action to undertake in willing to achieve our goal, as well as which "future action, unknown to us now" (TLP 5.1362), may eventually be needed. Concurrently, such lack of knowledge implies that it is impossible to antecedently lay out in thought all the factors that may determine the effectiveness of our actions and hence of the strategies these make up. Whether "the nail has been hit on the head" (TLP Pr ¶7) can be determined only after the event.

Now all this can be transposed, *mutatis mutandis,* to the *discursive* strategies deployed by the *Ethics* and the *Tractatus,* once we take into account that discursive strategies involve reasons rather than causes. Because language carries a performative function, propositions can affect us in various ways; they can, for example, produce changes in the ways we understand both ourselves and things at large. Hence a discursive strategy made of propositions formulated, interconnected, and ordered in particular ways can come to change things in us, and this change may, among other things, come to dispel confusions that were hindering our understanding of some matter before we encountered the propositions making up the discursive strategy. In other words, the discursive strategy sets up and articulates *reasons* for undergoing such a change, reasons that, if effective, will be seen as justifying the change. The strategy's success consists in this, though whether we can pin down why we underwent the change or even whether we are justified in having undergone it is a different matter.

Such similarities between discursive strategies and strategies deployed within the realm of things in nature (within the Attribute of Extension) seem to suggest that reasons and causes might be less foreign to each other than they are usually considered to be. Spinoza expressly holds that "the order and connection of ideas [and thus of reasons in general] is the same as the order and connection of things" (E IIp7) and thus of causes in general. The discussion in chapter 8 will show that Wittgenstein would not find much to object to here.

The tactical steps of the discursive strategy should obviously be ordered in accordance with their intended performative function, that is, in accordance with both the final goal of the strategy and their expected contribution to the successful execution of the strategy. Wittgenstein openly acknowledges this to be the case with regard to the strategy of the *Tractatus*: if the final goal of that strategy is indeed logical (the logical annihilation of the vantage point of God), then the ordering of the tactical steps (i.e., the constituent propositions) making up the strategy must proceed according to their "*logical* importance" (TLP 1n, emphasis added), exactly as Wittgenstein has it. The logical importance in question is of course identical to the *strategic* importance of these propositions. And since strategic importance in the case of discursive strategies is performative importance, the logical ordering in question is an ordering along the performative axis. Numbering these propositions according to their logical importance thus openly displays the sequence of the tactical steps making up the strategy of the *Tractatus* and so lays bare this strategy itself.

All this helps explain why purpose can be displayed only in action. In the kind of pedagogical contexts discussed previously, we saw that the instrument forwarding a strategy—and hence the associated purpose—at the linguistic level is telling nonsense. Once the requisite "good will" is at work, the teacher's use of telling nonsense constitutes an *action* to the pedagogical effect intended, one that employs linguistic terms to display both its character and the purpose driving it. Telling nonsense is thus the self-display of both action and purpose at the linguistic level. Telling nonsense is the means colloquial language supplies us so that we can handle action, strategy, and purpose at the linguistic level. And this is as it should be. Even if "colloquial language disguises thought" as such, "the purposes [*Zwecken*]" that drove the "construction of the external form of [the corresponding] clothes" (TLP 4.002) cannot differ greatly from all the other purposes that drive our activities— our actions and the strategies these form—hooking us onto the world. The "external clothes" in question could have been constructed only in a way that would have endowed us with the requisite means for managing purpose at the linguistic level, even if purpose may not inhabit the world or be a thought properly speaking.

My Strategy

The goal of Wittgenstein's strategy, then, is that of effectively demonstrating (*nachzuweisen* [TLP 6.53]), in a definitive and unassailable way, the *logical impossibility* of a vantage point outside thought, language, or the world. To achieve this goal is to demonstrate, in a way equally unassailable and definitive, that only the philosophical perspective of radical immanence is logically possible. It follows that philosophical activity can operate with logical consistency only "inside" this perspective. To the extent that it does not (but also, for the reasons indicated, to the extent that it does), the philosophical propositions advanced by this activity "are not false but nonsensical" (TLP 4.003). "The whole of philosophy is full of confusions" (TLP 3.324) precisely because philosophers have not realized that only the perspective of radical immanence can be logically possible and thus have failed to accept the consequences this has for the status of philosophical propositions generally.

By the same token, Wittgenstein's generic opponent is the philosopher who advances nonsensical propositions from "outside" the perspective of radical immanence. Since philosophy is "*full* of confusions," this category includes most philosophers. Most philosophers have not understood that they are advancing nonsensical propositions from outside the single logically possible philosophical perspective and are immersed in confusion for this reason. Since we have no reason to believe that philosophers are less intelligent than most people, it must be the perspective of radical immanence itself that impedes philosophers' access to it, which implies that this perspective differs radically from the points of view that most philosophers (and hence Wittgenstein's generic philosophical opponent) tend to espouse, as radically as, say, the STR differs from classical mechanics.

In the previous chapter I focused on scientific paradigm change precisely because I take the perspective of radical immanence to differ this radically from the perspective of Wittgenstein's generic philosophical opponent. The main task of that chapter was to clarify the notion of radical paradigm change though a conception of grammar foreign, at least in those terms, to the *Tractatus,* thereby providing a starting point for understanding that book's strategy. It is now time to draw the lessons of that examination. My strategy for understanding the strategy of the *Tractatus* starts by trying to carry the analogy with radical paradigm change in science as far as it can take us.

To exploit the analogy, I will initially place myself with respect to the *Tractatus* in more or less the same position the student eager to understand the STR places him- or herself with respect to a teacher of that theory. I will be taking TLP 6.54, the proposition summarizing Wittgenstein's description of his own strategy, as inviting us to read in the *Tractatus* recognizable forms of a telling nonsense that somehow adds up to an unassailable and definitive demonstration that the perspective of radical immanence is logically inescapable.

At the same time, however, the propositions I have formulated in trying to define the perspective of radical immanence are themselves nonsensical. An apparent proposition to the effect that the existence of a vantage point outside the world (or thought or language) is logically impossible could make sense only if it were possible to somehow *imagine* such a vantage point, even if only to rule it out immediately afterward. The reason is not hard to find. To claim, vaguely intimate, or just assume the impossibility of something is to endow this "something" with at least a modicum of imaginary consistency. This is the minimal consistency offering us the necessary foothold for distinguishing what is impossible from what is possible and thus for understanding why it is impossible in the first place, the minimal consistency permitting us to conceive of this impossibility by a stretch of the imagination. For example, a cat's smile without the cat does offer the kind of fleeting picture/no picture that can allow us to go on reading Carroll's novel with enjoyment. The same holds not just for chimeras, winged horses, and all kinds of science fiction constructs but also for twins that come to have different ages, for parallel lines that cross each other, for continuous curves with no tangents, and even for square circles.

By Wittgenstein's own lights, however, if the impossibility at issue is logical, it must envelop imagination as well; he maintains, "It is clear that however different from the real world an imagined world might be, it must have something in common—a form—with the real world" (TLP 2.022). Now since this shared form is nothing but the logical possibility of facts being structured in a certain way—"the form is the possibility of the structure" (TLP 2.033)—imagining the existence of this vantage point is by the same token *presupposing its logical possibility*. Wittgenstein expressly holds that the thinkable is what "we can imagine" (TLP 3.001) and that what is thus "thinkable is also [logically] possible" (TLP 3.02). Hence the quasi proposition at issue can make sense only by presupposing the logical possibility it rules out by what it upholds, while it simultaneously rules out by what it upholds the presupposition allowing it to make the kind of sense it does. Thus, although this may not be immediately apparent on the face of it, the putative proposition in question has a grammatical structure that looks identical to that of Weinberg's "definition": the proposition appears as self-destructive in a similar way, and so it cannot be a proper proposition.

We might be tempted to inspect the impossibility in question by taking our cue from TLP 3.032, which underlines the impossibility of imagining the geometrical coordinates of a nonexistent point, and conducting some kind of thought experiment to explore what such an external vantage point might possibly be and "where" it could be located. If we do so, however, we will soon discover that we unfailingly land in total incoherence and that our thoughts on the matter disintegrate, showing themselves to have not been thoughts at all. The propositions defining the perspective of radical immanence—and by the same token those defining the approach of Wittgenstein's generic philosophical opponent—appear incurably nonsensical in that they cannot be saved from self-destruction even in the wildest imaginative fancy.

We should, however, mistrust this sense of disintegration. Indeed, we should have avoided succumbing to the temptation leading to it, for the temptation muddles the issue by disregarding the peculiarities of the distinction between grammatical and logical impossibilities broached in the preceding chapter. In that chapter, I showed that it would be impossible to conceive of waves traveling in the absence of a material medium, for waves were by definition the propagation of disturbances in such a medium, or to think of how one might attend a ball while wearing and not wearing clothes *before* the novel thought put forth by the corresponding act of imagination has done its job. Trying to vaguely imagine ways of overcoming this analytic relation or that contradiction while confined in the old grammatical space, and thus still lacking the novel thought that could widen it, would make us experience a disintegration of our thoughts on matters of waves and clothing.

The lesson should be clear. Before a revolutionary grammatical leap comes to mark the distinction between the old grammatical impossibility and the novel logical possibility (by widening the grammatical space and reinstating logical consistency), there is simply no way for us to *generally* distinguish grammatical from logical impossibilities. Before the event, this distinction can be one of principle only, informed as to its own possibility by the radical grammatical changes that have occurred in history but remaining abstract, totally unspecified, and in this sense perfectly idle. Only grammatical leaps can so to speak materialize the distinction, but they do so only after the event. Succumbing to the temptation in question discounts the distinction between grammatical impossibility, which is by definition an ex post facto possibility, and logical impossibility, which is not, making us take them as of the same nature.

Nonetheless, Wittgenstein views grammatical possibilities or impossibilities as concerning how colloquial language disguises not logic as such but thought (TLP 4.002). For this reason, grammatical impossibilities *can* be overcome—only in principle, but *thoroughly*—through "the one and only complete analysis" (TLP 3.25), which brings the implicated propositions down to the level of Wittgenstein's "elementary propositions." At this level no conceptual or grammatical issue can arise, for all possible ambiguities in propositional content have already been clarified, all conceptual issues have already been resolved, and all possible grammatical leaps that might occur have already been addressed.

Let me elaborate. Since any imagined world will necessarily share "a form with the real world" (TLP 2.022), and sharing a form in this sense is sharing logical possibilities, the one and only complete analysis not only displays all logical possibilities but also takes care of all possible acts of imagination. Therefore at the "bottom" level reached by this analysis, everything imaginable has in principle been included, with nothing else left to imagine. It follows that to hold the impossibility of the external point to be logical rather than grammatical is to hold that entertaining the possibility of its existence is overstepping the limits of the imaginable. And since, on TLP 3.001, the imaginable coincides with the thinkable, this impossibility oversteps the limits of what can be thought at all. No act of imagination whatsoever can turn this impossibility into an ex post facto possibility.

It follows that the two ostensible definitions previously discussed, that of the perspective of radical immanence and Weinberg's of the electromagnetic field, although no different in grammatical structure and therefore apparently nonsensical for the same reason, have very different duties to perform: they are implicated in different contexts, they are advanced for different purposes, they aim at different targets, and thus they come to function overall in wholly different manners. On the one hand, Weinberg's definition is a self-destructive one of an actual something (the electromagnetic field) in the context of teaching; its purpose is to construct a rung of the ladder allowing someone learning the STR to conceive this something given the impossibility of conceiving it in the terms of the old grammatical space; its target is the "silent adjustments" at work in the old grammatical space; and its self-destructive character, which appears on its face, is instrumental to its success. On the other hand, my definition of the perspective of radical immanence is not one of a "something" at all but only of a logical impossibility in the context of philosophical activity; its purpose is to circumscribe a particular philosophical perspective; its target, if any, remains for the moment in the dark; and its self-destructive character does not appear directly on its face but is established only on the basis of a short argument. Yet this argument suffers from the same disease: in appealing to Wittgenstein's understanding of imagination, it employs the relevant all-encompassing terms and thereby assumes an external position of enunciation, the very position it is designed to preclude. The "argument" is thus self-destructive as well and appears in no position to establish anything.

What can we make of all this? At this stage, perhaps only that the considerations just adduced seem to imply that it is impossible to take the perspective of radical immanence as an object of scrutiny in the standard sense. What it is and entails can be worked out only from "within" it, not stated or argued for from its "outside." This remains to be clarified, but in any case, given the intimate connection of this perspective with nonsense, these considerations snowball into a serious warning: nonsense in the *Tractatus* is tricky to handle.

Be that as it may, becoming inextricably involved with nonsense at this relatively early stage of the discussion need not deter me from my strategy to understand the strategy of the *Tractatus*. If coming to understand a radical difference in philosophical perspectives sufficiently resembles coming to understand a radically novel paradigm in science, then the means for achieving such understanding *should* amount to the judicious employment of elucidatory (telling) nonsense. Defining the perspective of radical immanence and trying to work out what it amounts to in nonsensical terms may thus be seen simply as drawing the first lesson from the physics example: Weinberg was obviously and self-avowedly talking nonsense, but this was *the fundamental prerequisite* for the intended pedagogical success. Considering the "definition" in question as the first rung of the ladder I am trying to build in order to understand the strategy of the *Tractatus* might thus prove profitable after all.

Wittgenstein appears to side with me on the matter. In acknowledging that the way I define the perspective of radical immanence is nonsensical, I am concomitantly highlighting Wittgenstein's request (TLP 6.54) that readers of the *Tractatus* not to be unduly afraid of nonsense,[3] for, as he implies, nonsense is instrumental in understanding him. Hence in employing nonsense as the first rung of the ladder I am building to help understand the *Tractatus,* I am merely basing my strategy on what Wittgenstein himself appears to request expressly.

Further problems arise, however, for the grammatical structure of most propositions in the *Tractatus* differs from that of the one characterizing the perspective of radical immanence as well as from what my reading of TLP 6.54 (prematurely it seems) has suggested. Most of Wittgenstein's propositions appear to be grammatically legitimate, not lacking sense and arguable in one way or another. Likewise, the same appear as happily oblivious to the perspective I claim Wittgenstein to hold. For example, the propositions to which I have appealed in order to buttress my characterization of the perspective of radical immanence (e.g., TLP 2.022, 2.033, and 3.001), to say nothing of the one opening the *Tractatus*—"The world is everything that is the case" (TLP 1)—although obviously disallowed by the perspective of radical immanence, are put forth serenely and unconcernedly.

Wittgenstein thus does not appear to make a big deal of the perspective I claim he adopts. By the same token, I find myself incapable of supporting my claim by straightforwardly providing textual evidence to demonstrate that Wittgenstein indeed adopts it, as I would do for Weinberg if I had to explicate his pedagogical strategy more fully. Wittgenstein seems to expressly espouse the perspective I claim he espouses only in the few passages I indicated in chapter 1 and in some additional places I will mention later. For the rest, he appears to be unconcerned with that perspective altogether. At this point, then, my definition of the perspective of radical immanence is plainly nonsensical, while most of Wittgenstein's propositions seem not to be. While I am openly talking nonsense, Wittgenstein doesn't appear to do so, at least for the most part.

But then a question emerges naturally: if Wittgenstein indeed espouses the perspective of radical immanence, and if this perspective differs from that espoused by his generic philosophical opponent as radically as the STR differs from classical mechanics, why does he not proceed in the way I have just indicated? Why doesn't he follow Weinberg's example, ask for his reader's "good will," and go on from there? Why does he instead advance propositions that appear to make sense even as they are deemed nonsensical by the perspective of radical immanence? Is he or isn't he taking the same position with respect to his readers that Weinberg takes with respect to students? Does he uphold the perspective of radical immanence or does he not?

The short answer is that Wittgenstein does not proceed in this manner because doing so would be strategically ineffective. The longer answer involves the difference between the requirements for teaching a radically novel scientific theory and those for convincing anyone who opposes the philosophical perspective of radical immanence.

The Strategy of the *Tractatus*: Opening Moves

To understand this strategic ineffectiveness, note that, just as the teacher could not possibly convey the STR simply by expounding it in its own terms, perhaps Wittgenstein similarly cannot expound the perspective of radical immanence in its own terms so as to make his readers, and especially philosophical opponents, forgo their views and accept his. If Wittgenstein's philosophical perspective does indeed differ from that of his opponents as much as the conceptual system of the STR differs from that of classical mechanics, then to make such opponents come to realize that they are confused—as well as to recognize that Wittgenstein is compelled to advance nonsense (TLP 6.54) in countering such confusion—he has to build, rung by rung, the appropriate elucidatory ladder connecting his own philosophical perspective with theirs. What might sway such opponents in the direction of Wittgenstein's point of view and induce in them the eureka experience pertinent to the philosophical context at issue—the eureka experience making them "see the world rightly" (TLP 6.54)—is their coming to feel in their bodies the full force of Wittgenstein's exertions.

The ladder Wittgenstein must construct is one that will enable him to meet his opponents halfway and hence one whose creation will have already made the necessary concessions to the opponents' failures to understand him straightfor-wardly. Concomitantly, Wittgenstein will need to have already secured the requi-site "good will" from his opponent: they must have already "thought thoughts"—or at least believe themselves to have had thoughts—somehow "similar" to his (TLP Pr ¶1). Wittgenstein's strategy can prove effective only after these prerequisites have been satisfied. Independent of the toil necessary for establishing them, abstract pronouncements on the merits of the perspective of radical immanence—or more generally, self-enclosed presentations of a novel philosophical perspective—usually cannot engage opponents who are deeply convinced otherwise; such ways of pro-ceeding remain, as a rule, perfectly idle, if they do not incur downright philosophi-cal hostility. It follows that deploying his strategy by starting from the definition of the perspective of radical immanence would have not led Wittgenstein very far.

If such a straightforward strategy is indeed ineffective, then to understand the strategy Wittgenstein does deploy, as well as why this strategy *can* be effec-tive, we have to become clearer about the strategy's specific goal. Again, insofar as Wittgenstein espouses the perspective of radical immanence, his generic oppo-nent is the philosopher who takes as unproblematic any philosophical propositions tolerating, explicitly or implicitly, the existence of an external vantage point, and in most cases, those who do this build bodies of doctrine—that is, philosophical systems—wherein such propositions figure prominently. Wittgenstein's strategy, therefore, must aim at getting such opponents to see the nonsensical character of these propositions by penetrating their philosophical systems through the elucida-tory ladder he builds for the purpose, by opening this system up and by taking it from the inside.

As I demonstrated when discussing the grammar of paradigm change, our ability to distinguish sense from nonsense stems from the "enormously compli-cated silent adjustments" (TLP 4.002) on which our understating and expert usage of language rest. By the same token, unknowingly misjudging the leeway that these adjustments permit makes us mistakenly believe that we are proffering sense in a given context when we are not. Such unaware misjudging not only makes Witt-genstein opponents take it for granted that their propositions make sense but also builds up resistance against efforts to question the alleged sense. Accordingly, to achieve his goal, Wittgenstein has to elucidate to his opponents that the leeway to which they silently entitle themselves is not there.

To the extent that his strategic goal in espousing the perspective of radical immanence is to logically annihilate the vantage point of God, Wittgenstein must demonstrate once and for all that logic does not allow such a vantage point. But since logical propositions "treat of nothing" (TLP 6.124) and "say nothing" (TLP 5.43), demonstrating this cannot take place within logic. The demonstration can take place only in colloquial language[4] as "managed" from the background by the silent adjustments; that is, Wittgenstein's goal can be achieved *only grammatically*. This goal thus involves elucidating to his opponents that no matter how such ad-justments move about, the corresponding leeway cannot possibly allow the vantage point at issue. Such leeway is simply not there to begin with, and since the impossibility in question is *logical,* it cannot be arrived at, no matter how the grammatical space subtending colloquial language becomes widened. Any explicit or implicit entitlement to leeway of this sort can be only illusory.

The situation here is in a sense the inverse of that in expounding a novel scientific paradigm. In the latter case, the students come to the encounter making sense, but the teacher must speak nonsense to demonstrate the value of a new grammatical possibility. In the former, Wittgenstein's opponents (his students, as it were) come to the encounter advancing nonsense, though because they help themselves to an illusory grammatical entitlement, they believe themselves to be making sense. Thus Wittgenstein's task is to demonstrate to his opponents that they advance nonsense by elucidating the illusory nature of the entitlement in question, which is ruled out by logic. Whereas the teacher's task is to introduce a novel grammatical possibility, Wittgenstein's task is to elucidate that a grammatical possibility taken for granted is absolutely (i.e., logically) ruled out. To achieve this and thus convince his opponents, he must imitate the procedures of such opponents and thus appear to them as talking sense by seemingly helping himself to the same illusory grammatical entitlement.

Because Wittgenstein can circumscribe his goal in such a manner, he need not aim at the propositions of his opponents one by one, shooting them down in individual duels; the work to be done does not amount to an impossible infinite task. Since practically all propositions of his generic philosophical opponents, as they are interconnected in the bodies of doctrine they advance, help themselves to the same illusory grammatical entitlement, Wittgenstein need only set himself against all such propositions taken together. And this is *necessary* if Wittgenstein is to "solve the problems of philosophy" (TLP Pr ¶8) not one by one, writing the corresponding indefinitely large number of papers, but all together at one go. It follows that Wittgenstein has to *systematize his own* work by modeling it on the structure of the views against which he sets himself. By thus imitating his opponents' approach to philosophy, Wittgenstein makes the first significant concession to their failure to understand him, even though this initial strategic move requires him to adopt the habits of a chameleon or, better still, to act in full awareness as Woody Allen's Zelig.

In any case, all this clarifies why Wittgenstein had to give to the *Tractatus* the form of a complete logico-philosophical treatise. The work covers the whole of philosophy—there are two head propositions on the world, two on thought, and two on language, while their subdivisions cover all philosophical topics almost without exception—and it is set against the distilled compendium of all possible philosophical treatises by opponents, namely, those directly or indirectly hostile to the perspective of radical immanence.

Now that we understand Wittgenstein's goal and the opening moves of his strategy, we must ask how Wittgenstein can deploy this strategy. Again, for Wittgenstein, a "philosophical work consists essentially of elucidations" (TLP 4.12), the judicious use of telling nonsense to make a point go through. In the case at hand, then, Wittgenstein's use of telling nonsense, or equivalently, his construction of an elucidatory ladder, must be aimed at penetrating his opponents' philosophical systems, opening them up and taking them from the inside. But this is as far as the analogy with paradigm change in physics can take us, for the two ends of the ladder at issue are now very different.

In the case of paradigm change, the teacher already has the new paradigm while the students hold fast to the old, none of which is nonsensical; in the case of the *Tractatus,* however, Wittgenstein has no theory he can target or use to proceed with his work of elucidation. Since he views philosophy as "not a theory but an activity," there is no "number of philosophical propositions" (TLP 4.112), no body of doctrine whatsoever, at which he can aim or by which he can secure a stable anchor point for the ladder he must build if he is to lead his opponents out of the nonsense they are advancing. This is exactly the point, therefore, where the analogy with paradigm change encounters its limits: neither end of the ladder Wittgenstein has to build, the "good" end that will lead his opponents out of nonsense or the "bad" end that he is targeting, can be suspended on anything at all. Hence this a very strange kind of ladder, for it can join only the void with the void.

This double void marks the point where Wittgenstein's strategy must start to unfold. To see how it does unfold, we have to ask the question coming naturally to mind at this juncture: what sort of "stuff" can make up the ladder; what is the "stuff" on which Wittgenstein's opponents have to march in climbing up this ladder? Or, to repeat the term I employed in discussing paradigm change in physics, by what could Wittgenstein *flood* his opponents' philosophical system, given that he is not proposing any philosophical theory or any body of doctrine whatsoever? Answering this question will also clarify how Wittgenstein can allow himself to formulate philosophical propositions that are obviously nonsensical on the perspective of radical immanence while simultaneously holding fast to that perspective.

Working from Within

As it turns out, Wittgenstein can flood this system only with *nothing,* with the nothing constituting the perspective of radical immanence itself, as noted in chapter 2. Wittgenstein's strategy is to put this nothing to work, and the work this nothing performs is what makes the strategy effective. The nothing in question thus possesses a performative function, though one ultimately null and void for, at the end of the day (and as is to be expected from nothing), it leaves everything exactly as it was.

Recall that to hold the perspective of radical immanence is to claim that there is no external vantage point on which *any theory* about the world, thought, and language might rest. This claim is weird not only because it almost carries its nonsensical character on its face but also because it precludes any kind of assessment (of its truth, of its value, or of its effectiveness) issued from outside itself. The claim is not only self-avowedly nonsensical but also *self-justifying,* refusing by its very formulation to countenance any authority external to the perspective it defines by which the claim itself could possibly be assessed.

The reason should be obvious. Such an external element would by definition be one allowing the adjudication of claims such as "no theory external to the world and to thought and to language as a whole can be forthcoming." But if it permits this adjudication, it allows the possibility that such a claim might be wrong, and to allow that is to say that some theory external to the world and to thought and to language as a whole *can* be forthcoming. And anything that authorizes this possibility must itself occupy a position external to the world and to thought and to language, that is, a position that the claim defining the perspective of radical immanence rules out absolutely. Adjudicating the claim that defines the perspective of radical immanence is therefore impossible. By conceding *absolutely nothing* to any authority lying outside itself, this claim categorically denies the possibility of submitting it to any external valuation whatsoever.

This, then, is why the perspective of radical immanence cannot be stated or explicated with propositions not self-evidently nonsensical: such apparently bona fide propositions could come only from outside this perspective and thus would be disallowed by that perspective itself. Further, we cannot espouse this perspective by basing ourselves on some prior justification of its possible merits; any authority by which such merits could be described and assessed is disallowed by this perspective itself. Furthermore, the perspective of radical immanence by definition cannot provide any theory of the world, thought, or language of the general kind usually put forth within philosophy. In this sense, one who espouses this perspective has nothing substantial to argue *for,* something acceptable or not, with qualifications or not, in candid philosophical dialogue with a philosophical opponent. It follows that no one forwarding the perspective of radical immanence can appeal to *anything* that such opponents could readily recognize as relevant to their concerns. The idea that the perspective of radical immanence can flood such opponents' philosophical systems with *nothing* amounts precisely to this situation, which also explains why a straightforward presentation of the merits of the perspective of radical immanence would be strategically ineffective.

The only possibility left open for one espousing the perspective of radical immanence on Wittgenstein's terms is therefore that of taking the propositions *his opponents* advance and "merely" working on them. To pursue this strategy, one targets particular propositions or clusters of propositions referring to some more or less traditional philosophical subject; one then works in a particular way on these propositions to demonstrate to opponents that their propositions invoke, directly or indirectly, the external vantage point at issue. Of course, the result is by definition not amenable to any theory overarching *it.* Therefore, the corresponding strategy can be deployed only mutely, by itself and on its own, without the possibility of appealing to anything outside it; carrying it out is merely working out silently and to the end the destructive operation described. In short, this strategy amounts to exhaustively working out the perspective of radical immanence itself and by itself from within the opponents' system of thought. Since working from within in this manner is the *only logical possibility* left open to philosophical activity as such, this is all that philosophical activity can possibly do. Hence espousing the perspective

of radical immanence and considering philosophy as a thinking activity in the narrow sense, one that can neither base itself on any substantive theory (or thesis) nor
come up with any substantive theory (or thesis), are identical.

Alternatively, consider a metaphor gorier than that of the innocuous chameleon or the opportunistic Zelig but also more accurate, for it involves activity
whereas the more anodyne metaphors both involve passivity. Since the perspective
of radical immanence cannot exist by itself but must rest on its practitioners' feeding themselves with what their opponents put forth, Wittgenstein can be pictured
as a ferocious parasite so hungry for his opponents' propositions that he remains
totally indifferent to the fact that in eating them up, he is simultaneously ratifying
his own total annihilation as a philosopher. But all this is just another way of saying
that to espouse the perspective of radical immanence is simply to undertake the
only possible way of performing the elucidatory philosophical activity and thus the
only possible way for demonstrating the logical impossibility of an external vantage
point. At the same time, though, the use of the more violent metaphor to characterize the elucidatory and hence in principle commendable philosophical activity
nicely underscores Wittgenstein's radical opposition to "traditional" philosophy as
a whole, for most of which an external vantage point is quasi-constitutive.[5]

These considerations thus explain why Wittgenstein formulates propositions
that appear inconsistent with the philosophical perspective I claim him to hold. By
imitating the grammatical structure of the propositions his philosophical opponents put forth, Wittgenstein makes the necessary concessions to them, ones that
allow them to engage the *Tractatus* precisely as a (logico-)philosophical treatise
and become involved with it. But by the same token, Wittgenstein's propositions
fasten themselves on his opponents' propositions, just as a parasite would fasten
itself on its host, their author's strategic goal being to suck up these propositions in
his whole elucidatory movement, to eat them up and annihilate them all, together
with his own parasitical propositions, by the end of the *Tractatus*. To do this as effectively as possible, he has to select and order the propositions of the opponents
against which he will set himself; that is, he must order his own propositions according to their "logical importance."

The First Movement

But there is more. If they are to do their intended work, Wittgenstein's propositions can be, or rather should be, interpreted as bona fide philosophical propositions, the kind of propositions that have constituted philosophical activity since
its inception, wearing this guise all the way up to the reader's final, dramatic rec-
ognition of the author's purpose. Under that guise, and before the reader of the
Tractatus has managed to "climb out through them," they can and should be
treated according to the strictures of philosophical practice. They can and should
be argued for or against in the more or less typical philosophical manner; they can
and should be appealed to for buttressing a point of view; they can and should be
employed for elucidating this or that position (e.g., a paradigm change). Indeed,
the strategy of the *Tractatus* consists in this double movement: on the one hand,

earnestly partaking of the instituted philosophical activity; on the other hand, and concurrently, undermining the ostensible object of that activity so as ultimately to annihilate both object and way of treating it. The crucial point for understanding the strategy of the *Tractatus* is to get clear on what the corresponding double reading requires of us.

That the strategy of the *Tractatus* involves both these movements should come as no surprise. For one thing, no philosopher can come up with confusing nonsensical "propositions" without first engaging in philosophical activity. Since this activity merely aims at "the logical clarification of thought" (TLP 4.112), Wittgenstein's opponents must aim, in full awareness or not, at logically clarifying thoughts—their own first of all and in most cases those of other philosophers, too—whether or not they believe that their efforts are aimed at building an impossible philosophical theory. What is worthy in any philosophical work, however sharply it sets itself *against* the perspective of radical immanence, must "consist essentially of elucidations" (ibid.), no matter what its author might consider him- or herself to be doing or achieving. In a given philosophical work elucidations may well be covered under idle or pointless nonsense, but at least in some cases, the *genuine elucidatory fragments* a work might include can surface and prove helpful. Since language "disguises the thought" (TLP 4.002), and since no criterion overarching language as a whole can distinguish elucidatory from idle nonsense, a proposition that appears nonsensical may well express an elucidatory reminder—this was the intended function of the previously discussed propositions that carry their nonsensical character on their faces—while, conversely, a proposition that appears to be full of sense may well turn out to be pointlessly nonsensical.

The point that Wittgenstein's strategy necessarily comprises such a first movement allows us to start understanding the sense in which the *Tractatus* and the *Ethics* can be considered parallel philosophical undertakings. Given the previously discussed similarity, if not identity, of their purposes and ends and taking into account the historical difference in philosophical context, Spinoza's overall strategy might be judged to match the first movement of Wittgenstein's, even if the ostensible goal of Spinoza's strategy cannot be equated to the kind of logico-philosophical clarification that the first movement of Wittgenstein's strategy is intended to secure.

The first movement of his strategy fully licenses Wittgenstein to admit that, despite their having succumbed to idle or even confusing nonsense, Frege and Russell might have come up with important elucidatory achievements. Wittgenstein openly acknowledges that much in, for example, TLP 4.0031: "Russell's merit is to have shown [i.e., to have elucidated] that the apparent logical form of a proposition need not be its real form." This acknowledgment sets the stage for the kind of logico-philosophical clarification of language Wittgenstein seeks, that required for distinguishing elucidatory nonsense from idle nonsense in the context of philosophical activity. To bring such clarification to completion forms the goal of Wittgenstein's strategy along its first movement, up to the moment where this "positive" goal loses its point and becomes immersed in silence.

To forward his strategy, Wittgenstein combats the idle nonsense of his opponents' propositions through the telling nonsense of his own propositions. Ultimately annihilating both amounts to flooding his opponents' philosophical systems with the nothing of the perspective of radical immanence and thus leading those opponents to the nothing of philosophical silence. His strategy does, however, include the ethically commendable aim of showing his opponents that they have illicitly presumed a logically impossible external vantage point. To achieve this aim, Wittgenstein has to exhibit, to display, to put on view, to make manifest— in one word, *to show*—this logical impossibility by judiciously employing telling nonsense.

It follows that the nonsense Wittgenstein employs is not "profound" or "substantial" nonsense, supposedly pointing at some deep truths beyond words.[6] Although Wittgenstein's nonsense does have a point to make, which is to say that each of his propositions has particular work to accomplish and a particular task to perform, his "showing" has nothing to do with pointing at some allegedly ineffable content. Rather, it is always and only a displaying, a putting on view, intended to make his opponent come to see not some inexpressible truth but only some particular form or aspect of the logical impossibility of assuming, directly or indirectly, a vantage point external to the world, thought, and language. Only through such showing can Wittgenstein's opponents come to understand that they are talking nonsense and thus give up for good the illusory claim.

The goal of the first movement of Wittgenstein's strategy, as well as of the tactical steps making it up, is therefore genuine understanding. Attaining the goal here is intended to have the same kind of effect on his readers as attaining genuine understanding has in general and understanding the STR has in particular. Once we have felt the impact of Wittgenstein's propositions and, by activating our "good will" toward their author, have allowed them to affect us the way he intends, they manage to display what they cannot properly say or describe, namely, the logical impossibility in question. This is exactly what gets us "over them" (TLP 6.54) through the relevant eureka experience. And they achieve this through elucidatory effectiveness: they show by elucidating; they elucidate through showing. Thus showing and elucidating go hand in hand, just as elucidating and nonsense do.

Hence "what *can* be shown [through elucidating] *cannot* be said" (TLP 4.1212): to be capable of showing through elucidating means that we can resort to telling nonsense, and we resort to it because we are incapable of stating outright, of describing, or of proving—in one word, of *saying*—what we are after by our philosophical activity. No inkling of what commentators have called "substantial" nonsense is to be found in the *Tractatus*. From the perspective illuminated by the endpoint of his strategy, it becomes manifest that the nonsense Wittgenstein deploys, as well as the nonsense to which he explicitly refers, is always nonsense with no content of its own, *mere* nonsense, no better than the nonsense that Weinberg's definition and the claim defining the perspective of radical immanence must, by virtue of their grammatical structures, carry (or almost carry) on their surfaces.[7]

It appears, then, that telling nonsense, elucidating, and showing form the fundamental ingredients of Wittgenstein's strategy. The way they work together with respect to propositions put forth by his opponents is made clear by Wittgenstein himself.

For Wittgenstein, a "propositional sign," that is, the perceptible vehicle that expresses thought (TLP 3.1), performs its function only if it is itself "applied and *thought out [gedachte]*" (TLP 3.5, emphasis added). A propositional sign lacks sense—it remains idle—if our thinking has not endowed it with sense (TLP 5.4733). Now a propositional sign is made from parts (TLP 3.31). These are the "parts of the proposition characterizing its sense" that hook onto the world, acquiring thereby their own meaning (*Bedeutung*) only in the context of the full proposition (TLP 3.314). Wittgenstein refers to these "bits of meaning" as "expressions" or "symbols" (TLP 3.31). If a proposition is not nonsensical on its face, then it can lack sense only if some of the signs making it up have been given "no meaning" (TLP 6.53) and hence do not function therein as expressions or symbols. It follows that in any given context, a proposition incorrectly *appears* to us as having sense when we do not realize that we have failed to give meaning to some such partial signs (TLP 5.4733), when, that is, we are not aware that we have left them idle, exempt from the work they should have been doing within the proposition.

Since "an expression is the common characteristic mark of a class of propositions" (TLP 3.311), a nonsensical proposition will appear to have sense precisely when we have surreptitiously borrowed this "characteristic mark" outside the contexts where "it is used with sense" (TLP 3.326) and attributed it spontaneously and unawares, without thinking or being aware that thinking was required, to the corresponding partial sign. Like all bona fide propositions of colloquial language, this proposition, "just as it [is]," appears to be "logically completely in order" (TLP 5.5563), and this, together with the fact that the mark in question has been given sense again and again in various circumstances, blinds us by making us dumbly take for granted that the meaning of the corresponding expression or symbol applies as a matter of course to the proposition in question; at the same time, it is exactly this furtive loan that makes the whole proposition appear to have sense. This clandestine passage from contexts where an expression or symbol has meaning to contexts where it does not occurs because we have unheedingly taken the previously discussed "silent adjustments" as unproblematically allowing it. This surreptitious pseudoauthorization secures that passage and presents its outcome as always already having been accomplished, as always already having been an unquestionable *given*.

This clandestine passage prevents Wittgenstein's opponents from seeing that whatever proposition they were advancing can, despite its seeming innocence, differ markedly from apparently very similar propositions of colloquial language that, "just as they are, are logically completely in order" (TLP 5.5563). Getting them to see this in the context of philosophical activity requires work on Wittgenstein's part; specifically, he must carry out his strategy or, which amounts to the same, apply what he calls "the right method of philosophy" (TLP 6.53). This consists in "demonstrating" to an opponent facing him (*ihm nachzuweisen*) that he or she "had given no meaning [*Bedeutung*] to certain signs in [his or her] propositions" (ibid.).

Since "a proposition shows its sense" (4.022), it follows that Wittgenstein's opponents self-deceptively take themselves to see the sense their proposition allegedly shows, just as everybody should. But this is an illusion, for the sense at issue is purely and simply not there. Hence Wittgenstein's demonstration comprises two activities: *elucidating* how and why his opponents have fallen victim to this illusion and making them see that failure by *showing* where and how they failed. The whole tactical step, and thus all tactical steps of Wittgenstein's strategy, consists in displaying how his opponents' proposition exceeds the leeway allowed by the "silent adjustments" determining the sense of similarly structured propositions of colloquial language. On the perspective of radical immanence, this excess always arises from somehow accepting the logical possibility of an external, transcendent vantage point. This excess makes the proposition at issue land in the realm of pointless or idle nonsense, whereas Wittgenstein's means for showing all this is his judicious employment of telling nonsense.

It is perhaps worth clarifying that the "adjustments" the "riverbed" allows, thereby determining sense, should not be cashed out in terms of immutable logico-grammatical categories that allegedly underlie all meaningful uses of our colloquial language and whose boundaries would constitute the bounds of sense; nor should they be cashed out in terms of allegedly immutable logico-grammatical rules that organize colloquial language and that we should obey if we are to talk sense. As my discussion of paradigm change has shown, such categories and such rules can be neither a priori nor conventional, and hence using our language meaningfully involves neither conforming to the dictates of some pre-established linguistic harmony nor complying with linguistic laws of the sort we institute within human societies.[8]

The adjustments in question should not be cashed out in such terms because colloquial language is replete with telling nonsense, performatively effective nonsense, nonsense that affects us in various ways and cannot be placed under such immutable strictures. The inventiveness of colloquial language, which the riverbed adjustments permit, is inexhaustible, and both the context and the "good will" that allow us to recognize purpose are absolutely crucial for distinguishing telling nonsense from pointless or idle nonsense in our everyday dealings, where these adjustments leave all required leeway unproblematically free. Moreover, if "colloquial language is a part of the human organism" (TLP 4.002), making these adjustments be bodily functions, then the context and will in question can never be merely linguistic: the body itself typically supplies decisive elements for adjudicating sense. To cash out the adjustments in terms of either immutable grammatical categories or immutable grammatical rules would be to place grammar at a position overarching colloquial language as a whole, thereby elevating it to the dignified status of a secular God. This would be endorsing practically all that Wittgenstein opposes.[9]

Classifying Propositions

In deploying his strategy, Wittgenstein appears to proceed silently but manifestly to a tripartite classification of his propositions. Most belong to what we may call the first category. These are the propositions that line up along the first movement of Wittgenstein's strategy by presenting themselves as sincerely advanced within philosophy as generally practiced. They thus make up what we might call "partial arguments" (a partial set of elucidatory comments ordered according to their "logical importance" [TLP 1n]) within the body of the *Tractatus*. The propositions in this category are left to do their work on their own, as it were, in that their nonsensical elucidatory character is acknowledged only by the end of the work, at TLP 6.54. I mentioned TLP 1 in this respect and used many other such remarks in the course of my discussion.

The propositions in this first category appear to be bona fide philosophical propositions that tend to attract or repel readers of the *Tractatus* according to their philosophical commitments (Conant 2002). To the extent that some of the propositions strike a philosophical chord, the reader will engage with that work, trying to understand its specific arguments and general aim. Sooner or later, however, readers fall on propositions of a third category not yet discussed. These are dispersed throughout the *Tractatus* and talk about limits, what "cannot be said," what "shows itself," and other such odd things. Typically, readers cannot square these bizarre propositions with the rest, which not only appear to be straightforward philosophical propositions but seem to exalt the model of all clarity, logic itself. Thus engaged, a typical philosopher will want to make sense of the bizarre propositions and to fit them together with the others. This is why interpretations of the *Tractatus* differ as radically as they do.

To use another metaphor, such philosophers are typically caught in an intricate spiderweb. They take the propositions that initially engaged their attention to be philosophy as usual and nothing more. But as long as they innocently take this apparently obvious premise for granted, they cannot help getting inextricably entangled in the web. Their desperate efforts to free themselves simply suck them in deeper, for such efforts tend to entail formulating all kinds of fanciful interpretations that usually founder on other propositions of the *Tractatus*. They can save their self-esteem only by an act of "bad will," namely, putting the blame on Wittgenstein: it is *he* who is confused in this or that, it is *he* who overlooked the one or the other, it is *he* who got carried away by this or that. Conant and Diamond (2004) have aptly described the line of interpretation they put forth as "resolute," for it refuses to cut the *Tractatus* to measure in such self-serving ways.

The predicament is no fault of Wittgenstein's. Short of attaching to each one of his propositions the sign "Beware of the nonsense you are reading!" he has flashed all necessary warnings in his preface to the *Tractatus* and in many other places throughout that work, all well before the fatal proposition 6.54. Ignoring such warnings typifies readings that remain closed to the expressed intentions behind the text, leaving the relevant resistances intact. In those cases, readers do not take in what is printed on the page; rather, they unknowingly impose their own expectations on the text and hence read and confirm those expectations instead: they read in the text only what they themselves put there. The situation is all the more striking in the case at hand, because Wittgenstein's intention does not lie hidden somewhere deep in the folds of his text; it is put plainly on view on the very surface of the *Tractatus* by the kind of warnings just mentioned and presented as clearly as the overall strategy of that work can allow.[10]

The second category of Wittgenstein's propositions comprises those referring explicitly to Frege or Russell by directly engaging them either together or separately. For this reason, these propositions come out looking like philosophy as usual and hence are most likely to exacerbate the kind of misreading in question. In these cases, Wittgenstein appears to proceed as would virtually any other philosopher, advancing specific arguments to the effect that Russell and Frege commit various infelicities. For example, in speaking of Frege and Russell together, Wittgenstein uses phrases such as "having wrongly thought" or "having overlooked" (TLP 4.1272, 4.1273). Of Russell alone Wittgenstein speaks of "error" and points out specific instances where Russell was "not correct" (TLP 3.331, 5.525). To keep the balance, so to speak, Wittgenstein taxes Frege with "confusion" (TLP 5.02) and expresses surprise at a certain claim he makes (TLP 6.1271). Because Wittgenstein admits that he conceives of propositions much as Frege and Russell do (TLP 3.318), placing his endeavor squarely within the space opened by their work, the infelicities he adduces concern what falls explicitly within this space, particularly features of the logical symbolism that Frege and Russell had introduced. Wittgenstein says that this "symbolism does not exclude all errors" (TLP 3.325) and therefore proposes his own to replace it.

In proceeding in this way, Wittgenstein appears to apply himself directly to what he considers "the right method of philosophy" (TLP 6.54). He places Russell, Frege, or both in front of him, so to speak, and "demonstrates" directly to them the precise way they have "given no meaning to certain signs of their propositions" (TLP 5.4733). For example, the "laws of inference, which—as in Frege and Russell— are to justify the conclusions, have no sense [*sind sinnlos*] and would be [in the adequate symbolism Wittgenstein proposes] *superfluous*" (TLP 5.132, emphasis added). It follows that these laws of inference are in fact idle, exempt from any kind of real logico-linguistic work wherever they appear in the symbolism. More generally, "if a sign is *not necessary*, it is meaningless [*bedeutunglos*]" (TLP 3.328), remaining perfectly idle as well.

Compressing Wittgenstein's explicit criticism to its essentials, we might say that the symbolism of Frege and Russell requires drastic corrections because it does not clearly reveal the stark impossibility of speaking from a vantage point external to language (and thought and the world). For example, Russell must "speak about the things his signs mean" when he "draws up his symbolic rules" (TLP 3.331), which would require occupying a vantage point overarching both signs and things. Wittgenstein himself openly acknowledges that much: given that "a sign in its projection relation to the world is the proposition" (TLP 3.12), talking about the things that the signs mean would be "putting ourselves with the propositions outside logic, that is, outside the world" (TLP 4.12).

By correcting the symbolism of Frege and Russell from within the space their work had opened—with a proviso related to the third category of his propositions, to be discussed—Wittgenstein illustrates the first movement of his strategy in an exemplary fashion. Pointing out the infelicities of the symbolism in question constitutes the surplus elucidatory move making manifest to all those having uncritically espoused that symbolism (i.e., virtually all Wittgenstein's interlocutors) the stark logical impossibility of the external vantage point and their appeal to it. By the same token, this category of proposition reveals a fundamental prerequisite Wittgenstein must fulfill if he is to achieve his goal along the first movement of his strategy: a symbolism displaying clearly, if only silently, the logical impossibility of a vantage point external to the world, thought, and language is instrumental if the attendant logico-philosophical clarification is to be completed.

The third and last category of Wittgenstein's propositions comprises those presenting themselves as self-avowedly nonsensical. Such are the propositions in which Wittgenstein speaks about limits of which, according to these propositions themselves, we cannot speak, those disallowing language to speak about the logical form of picturing or of representation even as they speak about just that, and so on. Proposition 6.36 provides a particularly clear case: "If there were a law of causality, it might run: 'there are natural laws.' But that clearly cannot be said: it shows itself." This remark is obviously self-destructive, for as Russell remarks in his foreword, it manages to say exactly what it purports cannot be said, namely, that there are natural laws. As I have shown in the present chapter, the propositions of the *Tractatus* that are self-avowedly and self-evidently nonsensical constitute the main articulations of Wittgenstein's overall strategy, or equivalently, the main rungs of the ladder making up the *Tractatus*.

The reason for this should by now be clear. Because the grammatical structure of such propositions is identical to that of the claim defining the perspective of radical immanence, their roles should be strictly analogous: each engages this perspective from its own particular angle, that of the ethical, the linguistic, the scientific, and so forth; each appears to expressly involve, in one way or another, an external vantage point that each simultaneously precludes. Recall that telling nonsense is *performatively effective* nonsense. Here then are the propositions that, in carrying their nonsensical character on their faces within the context of the given philosophical debate, carry their elucidatory performative function in an equally open way, thus making this function manifest.

Through its self-avowedly nonsensical character, each one of these propositions performs the grammatical impossibility of advancing propositions that somehow rely on the existence of a vantage point external to the world, thought, and language. By their self-avowed nonsensicality, these propositions perform the grammatical destruction of this vantage point, thus coming to show (i.e., to make manifest) the logical impossibility of its existence and performing its logical annihilation, for imagination cannot come to the rescue here. Given their obvious strategic importance in this sense, it is no coincidence that these propositions in particular have principally sustained the misguided idea that the *Tractatus* may be pointing at some profound inexpressible "truths."

Performing Futility

I can now summarize all this. The goal of Wittgenstein's strategy is the logical annihilation of the position of God, that is, the foolproof demonstration of the logical impossibility of advancing philosophical propositions that draw, directly or indirectly, on the existence of a vantage point overarching the world, thought, and language. Insofar as this strategy accomplishes the purpose and achieves the goal for which it was designed, insofar as the reader of the *Tractatus*, together with its author, has climbed all the way up and off the ladder constituting that work, this vantage point is annihilated with the full rigor of logic. It is annihilated for good or, to use Spinoza's idiom, for all eternity. Showing or displaying this and putting into place a symbolism making this impossibility fully manifest is the positive goal of Wittgenstein's strategy, the endpoint of its first movement. Achieving this goal amounts to getting his readers, along with the author himself, to see "the world rightly."

To logically annihilate the transcendental vantage point is simultaneously to logically annihilate all philosophical discourse that directly or indirectly involves its existence and hence all philosophical points of view opposing the perspective of radical immanence. Since the philosophical activity achieving this goal does not rely on any theory but is "merely" the activity of demonstrating that this perspective is logically necessary and hence no philosophical perspective at all, carrying it out to its end annihilates that perspective, too. To put it succinctly, the perspective of radical immanence is annihilated by the philosophical activity of annihilating all philosophical views that, in opposing it, provide room for its existence. The completion of Wittgenstein's strategy annihilates all philosophy by demonstrating in an absolutely compelling—"unassailable and definitive"—manner that logic rules out any philosophical perspective whatsoever, those opposing the perspective of radical immanence together with that of radical immanence itself. The realm of philosophical silence is thereby attained in an unassailable and definitive manner.

Wittgenstein's strategy must be carried out along two concurrent movements simultaneously.[11] Along the first movement, the propositions of the *Tractatus* present themselves as typical philosophical propositions that can engage opponents. Insofar as they are confined within this first movement, these propositions appear to fall within traditional philosophical practice, and they can be employed for various elucidatory purposes not necessarily related to the *Tractatus*. The goal of this first movement is a complete logico-philosophical clarification, as this can be based on an adequate symbolism, perspicuously showing the logical impossibility of a vantage point overarching the world and thought and language.

Along the second movement of Wittgenstein's strategy, the opponents' propositions are shown to them as being idle or pointless nonsense. The means for showing or displaying this are the propositions of the *Tractatus* itself as they are logically ordered with respect to their tactical function. Insofar as the ladder constituting the *Tractatus* is actually being climbed to its last rung, these propositions function as telling (performatively effective) nonsense. Excepting their performative function when put to work, Wittgenstein's propositions are no better than the propositions of his opponents; in themselves, they belong to the realm of mere nonsense and they return to it after their work is done and the tasks of Wittgenstein's strategy have been accomplished. By the end, all propositions of Wittgenstein's generic opponent have been erased, the eraser being Wittgenstein's propositions, which becomes fully consumed in completing its work.[12] Thus the second movement of Wittgenstein's strategy subordinates the first: it is through this second movement that philosophical silence is reached.

Three additional points should be noted. First, the *Tractatus* and the strategy it deploys lack any philosophical foundation whatsoever. There is no philosophical theory on which Wittgenstein stands, no philosophical theory he aims to build, no philosophical theory he employs. He merely combats the idle nonsense of his opponents with his own telling nonsense—which in itself is no better—in order to flood their systems of thought with the nothing of the perspective of radical immanence and thus lead them to the nothing of philosophical silence.

Second, Wittgenstein does not achieve his goal by proving, *per impossibile,* the "truth" of the *Tractatus* or even by arguing straightforwardly in its favor. The demonstration carried out via the corresponding strategy in both its movements is deployed along the performative axis. Simply by performing this activity to its ultimate end, by performing, so to speak, *this futility itself,* Wittgenstein displays the futility of philosophical activity. In this sense, the outcome of his toil is immune to all philosophically substantive arguments excepting those that might concern effectiveness—that is, whether and to what extent the nail has been hit on the head.

The third point to note begins with recalling the general characteristics of philosophy discussed in chapter 3. Since proper philosophical understanding— amounting to mere elucidation—in principle involves understanding all "things" without exception, philosophy is all-pervasive. If we additionally take into account our tendency to try to provide reasons that go beyond our everyday practices, and if we realize concurrently that our efforts to properly understand (elucidate) some "thing" that has given us pause almost inevitably leads us to try to understand the next "thing" that appears as connected to the first, then the tendency to engage in philosophical activity becomes almost irresistible. Once we have embarked on it, we find it almost impossible either to do away with all the worries that have cropped up or to forget the occasional satisfactions we might have experienced in coming to some particular understanding. It seems, then, that we require well-crafted reasons for disengaging from philosophical activity.

To obtain such reasons, we need a vantage point from which we can comprehend philosophy as a whole. But then we fall on its all-pervasiveness. We discover that it is impossible to confront philosophy as an external subject matter and come up with reasons that might—by philosophical standards—objectively justify the disengagement. In fact, we discover that the all-pervasiveness of philosophy forbids us even to talk about it as a whole without employing the very resources it provides. When we try to do this, we invariably find ourselves caught within it, with no possibility of escape (Lecourt 1982). Trying to comprehend philosophy as a whole is thus like moving on a transparent Möbius strip: we think that we see philosophy lying at the opposite face and thus being external to us, allowing us to talk about it as a whole, but if we develop our philosophical strategies by moving far enough along the strip, we discover that we were always lying on the same face as did our supposedly external subject matter. The Möbius strip has but a single face although locally it has, and thus globally appears to have, two. In short, *no standard discursive means can be available for getting out of philosophy.* Performing philosophical activity to its ultimate end is the only way out.

This then is the strategy of the *Tractatus.* In composing his treatise, Wittgenstein thought that even if philosophy could not be comprehended as a whole from its outside, he could annihilate it performatively from the inside. Given philosophy's all-pervasiveness, this was the only therapy available for curing himself from the disease of philosophy—a disease to which, for the reasons given, he could not avoid succumbing—as well as the only therapy that could inoculate him against all possible recurrences of the same disease in other forms. This performative, if not properly homeopathic, therapy *required him to go through philosophy in its entirety.*

Even if the cure provided by such "therapy"—the "truth" of the *Tractatus*— is indeed "unassailable and definitive," even if Wittgenstein has indeed managed to annihilate all philosophy with no possibility of recovery or recombination, the question remains: exactly how did Wittgenstein manage to pass through philosophy in its entirety? An answer to this question will allow Spinoza to rejoin the conversation.

CHAPTER FIVE

Organizing Content

> In the actual world, in which everything is bound to and
> conditioned by everything else, to condemn and to think away
> anything means to condemn and to think away everything.
>
> —Friedrich Nietzsche, *Will to Power*, §584

IF THE FIRST MOVEMENT of Wittgenstein's strategy is subsumed under the second, then the propositions Wittgenstein advances along the first movement should be set forth in a way that makes them all self-destruct when the second movement is completed. Everything identifying and connecting those propositions within the body of that work—the way they are formulated, the way some invoke others, the way each exhibits its "logical importance" through the decimal figure numbering it, the way most function as "comments" to those preceding them in the order of logical importance—should make them all prone to such self-destruction. But the pattern delivering each to that fate need not be the same for all. I have distinguished three categories of propositions and tried to clarify their roles in bringing about this self-destruction and thus helping to achieve the overall aim of the *Tractatus*.

But what of Spinoza? If both Spinoza and Wittgenstein espouse the perspective of radical immanence; if Wittgenstein's procedure is indeed logically inescapable, as he takes it to be; if this inescapability entails that the propositions this perspective advances inevitably self-destruct and that the perspective itself inevitably consumes itself fully in completing its work; if Spinoza's propositions for the most part match Wittgenstein's propositions; and if Spinoza reasons with a formal rigor as implacable as Wittgenstein's—if all these are true, then we should expect a second movement of Spinoza's strategy to lurk in the folds of his text, a movement leading his own propositions to self-destruction, however much Spinoza himself recognized this. We thus need to examine whether and to what extent Spinoza's work takes into account the logical status and the ultimate destiny of his philosophical theory, sketching thereby—or at least leaving some room for—a second movement of his strategy.

Before arriving there, however, we must examine the nature and the extent of the match in question. If we give the label "philosophical content" to that which Spinoza's theory ascertains, as well as that which Wittgenstein advances along the first movement of his strategy, then this is a match of philosophical content. Yet philosophical content need not be confined to the idea or the thought a proposition or cluster of propositions is presumed to express, make manifest, or elucidate. I intend the term to further include the overall structure of each of the two works, the arrangement of the corresponding propositions within a unified body of text, the ways either author establishes his propositions—through explicit demonstrations and scholia in the case of Spinoza or through a particular way of formulation and articulation in the case of Wittgenstein—and all other such things. In short, I take "philosophical content" as referring to everything that helps make the *Ethics* and the *Tractatus* respectable philosophical treatises with a place in the history of philosophy.

In the present chapter, I begin to examine how the philosophical content of the *Ethics* matches that of the *Tractatus* and the constraints to which such a match must conform. My examination will largely suspend any consideration of the second movement of Wittgenstein's strategy; that is, it will unfold for the most part at the level of traditional philosophical activity. At the same time, however, I will try to gather evidence that Spinoza's propositions offer themselves to self-destruction, even if less explicitly than Wittgenstein's do.

To chart the terrain and provide an overview of the two works, I will begin by looking at the way each author organizes subject matter through the layout of his overall plan.

Plans

The plan of the *Ethics* is clear enough, at least on a first reading. The work is divided into five parts: part I, "Concerning God"; part II, "On the Nature and Origin of the Mind"; part III, "Concerning the Origin and the Nature of the Emotions"; part IV, "Of Human Bondage or the Nature of the Emotions"; and part V, "Of the Power of the Intellect or of Human Freedom." Thus the flow of the text goes from the general to the particular as well as from abstract theory to analyses of immediate application. All parts start from numbered definitions and axioms,[1] which are followed by numbered propositions and their demonstrations; propositions are sometimes preceded by lemmas and followed by corollaries. The last three parts of the *Ethics* are preceded by relatively lengthy prefaces; part II, by an insignificant short one; and part I, by none at all. Appendixes close some parts, and explications and scholia are dispersed throughout the text. The *Ethics* follows the geometrical order strictly, with Spinoza employing all the deductive machinery of Euclid to structure and organize the philosophical content he wants to establish and convey.

The philosophical content itself is brought forth in an order that, on the face of it, appears natural enough. Part I covers ontology (Substance, its Attributes, and its modes, both infinite and finite), and part II covers epistemology and the basics of the philosophy of mind as entailed by epistemology (the nature of bodies in general and of the human body in particular, ideas and their characteristics, the relation between body and mind, knowledge and its kinds, and so forth), while the last three parts cover the rest of what passes today as philosophy of mind, what one might call applied philosophical psychology. More specifically, part III—concerned with the "the origin and nature of the emotions" at the appropriate level of generality—closes with an exhaustive enumeration, conducted through definitions and explications, of all particular human emotions and culminates with a "general definition of emotions." Part IV then analyzes the role emotions play in perpetuating human constraint, while part V lays out the road for "human freedom" by setting out the ways in which the "power of the intellect" can harness the emotions.

Spinoza was fully justified in his choice of the treatise's title given that the systematic treatment of the emotions in the last three parts aim at rigorously presenting and bolstering a particular "road" (E V p42s) for carrying out one's life. As we saw in chapter 2, this is the road leading to "blessedness," amounting to the "intellectual love of God." Spinoza works through the emotions in all the practically relevant detail in order to free himself—as well as the readers who can be persuaded by the rigor of his demonstrations—from the corresponding bondage and thus attain the realm of freedom and bliss ensured by the intuitive "knowledge of the third kind," synonymous with the "intellectual love of God."

The first two parts of the *Ethics* set out the metaphysical basis of the approach (the perspective of radical immanence) and demonstrate its inescapability so as to place the ethical aim on an unshakable foundation. The backbone of the *Ethics* consists in the ethical aim that drives the whole approach; the use of ontology, epistemology, and philosophy of mind to ground the content of the aim; and the way that this content emerges and becomes clear from handling the emotions concretely. It gets its flesh from the effective treatment of those matters, which, to be compelling, must directly confront and respond to many other issues of typical philosophical concern. Thus Spinoza's position on what he does not explicitly countenance can often be inferred more or less directly from what he says. The *Ethics* forms a complete philosophical treatise for all these reasons.

This completeness is further attested by bibliographic facts. According to his biographers (e.g., Nadler 1999), Spinoza interrupted the composition of the *Ethics* after finishing its first two parts to write his *Theologico-Political Treatise*. He returned to the *Ethics* after (anonymously) publishing that work,[2] and its reception persuaded him that the prevailing political climate would not allow him to publish the *Ethics* during his lifetime. After finishing the *Ethics*, he started composing his *Political Treatise*, left unfinished by his death. These facts lend further credence to what I have already noted, namely, that the ethical aim of Spinoza's treatise does not refer exclusively to isolated human beings; the work equally concerns the body of individuals making up a political community. In this respect, the *Ethics* is also—

if not fundamentally—a work of political philosophy, a work intended to lay down the rigorous foundations of the method by which a community of people forming a social and political body can persevere happily in its own being as far as this can go.

Finally, the completeness of the *Ethics* as a philosophical treatise is borne out by the explicit connection it sets up with science. To repeat something noted in chapter 2, immediately after proposition 13 of part II, Spinoza introduces a digression (he calls it a "brief preface") on the "nature of bodies." This comprises a small set of axioms, lemmas, corollaries, and their demonstrations that together lead to a number of postulates regarding the human body. This digression forms the core of Spinoza's physics and presents what he needs to borrow from the science of his day to go on with his philosophical task. By the same token, the *Ethics* relegates to science what is not of proper philosophical concern. This gesture is structurally equivalent to the analogous gesture by Wittgenstein previously discussed.

As is the case for most philosophical works, one needs to read the *Ethics* meticulously at least twice. In the first pass, one proceeds from the beginning of the work to its final word so as to understand the ultimate conclusions and positions; the second reading, however, in a sense reverses direction: after one has ascertained exactly where the work intends to lead, one must read "backward" to determine whether the intended outcome is brought about by the means the author employed. Typically one locates various shortcomings and infelicities in this second reading, which thus generates the doubts, the critical objections, the disagreements, and the polemics feeding philosophy much of its daily bread.

In the case of the *Ethics,* however, a third reading seems necessary, one that follows a "vertical" track, viewing specific bits as falling under general aims. The scholia dispersed throughout the text—as well as the prefaces, the appendixes, and at least some of the explications—disrupt the flow of the geometrical order (as created by the definitions, axioms, lemmas, propositions, and corollaries together with the necessary demonstrations), calling readers to change the direction of their reading. They are invited to leave the rigorous geometrical order, momentarily forsaking it to take stock of the specific questions Spinoza was facing, of the particular philosophical context in which he was writing, of the wrong or confused answers that had been given to those questions, of the deep misunderstandings lying at the roots of these answers, and of the ways Spinoza would have replied *viva voce* to the expected objections. In short, the scholia, the prefaces, the appendixes, and some of the explications present the *motivations* that led Spinoza to the particular philosophical content his work presents and to his way of presenting it. Along this "vertical" direction, readers are thus invited to see and acknowledge the *intentions* presiding over the composition of the *Ethics,* that is, to understand the author of that work himself, at least as far as his work is concerned. It seems that understanding Spinoza's intentions with respect to his text is instrumental for understanding the *Ethics* and that Spinoza himself took this more or less for granted.

Nonetheless, since the scholia, prefaces, appendixes, and explications lie explicitly outside the geometrical order, we may infer that Spinoza considered them redundant with respect to his theses and their demonstrations and intended his readers to consider them precisely under this light. The fact that he nevertheless wrote them and let them stand implies that he considered them an indispensable aid to his readers in understanding the work. The existence of such material presents the clearest evidence that Spinoza intended his work to be a kind of textbook. Thus, while expressing the author's intentions and his own relation to his text, these elements both popularize the content of the *Ethics,* helping to render what its author wants to convey more clearly, and contextualize the work as a whole by relating it expressly to then prevailing philosophical views and placing it thus squarely within the ongoing philosophical discussion.

The rhythm of the text in the scholia, prefaces, appendixes, and some explications differs from that of the text presenting Spinoza's theses (definitions, axioms, propositions, etc.), as well as from that of the demonstrations establishing those theses (Deleuze 1981). The theses are written most succinctly, in a way that makes each individual word and each punctuation mark count; the demonstrations, albeit more discursive, are similarly terse and to the point, leaving no room for stylistic idiosyncrasies and even less for rhetorical flourish.[3] In other words, both theses and demonstrations are expressed in the austere and dispassionate voice expected from the geometrical order; their author and his own voice remain hidden behind this order.

Spinoza gives his own voice free rein, however, in the scholia, prefaces, appendixes, and explications, one altogether different from the impersonal voice expounding the geometrical order. More often than not, it is polemical, short-tempered, scornful, derisive, and even sarcastic. Spinoza sounds fully exasperated by the confusion reigning around him, by the failure of his contemporaries to see the truth he intends to convey, by their tendency to persist in their folly, and by their refusal to follow the road that would lead to their own happiness. He sounds fully aware that the task he has set himself and accomplished to his full satisfaction has almost no chance of reaching his readers and affecting them significantly. This is an angry voice addressed to all those whose attitudes forced him into isolation and solitude, a voice expressing Spinoza's will to persevere with his work tenaciously despite everything, his will to overcome even his own natural frustration.

Again, Wittgenstein did not intend the *Tractatus* to be a textbook (TLP Pr ¶1). Not only does he not aim to instruct his readers in any way, but he does not even expect this work to be understood by anybody who has not already entertained thoughts "at least similar" (ibid.) to those expressed in it. Scholia and other material that might have assisted his readers are totally missing: the logical order is left alone to speak for itself. Thus the purpose driving the writing of *Tractatus* differs in this respect from the purpose driving the writing of the *Ethics.* Nevertheless, many of the remarks just advanced with respect to the latter apply surprisingly well to the former, too.

For one thing, as both the first paragraph of the *Tractatus's* preface and all biographical accounts attest, Wittgenstein felt almost the same kind of exasperation with and anger toward his contemporaries that Spinoza felt in relation to his. Wittgenstein managed to get at best only a few people to see what he intended to convey, much less to accept it.[4] The unmistakable bitterness expressed in the paragraph indicates that Wittgenstein found himself obliged to pursue his self-assigned task in conditions of isolation and solitude closely resembling those facing Spinoza.

In leaving the logical order to speak for itself, Wittgenstein allows no room for formulating his exasperation and his anger in the body of his text. Nonetheless, despite his efforts to force them under that order, concealing them behind it, such feelings do manage to make an unequivocal appearance at the turn of this or that phrase. I noted the haughty tone of TLP 6.113, the high-handed dismissal of some of Russell's and Frege's views, the supercilious surprise Wittgenstein voices in discovering that even Frege succumbed to confusion (TLP 6.1271). Such turns of phrase ring a tone almost identical to the tone Spinoza lets resound from his scholia and other ancillary items.

Recall that in a letter to his prospective publisher, Wittgenstein explicitly states the aim of the *Tractatus* to be fundamentally ethical (in Monk 2005, 22–23). As he tries to make clear (no doubt ineffectually), reasons of consistency and stylistic economy forbade him to state this aim even in the preface of his work and much less to explicate it in the text itself other than with a few propositions concerning the ethical (discussed in chapter 2). To say more would unavoidably soil the austere purity of the logical order and merely contribute to what "many are babbling today" (ibid.). Thus Wittgenstein himself confirms that the *Tractatus* and the *Ethics* share virtually identical purposes and ends.

In refusing to provide scholia or other literary devices that could have assisted readers but would have disfigured the logical purity of his text, Wittgenstein took himself to be meeting a strictly ethical demand. Scholars have highlighted that in the time and place Wittgenstein worked and within the theoretical and ideological context that determined how he conceived his task, uncompromising austerity in expression was considered an absolute requirement for speaking about matters of importance.[5] Literary crutches, ornaments of speech, rhetorical embellishments, and all expressions of feelings were taken as indications of degeneration, as devices that deepened the reigning confusion and added to the generalized "babbling," as the enemy. In this context, and given that logic constituted the model of all austerity, Wittgenstein was compelled to hold that the austerity of the logical order should come out at its purest.

No such ethical demand constrained Spinoza. In the context in which he wrote, freedom for expressing one's ideas, whatever these were and however they might be formulated, was the overarching demand. Given this context, one might perhaps go as far as to say that Spinoza regarded it his *duty* to vent his mind-set and hence to state and expound his criticism in the terms he found most appropriate. For example, in expressing rhetorical surprise at the fact that the philosopher who had broken the new philosophical ground, the "renowned Descartes" (E III Pr), had succumbed to confusion, Spinoza uses terms much harsher than those Wittgenstein employs with respect to Frege or Russell, who, as Wittgenstein concedes (TLP Pr ¶6), had opened for him the corresponding ground. Thus Spinoza allows himself to write: "I could scarcely have believed [that this view was] put forward by such a great man. . . . Indeed I am lost in wonder that a philosopher who had strictly resolved to deduce nothing except by self-evident bases and to affirm nothing that he had not clearly and distinctly perceived, who had often censured the Scholastics for seeking to explain obscurities through occult qualities, should adopt a theory more occult than any occult quality. What, I ask, does he understand by the union of mind and body?" (E V Pr). Once this disparity in historical context is taken into account, the similarity between, if not identity of, the voice of Spinoza and that of Wittgenstein comes out even more conspicuously.

This difference in expression of personal voice aside, Wittgenstein does nevertheless hold that his readers will be in a position to understand the propositions making up his book only after they have understood its author (TLP 6.54). To understand the author in this sense is to understand his motivations for writing his work as he did, to understand him in relation to that work. Understanding the book's propositions per se must follow from, or be subsumed under, understanding the intentions dictating the formulation of these propositions. Diamond (2000b) has explained how coming to understand Wittgenstein's intentions in this manner is instrumental in understanding the point of the *Tractatus*. It follows, then, that Wittgenstein and Spinoza each considers a fundamental prerequisite for understanding his work to be understanding the intentions presiding over its composition, even if the two men employ markedly different means for manifesting those intentions. Once again the similarity between the two approaches proves striking.

In chapter 4 I hinted at the reasons this fundamental prerequisite for understanding the *Tractatus* comes out explicitly only at its penultimate proposition. I can now state these reasons more amply. First, in TLP 6.54 Wittgenstein openly acknowledges that his strategy has incorporated a second movement, one that should have led the reader to appraise the *Tractatus* not merely as a philosophical treatise of the usual sort but also as a "ladder" of performatively effective nonsensical propositions whose final goal is the annihilation of philosophy in its entirety. The wording of the remark implies that the second movement subsumes the first, or equivalently, that the figure of the ladder subsumes that of the treatise. The text at TLP 6.54 forms the juncture where the two movements of Wittgenstein's strategy culminate and combine their effects so as to lead the reader to the final "thesis" of the treatise and the final rung of the ladder, out of the *Tractatus* and thence out of philosophy altogether.

To understand Wittgenstein's intentions in writing the *Tractatus* is to understand *that* his strategy consists of two movements unfurling together and *how* they unfurl together, as well as *why* this double movement constitutes the only rigorous way for demonstrating the futility of philosophical activity in its entirety. But the rigor of Wittgenstein's demonstration required that the proposition displaying the strategic importance of understanding his intentions be placed at the juncture that completes this demonstration through the culmination of both movements of the strategy. It is only *after having gone through philosophical activity in its entirety* that readers can reach a position permitting them to assess the futility of that activity with the rigor required to justifiably make them forsake it for good.

If any philosophical work generally requires two readings, it is the *Tractatus*, for TLP 6.54 compels its readers to carry out a second reading radically different from all its typical namesakes. This second reading should reveal that a second movement of Wittgenstein's strategy is at work within his text; that this second movement erases the philosophical content the first appeared to have advanced, thus subsuming this first movement under it; that this second movement can indeed take the figure of the ladder, in contrast to the first, which takes the figure of the treatise; and so on. This second reading should also let readers locate the major articulations of the second movement; assess the logical importance of the *Tractatus*'s propositions along the performative axis; understand the particular ways by which Wittgenstein manages to do away with the philosophical content of each particular philosophical subject; appreciate how all philosophical subjects hang together for him; and finally, realize how this method of proceeding from philosophical subject to philosophical subject does away with the whole of philosophy. Only after this particular second reading has been thoroughly carried out can the more typical sort—checking for possible shortcomings, infelicities, and so on—be undertaken responsibly and in good conscience.

Many commentators have not acknowledged the second movement of Wittgenstein's strategy in the *Tractatus*. They have taken the work to be of a piece with philosophical activity as traditionally practiced and have thus considered the major articulations of the second movement to be either gestures toward some deep philosophical "truths" beyond words (which are neither properly ineffable—for the commentators in question manage to state them quite well—nor significantly different from the results of typical philosophical activity) or mere infelicities of the first movement alone. In other words, they have confounded the second reading required by any philosophical work with the very particular second reading required by the *Tractatus,* leading them to fault Wittgenstein for their own inability to understand his text. Nonetheless, disentangling the two movements of Wittgenstein's strategy and thereby coming to realize the number and kind of readings the *Tractatus* requires is necessary if we are to make headway in understanding what Wittgenstein might be doing with respect to philosophical content.

Doubtless, this simultaneous double movement in Wittgenstein's strategy makes the effective plan of the *Tractatus* more complicated than the plan of the *Ethics*. In fact, two plans with different joints are superimposed, one atop the other. The first plan lays out the first movement of Wittgenstein's strategy while presenting the *Tractatus* as a philosophical treatise of a traditional cast. The numbering system adopted displays this plan concretely by ordering the constituent propositions according to their logical importance. The second plan organizes the second movement and remains more or less hidden on the first reading. It appears only after the reader has fully grasped what TLP 6.54 intends to convey, and hence it can be acknowledged and identified only in the course of the second reading.

The main joints of this second plan consist in the propositions of the third category I distinguished in chapter 4. These concern different philosophical subjects and talk about limits, about what cannot be said but nevertheless shows itself, and about other such things; by the same token, they present a grammatical structure that, as does that of Weinberg's "definition," makes them carry their nonsensical character on their faces. The second reading makes manifest that these propositions were preparing the reader for the shock that TLP 6.54 invariably produces in the first reading—and again, the logical austerity by which Wittgenstein abided prevented him from employing other means for warning readers that they should expect such a shock.

The role each such proposition plays in erasing the philosophical content of its corresponding philosophical subject can be proper only to that subject. Accordingly, the patterns of this erasing need not be the same for all philosophical subjects, and hence no specific order of logical importance need distinguish the propositions of one category from those in other categories: no particular configuration of Wittgenstein's decimal numbers can single them out. The major articulations of the second movement of Wittgenstein's strategy presumably might be characterized by any decimal number, so that the main joints of the two plans can be very different. Since the main rungs of Wittgenstein's ladder thus need not coincide with the main "theses" of his treatise, the main joints of the second plan cannot be read off from the face of the text, and hence that plan itself is not easily graspable as a whole. The only readily visible plan is that of the first movement of Wittgenstein's strategy.

Structures

Within its geometrical order, the text of the *Ethics* distinguishes propositions, lemmas, and corollaries from their demonstrations. Although all propositions are numbered consecutively, it is clear that some are strategically more important than others. Some receive support from more than one demonstration, and some scholia are richer and lengthier than others. In many cases, and as one would expect from the geometrical order, demonstrations refer backward to propositions already demonstrated and sometimes forward to results to be demonstrated later. All these features structure the *Ethics* in complex ways, all of them essential for understanding the work. To experience the compelling power of Spinoza's treatise, readers must follow a meticulous reading that will allow them to identify all these structural futures, to assess the role each plays in establishing philosophical content, and to realize how they are all interrelated within the text.

On the face of it, the structure of the *Tractatus* appears much simpler. Scholia, prefaces, appendices, and explications are absent, while demonstrations are not distinguished from propositions. In place of demonstrations the *Tractatus* offers elucidations, though the series of propositions that constitute "comments" on a given superordinate proposition, as ordered according to their logical importance, could pass for a demonstration of that proposition.[6]

All such comments are brought up to the seven head propositions of the *Tractatus,* each labeled with an integer. All other propositions ultimately refer to one or another of these head propositions, but each subordinate proposition refers as well to all the ones in the sequence lying above it in the order of logical importance. Thus the numbering system of the *Tractatus* not only orders its propositions but also interconnects them within a tightly knit structure. The fact that some propositions appeal indirectly to propositions not numerically connected to them tightens this structure even further. I should underline that all my remarks here concern the first movement of Wittgenstein's strategy. Comments, connections among comments, order, and logical importance are all part and parcel of standard philosophical activity and hence should be taken as referring to the structure of the *Tractatus* as a (logico-)philosophical treatise participating in that activity.

The particulars of the structure exhibited by Wittgenstein's work determine the conditions for the best way to conduct the first reading along the first movement of that work's strategy. This reading starts with an acknowledgment that there are different *layers* to the text under each of its head propositions (except the seventh, which has none)—layers distinguished by the number of decimal places in the numeral naming each proposition—and then goes on by driving a course that takes full account of all the connections among these layers. This is a course of reading proposed by Aenishänslin (1993), whose principle, once we allow for repetitions, leads readers along loops, making them pass seamlessly through all successive layers while fully respecting the logical importance of each proposition.

The principle runs as follows: Start from any proposition named by a given numeral and read on until you reach a proposition named by a numeral with an additional decimal place. At this point, go to this deeper layer and read all propositions named by numerals with the additional decimal place; after you complete that series, loop back and reread the proposition that led you into this deeper layer in the first place (before looping back you may have to repeat the procedure as many times as an additional decimal place crops up within this deeper layer). Once you emerge from all loops underlying the proposition from which you started, reread that proposition and then go on to the next proposition having the same logical order (viz., the same number of decimal places) as the initial proposition. Finally, repeat this procedure as many times as required.

Here is a random example: Start, say, at TLP 3 and then read 3.1 through 3.14. At this point, start the first loop, which runs from TLP 3.141 to 3.144, and then start the second loop by reading 3.1441 and 3.1442. Emerge from this second loop by repeating TLP 3.144, which exhausts the layer of three decimal places, and repeat TLP 3.14 to leave the first loop. Since no proposition 3.15 exists, repeat TLP 3.1 and then proceed to TLP 3.2, repeating the procedure until you exhaust the layer with one decimal place. Now continue with TLP 4 in the same way. The *Tractatus* goes to a depth of six layers (its propositions are labeled by figures with up to five decimal places), but in principle the number of possible layers has no limit. Wittgenstein employs only five levels, presumably from reasons of stylistic economy; he considered this number of layers as sufficient for conveying what he intended to the kind of reader he had in mind.[7]

This reading principle can readily accommodate and explain the fact that Wittgenstein includes zero in his numbering system. While the layered structure of the *Tractatus* shows that the natural flow of the text leads from, say, TLP 5.1 to both 5.11 and 5.2, Wittgenstein inserts proposition 5.101 to clarify the intended sense for 5.1 before letting the reader go on with that natural flow. If a series contains more than one proposition whose designation includes a zero—or if one designation includes more than one zero—the series that the corresponding propositions constitute is also ordered according to logical importance.

This scheme provides a best first reading, for it follows the first movement of Wittgenstein's strategy as closely as possible. It goes through all the steps of the plan he expressly laid out for this movement and fully respects the *Tractatus*'s organization as a treatise by paying all due attention to the hierarchical structure of logical importance. For this reason, and once the reader has given the text the necessary amount of thought, this reading can help clarify exactly how the propositions connect with one another and why their logical importance is graded the way it is, as well as how, both separately and in concert, they can function as elucidatory comments on the propositions preceding them in the order of logical importance. Once the first reading is thoroughly carried out in this manner, and once the existence of a second movement of Wittgenstein's strategy is taken into account, the reader is in a much better position to discern which propositions constitute the major articulations of the second movement and how the connections these bear with the others can bring about the annihilation of each philosophical subject in particular and of all of them together.

Again, the readily visible plan structuring the *Tractatus* lays out the first movement of Wittgenstein's strategy. The plan itself is clear enough. The seven head propositions may be read as chapter titles indicating the contents of the treatise as a whole, while the propositions following each head proposition may be viewed as the content of the corresponding chapter. These chapters have differing lengths; for example, "chapter 7" has no proposition following it and hence no content, while "chapter 1" is very short. Here are the head propositions as running text:

> (1) The world is all that is the case. (2) What is the case—a fact—is the existence of atomic facts. (3) A logical picture of facts is the thought. (4) A thought is a proposition with sense. (5) A proposition is a truth function of elementary propositions (an elementary proposition is a truth function of itself). (6) The general form of a truth function is $[\bar{p},\bar{\xi},N(\bar{\xi})]$. This is the general form of a proposition. (7) What we cannot speak about we must pass over in silence.

Several remarks are in order here. First, with respect to the organization of philosophical content, note that the two initial propositions concern the world and therefore ontology; the second two concern thought in its relation both to the world and to language, which is to say, epistemology and the basics of the philosophy of language; the next two concern language itself and its workings; and the seventh, having nothing beneath it in the order of logical importance and bearing an air of tautology, leads us to exit the work. Given all this, we can see that the *Ethics* and the *Tractatus* organize philosophical content in almost exactly the same manner. Both start from ontology, and both go on to epistemology, differing only in their respective third parts.[8] Spinoza—having included his conception of language in the first two parts of his treatise—discusses in the three remaining parts what I have called applied philosophy of mind, whereas Wittgenstein devotes his third part to the philosophy of language. This difference is readily attributable to factors already discussed. On the one hand, the historical context in which Wittgenstein worked had elevated philosophy of language to the rank of *prima philosophia,* a situation having markedly influenced his views and compelling him to offer language a prominent treatment by addressing it at length. In fact, "chapters" 5 and 6 make up more than half of the *Tractatus.* On the other hand, applied philosophy of mind was of minor concern to him, for psychology had already broken many of its ties with philosophy to become an independent discipline in its own right.

Second, presenting the head propositions consecutively shows them capable of forming a discursively articulated piece of writing constituting a succinct and self-sufficient summary of the whole of the *Tractatus.*[9] For conclusive evidence, it suffices to replace the numbers naming the propositions with appropriate conjunctions and to provisionally forget that these propositions function as chapter titles as well. Indeed, once the symbolization in TLP 6 is understood, the summary constructed in this way allows one to acquire at a glance an exact overview of the philosophical content Wittgenstein advances along the first movement of his strategy. A reader who finds this overview sufficient for assessing its correctness need go no further. As this is highly unlikely, however, Wittgenstein has striven to provide that reader with a set of logically ordered comments that should do the job. The ideal of austerity and perhaps respect for his readers' thinking abilities[10] prohibited him from adding comments that would explain his intentions, that is, explain himself (TLP 4.026), any further.

So even though Wittgenstein structured his treatise in the form of distinct numbered propositions, we can nevertheless read his work discursively, layer by layer. Once the propositions of any given layer in any chapter are laid out in the form of running text, once the numbers are suppressed and the necessary conjunctions added, and once a few short explications, provided by the layers beneath, are considered, these propositions would add up to a discursively articulated, self-sufficient text, just as the head propositions do. This principle may not apply with equal literary felicity to all layers, but it applies so well to so many that it almost forces us to believe that Wittgenstein intended it. In any case, as the length of some of the work's propositions attests, Wittgenstein had no objection to discursive text as a matter of principle. On the contrary, he may well have ordered the propositions as he did, along successive layers, precisely because this particular ordering offered them to such a discursive reading.

For a purpose I will state in a moment, allow me to run this experiment for the first layer under TLP 6.

> (6.1) The propositions of logic are tautologies. (6.2) Mathematics is a logical method. (6.3) Logical research means the investigation of all regularity. And outside logic all is accident. (6.4) All propositions are of equal value. (6.5) For an answer which cannot be expressed, the question too cannot be expressed.

The riddle does not exist. If a question can be put at all, then it *can* also be answered.

The technique works here, too. This layer succinctly summarizes how Wittgenstein conceives different kinds of propositions and forms a more or less felicitous discursively articulated piece of writing. At the same time, if these propositions are considered separately, as Wittgenstein laid them down, they constitute a table of contents with respect to the layers lying underneath the particular layer they form.

Thus, after having put forth the "general form of the proposition" in TLP 6, Wittgenstein immediately goes on to distinguish "kinds" of propositions, starting as he should—since his treatise is a *logico*-philosophical one—from the propositions of logic. He then moves to the propositions of mathematics and from them to the vantage point from which he will consider the propositions of natural science (logical research into "regularity"). The relevant subject matters here are induction (TLP 6.31); the "law of causality" (TLP 6.32); the a priori as it works in science (TLP 6.33); natural laws (TLP 6.34); the geometrical character of picturing physical theories as he does in the propositions immediately above (TLP 6.35); causality in its relation to time, space, events, and inductive inference (TLP 6.36); and finally, necessity and its relation to the will (TLP 6.37). Note that Wittgenstein distinguishes psychology and the theory of evolution from philosophy and equates them to one or another "hypothesis of natural science" in earlier passages (TLP 4.1121 and 4.1122, respectively).

After having done away with natural science in the layer of TLP 6.3, Wittgenstein proceeds to ethical propositions through the introduction of the fundamental ethical term "value" (TLP 6.4) and finally to philosophical propositions to the extent that these are not treated in the other chapters. Thus TLP 6.51 considers skepticism; TLP 6.52, together with the layer lying underneath it, the "problem of life"; TLP 6.53, philosophical method; and TLP 6.54, the character of Wittgenstein's own philosophical propositions. All these philosophical subjects are treated through the lens of logic and on the basis of the relevant kinds of propositions, which is only to be expected given that the *Tractatus* presents itself explicitly as a *logico*-philosophical treatise.

The reason I chose this layer should be obvious. If we relate the table of contents just canvassed to the table of contents provided by the head propositions laid out above, and keep the ethical overall aim of Wittgenstein's work in consideration, the completeness of the *Tractatus* as a (logico-)philosophical treatise becomes perfectly visible. Almost all subjects of philosophical concern are covered in Wittgenstein's particular way, meaning that Wittgenstein does indeed pass through philosophy in its entirety so as to annihilate it altogether in the second movement of his strategy.

Beginnings

Again, Wittgenstein starts his treatise from two "chapters" concerning the world, goes on to two "chapters" concerning thought, and from there proceeds to two "chapters" concerning language so as to complete his work with silence. I noted earlier (note 29 to chapter 1) that this sequence reverses the order his contemporaries would have expected. Many would agree that the formulation of traditional philosophical problems "rests on the misunderstanding of the logic of our language," inferring that the analysis of language, as carried out with the new means logic then appeared to provide, should make these problems dissolve. But they would further maintain that only logic and language can form the proper subject matter of a philosophical treatise and hence that the two first parts of the *Tractatus* should have followed the third as its mere applications. Russell's proposal that Wittgenstein's work should be titled "philosophical logic" finds its roots here.

I have also explored why espousing the perspective of radical immanence led Wittgenstein to a different road. On the face of it, the bare formulation of his final goal might well have been acceptable to his contemporaries, but he could reach that goal only by passing from a first movement of his strategy along which traditional philosophical problems had to be engaged seriously. And I showed how the plan of the *Tractatus,* as well as many aspects of the structure organizing that work, form part and parcel of this first movement. Once the requirements of the second movement of Wittgenstein's strategy are taken into account, we can see that the sequence in which his treatise addresses philosophical subjects had to be the affair of the first movement.

Given this, Wittgenstein's course of beginning the first movement of his strategy, and hence his treatise as a whole, from the world is perfectly natural, if not compelled by the logical order itself. Since language speaks about the world, one is not expected to examine language before examining that of which language speaks—before, that is, examining how the world offers itself to its linguistic appropriation. The same consideration holds for thought. If thought is the logically articulated "picture of the facts" (TLP 3) making up the world (TLP 1.1), then it is again only natural to examine what is pictured before examining the ways of picturing it. This should be the case even if it is impossible to access the pictured independently of its picture, and hence the philosophical treatment of the world "in itself" can do nothing more than mark this kind of logical precedence. By the same token, if the propositions of language are pictures of reality (TLP 4.01), while "the total reality is the world" (TLP 2.063), then the examination of language and its propositions cannot precede the examination of the world in the logical order; moreover, if propositions are there to express thoughts in a way perceptible by the senses, as Wittgenstein holds (TLP 3.1), then the examination of language cannot precede the examination of thought.

Again, Wittgenstein's overall aim is to see (and help his readers to see) the world rightly (TLP 6.54). That the last (actually, the penultimate) word of the *Tractatus* before letting us out to silence is identical to its first confirms that the best way of reading that work is the one I have been suggesting. Coming to see the world rightly after the exhaustive first reading is coming full circle to see that the world is indeed "everything that is the case" (TLP 1). But in light of TLP 6.54, readers are now in a position of reassessing TLP 1 and realizing that it is not intended to be a disputable philosophical thesis, as they unavoidably considered it to be on starting the first reading. By having thus taken stock of Wittgenstein's intentions, readers can carry out the second reading, which will help them realize that all the book's propositions should finally be considered as nonsensical while forcing them to acknowledge the necessity compelling Wittgenstein to proceed to his demonstration along two simultaneously unfurling movements. Coming to see the world rightly following this second reading is thus simply coming to see it as it really is—as indeed everything that is the case—with no philosophical worries concerning it and with no need, wish, or temptation for indulging in philosophical activity, now seen as futile. At the close of the second reading, the claim that the world is everything that is the case should be taken for granted just as everybody *not* involved in philosophical activity does in everyday dealings: "there is then no question left, and just this is the answer" (TLP 6.52).[11] That the first and last word of the *Tractatus* resonate in such a way is not only logically dictated; it is also an apt, felicitous, and perfectly innocent effect of literary style.

As we saw, Spinoza too starts from the world, that is, from ontology. And I noted that such a beginning is surprising given the historical context in which he wrote, for one of the main moves made by Descartes, the "great man" (E V Pr) who had opened the new philosophical terrain where Spinoza himself worked, was to subsume ontology to epistemology. I briefly discussed the sense of this move in chapter 1: Descartes deploys a strategy of hyperbolic doubt that questions everything until reaching the rock-bottom certainty of the *cogito*, which will allow him to erect his system on an unshakable foundation. How could Spinoza come to occupy the philosophical terrain opened by this gesture while reversing the gesture itself? To understand why Spinoza begins the *Ethics* as he does, we must try to answer this question.

First of all, recall that a fundamental move in Descartes's overall approach was to place the solitary ego at the center of his system while endowing it with an unlimited power of reason. This is a power capable of establishing the certainty of the ego's own existence as a thinking thing and, on this basis, of proving God's existence by relying solely on itself. Placing the ego of the philosopher enunciating the proof—understood, as I said, to represent the person in the street[12]—at the center while endowing it with such unlimited power constituted a revolutionary gesture with far-reaching consequences. Making God's existence follow epistemically from the existence of any ego deploying Descartes's discourse removes God from the foreground and turns Him into an object of reasoning for anyone who exercises that power effectively. Organized religion thus loses one of its main ideological grips as God becomes, in this particular sense, subject to that power. Concomitantly, His epistemic role is demoted to that of simply guaranteeing the truth of clear and distinct ideas, ones the ego is able to entertain by itself. True, the argument establishing this role depends on God's perfection, and God subsequently emerges as *ontologically* prior, since He remains the Creator of the world and of humankind, but this barely moderates the gesture's arrogance. The power and self-sufficiency of human reason are established, ontology is subsumed under epistemology, and modernity is ushered in.

Spinoza was certainly impressed but perhaps not fully convinced. He did not like skeptical arguments (Popkin 1979, chap. 12), and hence he might well have viewed Descartes's employment of hyperbolic doubt to attain a presuppositionless certainty as a merely rhetorical maneuver. In particular, Descartes's proof for the existence of God and the place his system carved out for Him likely left Spinoza deeply unsatisfied.[13]

For one thing, the proof in question makes it evident that God's place within the Cartesian edifice enables Him to keep most of the anthropomorphic characteristics attributed to Him by the Scholastics. Yet these characteristics undermine Descartes's radical proposal on the self-sufficiency of human reason. By guaranteeing the truth of clear and distinct ideas, God comes to occupy the place of a benevolent, omniscient, and absolutely free arbiter who can intervene in anything according to His will. This makes human reason dependent, even if only in the last analysis, on an authority superior and external to it, and such a dependence vitiates the self-sufficiency with which Descartes himself had endowed it. To the extent that it allows God to intervene from the outside, Descartes's system regresses to the Middle Ages, allowing the Scholastic tradition to reenter through the back door with disastrous consequences for the revolutionary thrust of that system. To put it succinctly, Descartes's approach is not radical enough: it concedes too much to the prevalent ideas of his day.

Nevertheless, Spinoza understood that the need for such concessions is rooted in the way the relative positions of God and humankind emerge from Descartes's philosophical strategy. First, if human reason is indeed self-sufficient, as Descartes claims and as Spinoza would agree, the truth of clear and distinct ideas should be in no need of external guarantees. Spinoza is perfectly explicit on this. As we saw in chapter 1, from the *Treatise on the Emendation of the Intellect* on, he firmly held that only truth can guarantee truth, its measure being a truth we already possess. For Spinoza, however, the self-sufficiency of human reason does not imply that humans have been created in the image and likeness of God, almost as small gods in their own right, placed at the epicenter of the world and thereby capable of surveying it as a whole, if only from a lower position. Rather, each human being and his or her reasoning power should be considered as remaining thoroughly *subject to* rather than as *standing at* epicenter of the order of things, even of things epistemic, for humanity is part of nature and not a "kingdom within a kingdom" (E III Pr), playing the role of God—being king—at a smaller scale. Yet Descartes's strategy goes in precisely this direction. Although that strategy reserves for God the ontological place of honor as Creator and offers Him the role of guaranteeing truth, by turning the human ego into the philosophical protagonist, it inflates human arrogance and belittles God (Nature) correspondingly. Thus this approach is in a sense also too radical: even if God is allowed to retain the high supervision of all His creations, the one among them privileged to bear the image and likeness of the Creator becomes elevated to mastery over the rest of nature. (We are still in the grips of this arrogance and in the process of paying the consequences.)

The crucial point of concentration for all these issues lies in the exact characterization of God and His position with respect to both the world and humanity. One may argue that offering God the role of guaranteeing truth, conceiving Him as the Creator of the world and of humankind, endowing Him with free will, attributing to Him the best human characteristics elevated to the superlative, *and* taking Him to occupy a position external to the world go together as a matter of logic. Hence to deny the possibility that any being, however exalted, and any entity, however supreme, could come to occupy such an external position would be to strip God of all these features as inseparably connected to that external position.

Concomitantly, nobody could deny that the world goes its way independent of human will and that we are subject to all kinds of worldly forces going far beyond our control. God should thus be maintained as simply the *power* determining everything that happens or could possibly happen to the world and to humanity, the power impressing the requirement for the attendant humility. But what happens or could possibly happen is determined by the laws of nature, which science was then in the process of establishing. Thus, if God cannot inhabit any external position whatsoever, so that His power cannot emanate from any position outside nature, that power must be identical to nature's own power, making God identical to Nature. All the elements allowing the formulation of the perspective of radical immanence are thus set into place. Bringing together these elements in the form of a rigorous philosophical theory required somebody who could reason with relentless rigor and who would not be afraid of drawing the consequences unequivocally. Spinoza was precisely such a one.

This explains why Spinoza begins the *Ethics* with ontology. Engaging the novel philosophical terrain as forcefully as the situation required hinged on understanding the character and the position of God. For all kinds of reasons—metaphysical, scientific, ethical, political, and ideological—this was the crucial issue, one that even the "great thinker" Descartes had found impossible to disentangle. Moreover, if God is indeed the power determining everything and if Spinoza could not allow himself anything less than a demonstration from first principles, he was compelled to begin his treatise with God considered as identical to that power and hence as the fundament of all ontology. In addition, initially reversing the very gesture that had opened the novel philosophical terrain marked a point of central importance for Spinoza: this was the first installment of the rigor he sought to use in executing the task he had assigned himself.

So Spinoza, too, unfolds his work by reversing the sequence his contemporaries would have expected him to pursue. This marks yet another important similarity between his approach and that of Wittgenstein, one that has to do less with matters of personal idiosyncrasy or literary style than with historical context and philosophical content. Most of their contemporaries failed to fathom why either author proceeded as he did, and this failure contributed to the isolation of both and to the solitude in which each was obliged to carry out his self-assigned task.

Having thus seen why both Spinoza and Wittgenstein open their works from what seems to be the same starting point, namely, ontology, we should go on by examining whether and to what extent the ontological content advanced by the *Ethics* can measure up to the ontological content advanced by the first movement of the *Tractatus*. But we are not there yet. The differences in plan and structure, as well as the fact that Wittgenstein invites us to consider all propositions of the *Tractatus* as nonsensical, oblige us first to consider the constraints to which such a match should be expected to conform.

Matching Constraints

The philosophical content that Wittgenstein advances along the first movement of his strategy should be capable of bearing out the demonstration (always in Wittgenstein's sense) that, at the end of the day, no philosophical issue whatsoever is to worry the reader of the *Tractatus*. But what can this philosophical content possibly be? What kind of propositions should Wittgenstein set forth along the first movement of his strategy?

Let me try to clarify this question by focusing on the beginning of the *Tractatus*. According to Wittgenstein, "The world is everything that is the case" (TLP 1), and what is the case can be only a fact. Hence "the world is the totality of facts, not of things" (TLP 1.1). Although this bit of reasoning is not self-evidently nonsensical, both its starting point (TLP 1) and its outcome (TLP 1.1) are obviously ruled out by the perspective of radical immanence, since both invoke a vantage point overarching the world. This is why TLP 6.54 deems nonsensical all such apparently innocent propositions of the *Tractatus*, including TLP 1 and 1.1.

Yet a contrary proposition to the effect that the world is the totality of things, not of facts, seems to share the same grammatical characteristics: it too does not carry its nonsensical character on its face, it too presumes the same external vantage point, it too is ruled out by the perspective of radical immanence, and it too is ruled nonsensical by TLP 6.54. Moreover, both TLP 1.1 and the contrary proposition that the world is the totality of things, not of facts, appear to be fully legitimate philosophical propositions, and either might in principle form the starting point— or a major articulation—of a bona fide philosophical system. If both are equally nonsensical, why does Wittgenstein choose the one and not the other?

At least one of reasons determining Wittgenstein's choice must be that TLP 1.1, although appearing to rely on an external vantage point, lends itself to becoming a rung of the ladder—a step of the demonstration—that will eventually lead out of philosophy altogether and hence to its own erasure as a philosophical proposition, while the contrary proposition does not offer itself to such a role. Nonetheless, in coming to erase TLP 1.1, the demonstration would come to erase the contrary proposition, for demonstrating the nonsensical character of TLP 1.1 will by the same token demonstrate the nonsensical character of its contrary. It seems therefore that nonsense cannot be the only issue here, and we must try to become clear on what more is involved. On the evidence provided by the *Tractatus*, the difference between TLP 1.1 and its contrary is that the order of logical importance connects TLP 6.54 to the assertion of the one and not of the other.

Now, that TLP 1.1 is connected to TLP 6.54, that TLP 1.1 comes after TLP 1 in the order of logical importance, and that there are other propositions following it in the same order imply that TLP 1.1 should not be considered solely by itself, as a thesis whose philosophical content could be supported or refuted on its own. It seems that no answer to our question can be forthcoming unless we take into account all such connections, for it is only through them that we might come to understand how Wittgenstein intends the content of TLP 1.1 to be taken. We should deal with it in more or less the way we would deal with a proposition of any philosophical treatise whose content becomes specified by analogous connections to other propositions. This is just to reiterate that the first movement of Wittgenstein's strategy involves reading the work as a typical exercise in philosophy: every proposition of the *Tractatus* constitutes an integral part of that work, embedded in its overall structure, articulated with all its other propositions in the appropriate logical order, and bearing a content intimately related to the content of all these other propositions. It follows that the demonstration leading to the erasure of TLP 1.1 (and its contrary) by the second movement of Wittgenstein's strategy can be effected only in the very particular context the first movement establishes.

In the preceding chapter I sought to show how Wittgenstein carries out this demonstration by working from within his opponents' system of thought, wherein propositions such as TLP 1.1 and its contrary appear as unproblematic, and by employing his own propositions as elucidatory or telling nonsense. So now we should ask how TLP 1.1 can function as elucidatory nonsense in the context established by the *Tractatus* while its contrary apparently cannot. To answer this question, we need to consider not just the context established by the *Tractatus* as a whole but also the invitation made in TLP 6.54—the proposition where the two movements of Wittgenstein's strategy culminate and combine their effects—namely, to consider Wittgenstein's purpose as a prerequisite for understanding the point of the propositions he advances.

So what does taking Wittgenstein's purpose as such a prerequisite involve? Diamond's (2000b) observations suggest a few answers. First, by inviting his readers to recognize his purpose in advancing his propositions and thus to enter in communion with that purpose, Wittgenstein asks them to recognize and foreground their own purposes in reading his work and thus to open themselves to the text they are reading. The request for such a communion—for such a "meeting of minds," as Frege calls it (in J. Weiner 1990, 230)—need not have anything mystical about it.[14] As we saw in the preceding chapter, it boils down to the necessary prerequisite for any adequate reading, namely, that a reader activate the requisite "good will" with respect to the text he or she engages.[15]

Second, recognizing Wittgenstein's purpose in such a fashion is equivalent to coming to realize that the purpose in question *animates* TLP 1.1 (and all propositions of the *Tractatus*) almost in the literal sense. Instead of being frozen and impersonal pieces of writing passively awaiting their impartial evaluation, TLP 1.1 and each other proposition of the work are caught in the performative movement constituting the *Tractatus,* and in contributing its own part to the deployment of this movement, each acquires a life of its own following the corresponding cycle to the end: it is born as a philosophical proposition under a certain grammatical form, it matures through its connections with the text's other propositions, and it finally expires together with them all at the close of the work.[16] In coming to realize this shifting character of these propositions, readers experience their lifecycles and the final outcome with their bodies by undergoing the relevant eureka experience. We saw how completing the second reading of Wittgenstein's work does away with the *philosophical* content of TLP 1 while making the world appear as indeed everything that is the case, without any accompanying worries.

In discussing Wittgenstein's strategy, I pointed out that we have no over-arching rules, norms, or criteria by which to arbitrate between telling and point-less nonsense once and for all in the corresponding general terms; only context and purpose can do the job. Recognizing Wittgenstein's purpose in composing the *Tractatus* and taking the context that his work establishes fully into account are thus the two fundamental prerequisites for coming to understand how TLP 1.1 can function as telling nonsense while its contrary cannot: Wittgenstein's purpose animates TLP 1.1—and not its contrary—in that the apparent sense of the proposi-tion (the philosophical content that it appears to forward) is used as an instrument in Wittgenstein's demonstration that all propositions of that work, including TLP 1.1, are ultimately no better than pointless nonsense. The demonstration is carried out within the context established by the *Tractatus* and with the means (i.e., the propositions) this context provides, while from the perspective of the demonstra-tion's endpoint, TLP 1.1 and all the other propositions appear as having participated in the form of telling nonsense before coming to die out as pointless nonsense. It is the apparent sense (the philosophical content) of TLP 1.1 that functions in that context as telling—performatively effective—nonsense, while the apparent sense of the proposition contrary to TLP 1.1, what Wittgenstein calls the "reverse" (TLP 5.2341) or the "opposite sense" (TLP 4.0621) of a bona fide proposition, seems un-able to function in that capacity. Both TLP 1.1 and its opposite sense are, of course, no better than pointless nonsense if considered as philosophical propositions in themselves.

To explicate the role of apparent sense and "reverse (apparent) sense" in Witt-genstein's demonstration—or equivalently, the live, shifting character of Witt-genstein's propositions—it is helpful to start by elaborating how the absence of overarching rules, norms, or criteria arbitrating between telling (elucidatory) and pointless nonsense may allow propositions to assume radically different functions even when sharing an identical grammatical structure, at least on the surface. To that end we might appeal once again to Weinberg's characterization of an electro-magnetic field, which is perhaps easier to handle for the purpose at hand since it is self-evidently nonsensical and hence seems to present no content or apparent sense.

Thus not only is Weinberg's "definition" to the effect that "the electromag-netic field is a taut membrane without the membrane" just as nonsensical as is the proposition, say, "the electromagnetic field is a cat's smile without the cat," but the two share the same grammatical structure and hence are nonsensical in exactly the same manner. Nonetheless, everybody somewhat versed in physics would pre-sumably feel that a world of difference separates the two propositions. The issue of distinguishing telling nonsense from pointless nonsense reduces here to the issue of determining how the two propositions can induce such a feeling of essential difference despite their grammatical congruence.

Given the previous discussions, the answer is not difficult to find. As the preceding chapter suggests, all the "expressions" or "symbols" (TLP 3.31) that Weinberg's definition brings together—"electromagnetic field," "taut membrane," "without the membrane"—retain a connection with the contexts in physics where they function straightforwardly in their standard senses, so that they are spontaneously taken as bringing those contexts into play as their background. Weinberg's students will acknowledge this straightaway, with no need to be told. Concomitantly, the teaching context where both Weinberg and his students reside and where Weinberg's definition is advanced makes manifest Weinberg's purpose for combining those expressions in a patently nonsensical way: as I explicated, that purpose is to make his students come to understand the STR *while presupposing*—and openly displaying in the wording of the definition—that classical mechanics cannot provide a conceptual basis for entering the conceptual system of that theory.

In spontaneously acknowledging such standard contexts in conjunction with immediately recognizing that purpose, Weinberg's students are almost compelled to open themselves to and enter in communion with what Weinberg had in mind in advancing his pseudo-definition and thus to activate their good will with respect to the nonsense confronting them. We may say, therefore, that although the expressions in question are combined in an ostensible definition that is patently nonsensical and instantly recognizable as such, the "propositional sign" (TLP 3.1) that this definition forms is nevertheless "applied and thought out" (TLP 3.5)—this is what makes it part of Weinberg's thinking strategy—in a way that fulfills its intended aim in the corresponding teaching context. This is why it functions effectively as telling nonsense in that context.[17]

The proposition "the electromagnetic field is a cat's smile without the cat," however, strikes us as pointlessly nonsensical because there is no readily imaginable context that could provide some more or less natural accommodation for all the expressions or symbols in it: cats and their smiles, whatever these might be, do not dwell naturally in contexts where electromagnetic fields dwell. Of course, nothing in principle prevents that proposition, or even the proposition "the electromagnetic field is mefkumsunus," from functioning effectively as telling nonsense that might serve one or another purpose in some context.[18] Alternatively, Weinberg's pseudo-definition would function as pointless nonsense for someone wholly unfamiliar with physics and its teaching contexts.

To go back to TLP 1.1 and its contrary, recall the distinction between Weinberg's position with respect to his students and Wittgenstein's with respect to his readers. In proposing his definition, Weinberg advances openly telling nonsense to help students understand the STR; in formulating TLP 1.1 (and the other propositions of the *Tractatus*), however, Wittgenstein advances, in the proper logical order (i.e., in the way systematically establishing the context of the *Tractatus*), apparent sense to show his reader that all philosophical propositions (including Wittgenstein's own and hence TLP 1.1) ultimately amount to pointless nonsense. For the demonstration to go through, TLP 1.1 and all companion propositions *should* appear to have sense, while it seems that the corresponding "reverse (apparent) sense" cannot do the job.

Since TLP 1.1 appears to be grammatically identical to the contrary proposition, and at least the text's other not explicitly nonsensical propositions are grammatically identical to propositions contrary to them,[19] there is nothing that could distinguish these Wittgensteinian propositions from their contraries—and, by the same token, from the corresponding propositions of Wittgenstein's generic philosophical opponent—other than what they appear to assert while participating in ordinary philosophical activity. What carries Wittgenstein's demonstration through is not just the apparent sense of TLP 1.1 (and the book's analogous propositions) but also the fact that Wittgenstein himself asserts that apparent sense and concurrently denies the "reverse (apparent) sense."[20] Deliberately asserting this apparent sense (and concurrently denying the "reverse" sense) is equivalent to "applying and thinking out" (TLP 3.5) the corresponding "propositional sign" (i.e., TLP 1.1 itself) in the context at hand. Obviously, the apparent sense deliberately asserted for all propositions in the *Tractatus* makes up the philosophical content that the work advances along the first movement of its strategy.

Consequently, given that the *only* difference between TLP 1.1 and its contrary is that the latter denies the philosophical content the former asserts, it follows that the philosophical content asserted by TLP 1.1—and by obvious extension, the content of all propositions in the *Tractatus* of at least the first category—plays a determinative role in demonstrating the logical inescapability of the perspective of radical immanence. Thus to say that Wittgenstein's strategy comprises a first movement, that this movement participates in traditional philosophical activity by advancing a particular philosophical content, and that both this first movement and that content should be considered seriously amounts precisely to saying that the philosophical content plays this determinative role. By the same token, the philosophical content of the proposition contrary to TLP 1.1 and thus the philosophical content of the propositions advanced by Wittgenstein's generic philosophical opponent would not allow the intended demonstration to go through and could not establish the inescapability of the perspective of radical immanence; more strongly put, they would counter this perspective. In sum, the *Tractatus*'s propositions can play their elucidatory role as telling nonsense and contribute to demonstrating the futility of all philosophical activity only if Wittgenstein deliberately asserts their philosophical content as taken straightforwardly.

So coming to understand Wittgenstein's demonstration fully means coming to realize how TLP 1.1 (and not its contrary) functions as a rung of the ladder leading out of philosophy altogether as well as how the *Tractatus*'s propositions, as laid down according to the order of logical importance, function in the same manner both individually and in concert. To realize this is also to realize how each of these propositions—or, more realistically, each significant cluster of them—comes to erase the propositions of Wittgenstein's generic philosophical opponent, that is, anyone who somehow counters the perspective of radical immanence. The final outcome consists in coming to fully understand how all of philosophy is entirely erased in the end.

To arrive at such full understanding, we would have to reenact the entire demonstration constituting the *Tractatus*. Herculean as this task might be,[21] bringing it to completion would constitute the royal road for coming to understand what the philosophical content of the *Tractatus* involves, why Wittgenstein asserts this particular content along the first movement of his strategy, and why it is this content that is determinative for demonstrating the logical inescapability of the perspective of radical immanence.[22] In short, this is the royal road for coming to understand the philosophical content of the treatise constituting the *Tractatus* as well as the reason this content forms the ordered rungs of a ladder leading out of philosophy altogether.

The demonstration in question annihilates philosophy in its entirety for it destroys every rung of the ladder as it climbs inexorably from each to the next, leaving nothing behind after it has scaled the last rung. This is the *self*-annihilation of philosophy, because it is the philosophical content Wittgenstein asserts that does the job; concomitantly, the ladder and its rungs are *self*-destroyed because the content achieving the annihilation is similarly destroyed in the process. To reenact this demonstration would be to painstakingly work through all those rungs in all their interconnections (something like strings attaching them together and thus forming a firm ladder) so as to understand how they manage, both separately and together, to attain total annihilation.[23]

Thus if philosophical content is determinative for Wittgenstein's demonstration, then it is the content that Wittgenstein asserts along the first movement of his strategy that permits a match with the philosophical content Spinoza advances in the *Ethics*. The previous discussion renders perspicuous why this match is possible in the first place and contributes to setting out the constraints framing that match and thus allowing us to take up its examination on the broader grounds pertinent to "normal" philosophy.

To pinpoint this frame of constraints and to answer the question with which I began the present section, I propose an argument to explain why the propositions Wittgenstein advances along the first movement of his strategy should be fundamentally the same as Spinoza's propositions. This is a kind of converse of the argument I forwarded at the beginning of the present chapter when I suggested that a second movement of Spinoza's strategy might be prowling within the folds of the *Ethics*.

This parallel will rest on the satisfaction of several conditionals. If Wittgenstein is compelled to deploy his strategy along two movements at the same time, with the first appearing to engage philosophy as usually construed; if Wittgenstein's *whole* procedure (i.e., both the movements in his strategy) is logically inescapable, as he takes it to be; if Spinoza, too, espouses the perspective of radical immanence; if historical context allowed Spinoza—at least in full awareness—only one movement for his strategy, leading to the construction of a philosophical theory; if, insofar as we consider the perspective of radical immanence to be a bona fide philosophical perspective, there is a *unique gist of it*;[24] and finally, if Spinoza reasons as rigorously as does Wittgenstein—if all these conditions obtain, then Spinoza's philosophical theory is the closest one can get[25] to the substance and point of the perspective of radical immanence, barring the theory's self-destruction. It follows that Spinoza's propositions best lend themselves to their own self-destruction in the way intended by Wittgenstein, so that Spinoza should have adopted precisely these along the first movement of his strategy. Therefore Spinoza's propositions *should* fundamentally match the propositions that Wittgenstein appears to advance along that first movement.

To put it in more succinct terms, given that the philosophical content of the *Ethics* matches the content Wittgenstein advances in the first strategic movement of the *Tractatus,* then, had the latter work been limited to its first movement, the two works should have resulted in the same theory. It follows as a corollary of such identity that Spinoza would have wholeheartedly rejected the proposition "the world is made up of things, not of facts"—as well as all its metaphysical presuppositions and consequences—once the necessary translation into his philosophical vocabulary had been effected. Showing this to be the case forms an object of the ensuing discussion.

Having shown how Spinoza and Wittgenstein organize the philosophical content of their work, we can embark on the examination of the corresponding match. This examination, moreover, will begin where they both begin: ontology in its connection to epistemology, or in a word, metaphysics.

CHAPTER SIX

Metaphysics

> What is needed above all is an absolute skepticism toward all
> our inherited concepts.
> —Friedrich Nietzsche, *Will to Power*, §409

HAVING SEEN WHY BOTH Spinoza and Wittgenstein begin their treatises
from ontology, one may feel compelled to ask about the specifics of this beginning.
What does ontology amount to for each author? How does each understand and
use the metaphysics supporting and structuring this ontology?

To answer these questions, recall how the scientific upheavals of his day determined how each man set out to accomplish his task. My discussion of this, however,
was restricted and in a sense anachronistic: it highlighted only the conceptual and
hence grammatical aspects of those upheavals and aimed only at displaying how
awareness—sharp in the case of Wittgenstein, diffuse in the case of Spinoza—of the
corresponding grammatical changes were instrumental in the way each conceived
his overall project. I must now adjust the historical perspective by remarking that
the relevant upheavals in both periods involved entities at least as much as they
did concepts and their grammar. These upheavals were ruthlessly destroying long-
standing ontological distinctions and wiping out whole categories of entities and
processes while bringing forth novel categories of entities, processes, and properties of both.

In Spinoza's time, the heavenly and earthly realms, hitherto held as radically
distinct, were coming to be seen as a single boundless universe that was proving
amenable to full description in the "language" of geometry. Concomitantly, the
state of rest ceased to be viewed as the "natural" end of motion, for uniform rec-
tilinear movement could go on indefinitely, and the distinction between "natural"
and "violent" motion started losing its grip on learned discourse. Simple mechanical causes came to replace the complex arsenal of Aristotelian causation, doing

away with teleology and inherent design,[1] and instead of being viewed as composed of different proportions of the age-old four elements, things were now considered merely as extended bits of matter. Quality retreated before quantity and was left only to characterize "secondary" properties, allowing our senses to distinguish one thing from another. The whole received ontological underpinning of the world was changing drastically, with a wholly novel class of entities and processes inexorably coming to the fore.

In Wittgenstein's day, the ontological commotion was perhaps even more dramatic. A single spatiotemporal continuum came to replace separate space and time in the discourse of physics; a newly recognized finite upper limit to velocity endowed that limit with the novel ontological status of a universal constant; the equivalence of mass and energy rendered materiality itself elusive. In addition, novel kinds of entities, such as the electromagnetic field, came to acquire ontological pride of place, while not just atoms but their unperceivable constituents in all their weird properties were coming to be viewed as the building blocks of the universe. Concurrently, energy was now seen as chunky; particles and waves started appearing to be facets of the same entity, thereby blowing up the distinction between what is localized and what is indefinitely extended; and people began speaking of novel entities and properties such as quantum objects and stationary and excited states. Into the bargain, the possibility merely of measuring properties of things was seen to smack against limits strangely inherent in nature but nevertheless structuring it at a deeper ontological level.

When Spinoza was writing, the ontological upheaval of his time had not yet settled down, and the ontology going with it had not become secure; novel surprises might well be in the offing. Accordingly, philosophical wisdom dictated that his philosophical theory should not take a definitive stand on what there is. To employ a mildly anachronistic formulation, his system had to consign all attendant concerns to science, leaving room for what might be coming from that direction. This system should borrow from science only the most fundamental concepts, those that were indispensable to the upheaval in question. Spinoza thus needed to limit his ontology to first principles and to what, from the science of his day, could be elevated to that rank, programmatically leaving out most ontological details. By the same token, his metaphysics had to remain minimal, even if the vocabulary he used to express it sounds rather baroque today. Spinoza starts from God and develops only what can be directly inferred from that basis in this spirit and for fundamentally this reason.

Nonetheless, the major upheaval in the sciences of Wittgenstein's day constrained his overall approach even more stringently. No entity could become fully secured against future upheavals, and hence, to the extent that the first movement of his strategy compelled him to speak of ontology and countenance metaphysics, this ontology had to be limited to principles and the metaphysics had to remain minimal all the more.

But what can this ontology of principles and this minimalist metaphysics amount to for the two authors? How can they draw a line between what they should borrow from the science of their day and what they should safely leave to it? What kind of conception of the world can such an attitude imply for those espousing the perspective of radical immanence? To address these questions, we have to enter the matter directly and start from the place where both Spinoza and Wittgenstein begin developing their own answers. This is the issue of the world or, to say much the same in the more consecrated philosophical vocabulary, the issue of substance.

Substance I

Part I of the *Ethics* is devoted to God. It begins abruptly, with no preface, introduction, or rubric that might prepare the reader for what is coming, by laying out eight definitions. They characterize, in order, the senses Spinoza will give to (1) the self-caused; (2) the finite; (3–5) substance, attribute, and mode, the three items exhausting his ontology; (6) God; (7) a free thing; and (8) eternity. A slight differentiation in formulation is to be noted: all definitions are preceded by the first-person clause "I mean" except definitions 2 and 7, which open with the impersonal "it is said to be." Presumably, Spinoza intends to distinguish formulations for which he assumes full responsibility from those he borrows from the common philosophical stock and, in borrowing, ratifies. All other definitions in all parts of the *Ethics* are preceded by "I mean," "I understand," or "I call," first-person clauses that appear in most of the scholia, prefaces, appendixes, and explications as well.

Using these definitions, Spinoza submits seven axioms, thereby launching all the Euclidean gears to demonstrate the thirty-six propositions making up part I of the *Ethics*. As the scholia and the lengthy appendix attest, this part carries an unmistakable polemical thrust whose main target is Descartes. Curley (1988) has argued convincingly that Spinoza's strategy is to start from definitions and axioms that his opponents would have allowed but then turn them around, so to speak, to establish what such opponents could hardly consider, let alone accept. Curley contends that this form of confrontation is typical of Spinoza, and it is worthwhile to join Ostrow (2002, 84) in noting that "Wittgenstein, in his characteristic way of responding to his predecessors, must be understood essentially as *co-opting* the Fregean insight, taking what he sees as important in Frege's approach and using it toward a very un-Fregean end." Spinoza and Wittgenstein thus seem to adopt the same attitude toward their teachers, one obviously related to what I have been calling working from within the opponent's system of thought.

Spinoza starts from the mathematically flavored and largely acceptable definition of God as "an absolutely infinite being [that is] substance consisting of infinite attributes, each of which expresses eternal and infinite essence" (E I def 6), and then proves in a single breath, as it were, that "there can be or be conceived" (E I p14) no substance other than Substance, that Substance necessarily exists, and that it is identical to God: "God or substance consisting of infinite attributes, each of which expresses eternal and infinite essence, necessarily exists" (E I p11). The necessary existence of the single Substance is demonstrated by three distinct proofs, all following from the same—again, more or less indisputable—definitions and axioms; the overkill, especially the third proof, is expressly intended for those who might not "easily perceive" (E I p11s) how the first, concise proof works.

This extremely dense piece of text culminates in the strict identification of God with Substance: "There can be, or can be conceived, no other substance than God" (E I p14). Spinoza starts from God and equates Him with Substance, taking no hostages. Having covered less than one-third of only the first part of his treatise, he lays to rest all interminable cogitations on what God is and on how His existence can be proved while simultaneously providing the foundation of his whole ontology.

The next proposition states the converse, so to speak, of the one just quoted: "Whatever is, is in God, and nothing can be or be conceived without God" (E I p15). This proposition sets the fundamental ontological and epistemological coordinates of the perspective of radical immanence. Since nothing can be or be conceived outside God or Substance, God or Substance can have no exterior providing a point from which anyone or anything, however exalted, could relate to the world as a whole. God, by being identified with Substance, *is* the world.

Starting from E I p16, initiating the discussion of what "follows from the necessity of the divine nature," and up to the end of part I, Spinoza demonstrates the kind of causal relations entailed by his axioms and definitions.[2] In his typically implacable and thoroughgoing manner, he presents us with a causal nexus wherein everything not only has a cause (E I def6) but is the cause of something else, thus bearing effects of its own: "Nothing exists from whose nature an effect does not follow" (E I p36). The nexus encompasses even God, who is self-caused. Thus God/Substance becomes equated with the self-caused thing, a gesture providing Spinoza's initially abstract definitions both with content and grounds for justification. In tune with the science of his day, Spinoza begins by intimating that he seeks to establish no less than the metaphysical foundations of the world's causal constitution.

This causal constitution presents a hierarchical structure. All causes are brought up to God, which is the "absolutely first cause" (E I p16c3) as well as "the efficient cause of all things that can come within the scope of infinite intellect" (E I p16c1). The scope of the infinite intellect presumably excludes things that would be self-contradictory or ruled out by the "laws of God's own nature" (E I p17),[3] and this nature is governed by the strictest necessity: "Nothing in nature is contingent, but all things are from the necessity of the divine nature determined to exist and to act in a definite way" (E I p29). The binding character of this necessity is further specified, if not strengthened, by Spinoza's demonstration that "things could not have been produced by God in any other way or in any other order than is the case" (E I p33). In thus ruthlessly depriving God of any freedom of intervention,[4] this is one of Spinoza's propositions that his contemporaries would have found hardest to swallow. Its author was acutely aware of this difficulty and so felt obliged to forward two lengthy scholia to give his proposition a chance at acceptance. More important here, however, is that this numbers among the propositions that make Spinoza's naturalism come out at its purest while rendering the radical character of his approach perspicuous.

The claim that all things are "produced" by God in the way and the order "that is the case" identifies the whole causal machinery with God's power. In other words, God is all power and nothing but power, for "God's power is his very essence" (E I p34). The deployment of this power, that is, the way God acts, inexorably follows "solely from the laws of [God's] own nature" (E I p17). Since there can be or be conceived nothing outside God/Substance, God's actions cannot be "constrained" (ibid.) by anything other than His power. The issue, then, is that of the laws of God's nature, and those laws alone, while the causal machinery in question must be strictly identical to those laws.

On the one hand, Spinoza's philosophical acumen forbids him from directly identifying the laws of God's nature with the natural laws that the science of his day was then establishing. Science had not yet arrived at the definitive formulation of these laws, and Spinoza sought to provide only their overall metaphysical underpinning. On the other hand, it was impossible not to transfer the absolutely infinite character of God to the laws of His nature in some way, thus making them appear somehow at odds with the finite causal relations holding among the finite things then concentrating the attention of science. Accordingly, Spinoza's overall metaphysical underpinning had to reconcile the infinite character of God's causal power with such finite causes. To achieve this reconciliation, Spinoza first openly assumes the distinction between what is produced directly by the absolute power of God and what is produced by an "ordinary" finite cause (E I p29d). Second, he secures the link between the two by the causal aspect of God's radical immanence, namely, by holding that "God is the immanent, not the transitive, cause of all things" (E I p18).

Spinoza secures this link in the following way. First, God's status as immanent cause implies that His power continues to dwell in the causal relations between any finite thing A and another finite thing B; God remains the efficient cause of all things, infinite as well as finite (E I p16). Second, the infinite character of His causal power is preserved because the causal relation between the two finite things A and B are for Spinoza not isolated: the finite thing A, which has caused the finite thing B, has been caused in its turn by another finite thing A_1, and this by another finite thing A_2, and so on, ad infinitum. An infinitely long causal chain is implicated in the causal production of any single finite thing. This causal infinity manifests the particular "fraction" of God's infinite causal power involving the finite thing B, while the infinitely many infinite causal chains implicated in the causal production of all things fill in the other fractions. God's ubiquitous causal presence is thereby ensured. Spinoza states it thus: "Every individual thing, i.e., anything whatever which is finite and has a determined existence, cannot exist or be determined to act unless it be determined to exist and to act by another cause which is also finite and has a determined existence, and this cause again cannot exist or be determined to act unless it be determined to exist and to act by another cause which also is finite and has a determined existence, and so on ad infinitum" (E I p28).

Spinoza tries to make the connection between God's infinite causal power and the form of causality involving finite things more detailed by bringing in what he calls the "infinite immediate mode" and the "infinite mediate mode."[5] Moreover, he introduces a partition between *"Natura naturans,"* or Nature naturing, and *"Natura naturata,"* or Nature natured (E I p29s), which helps spell out both the distinction and the connection between cause and effect from the somewhat different angle of the distinction and connection between the part of Nature that "natures," or produces, and the part of Nature that is being "natured," or produced. It is this separation between cause and effect as it accompanies the unbreakable bond between the two—required by God's immanence and ensuring Nature's unity—that allows the grand identification I mentioned in chapter 1, namely, that between God and Nature (E IV Pr).

By making God the immanent rather than the transitive cause of all things (E I p18), Spinoza excludes, *from the decisive causal angle,* the possibility that even the most powerful causal agency could possibly occupy a position external to the world, allowing it to intervene in the world's workings. For although he has already ruled out final causes as "mere figments of the imagination" (E I Ap) by holding God to be the efficient cause of all things (E I p16c1), continuing to consider efficient causation as transitive would have allowed the possibility of an external position from which some causal power could somehow get in the way of the world as a whole.[6]

By characterizing God as the immanent cause of all things, Spinoza finishes setting the coordinates of the perspective of radical immanence while manifesting a drive to go to the roots of things with a relentlessness almost unthinkable in his time.[7] The fact that Spinoza's first definition involves cause, as well as that he rounds off the first part of his treatise with the analysis of causation, marks how steadfastly his eye was turned toward natural science, then in the process of jettisoning the Aristotelian conception of causation (Carriero 2005). Spinoza's definition of God on the basis of infinity points at the same direction, for it pays due respect to the constitutively mathematical character of the new science (Mason 1999, 27). All this, then, explains why Spinoza concentrated his polemics on causation.

Spinoza's Grammar

Allow me to summarize what Spinoza has established thus far. Respecting his mathematical turn of mind, one might compress the philosophical content he advances in one grand formula: God = infinitely many infinite Attributes expressing infinite essence = Substance = the self-caused thing = Nature. I deem the equation between the two ends, namely, God = Nature, grand not only because I want to stress Spinoza's radicalism but also because this equation translates directly into the language delineating our (and Wittgenstein's) philosophical landscape. By this equation, God's existence, God's all-inclusiveness, and God's causal self-sufficiency become Nature's existence, Nature's all-inclusiveness, and Nature's causal self-sufficiency.

To put it in contemporary vocabulary, then, the philosophical content Spinoza considers himself to have established is that Nature exists, Nature is all that exists, and Nature is causally self-sufficient. Establishing such philosophical content, however, hardly offers what one traditionally would have expected. Most standard ontological issues are left to science,[8] which vindicates my claim that Spinoza's ontology should remain at the level of principle. The content in question is just a way of expressing the perspective of radical immanence, which implies that the grammatical status of the corresponding demonstrations cannot be very different from an ordered set of elucidatory comments.[9] If it is difficult to say that they constitute a Wittgensteinian ladder that readers should scale, leave, and then discard—for the structure of the *Ethics* at least appears to differ markedly from that of the *Tractatus*—we might propose instead that the *Ethics* presents itself thus far as a quasi-Wittgensteinian elucidatory *scaffold* seemingly erected around Nature from the outside for the purpose of "seeing the world (Nature) rightly": once Spinoza's readers have managed to understand the relevant demonstrations and thus climb arduously (rigorously) on the scaffold, they will see that Nature exists, that Nature is all that exists (and thus includes themselves), and that Nature is causally self-sufficient. In providing his minimal metaphysical underpinnings, Spinoza has been building just such a scaffold. Moreover, once readers have attained the view that the scaffold offers, the scaffold and its planks become dispensable.

But of course it is we who can perceive the *Ethics* in this light and not necessarily Spinoza. Elucidation does not seem to enter his text, and telling nonsense appears totally foreign to him. Hence the question inevitably surfaces: given that Spinoza explicitly and unwaveringly holds to the perspective of radical immanence, where does he take himself to be standing? In speaking about Nature as a whole, what position of enunciation does he take himself to occupy? Given the conclusions for which Spinoza explicitly argues, does he or does he not realize that even beginning to carry out his demonstrations requires him to assume a logically impossible position ruled out by the very thing he is trying to establish? Is he or is he not aware that, in trying to set up the ontological bases of Nature, his discourse must ultimately self-destruct?[10]

Feeling somewhat embarrassed on Spinoza's behalf, we might try dodging the question by adverting to historical context. In our philosophically prosaic days, we seem hardly to require a direct proof—such as the medieval proofs for the existence of God—to establish that Nature is all that exists and causally self-sufficient. This seems too obvious to require justification, let alone proof. By contrast, such proofs attracted much of the philosophical attention in Spinoza's times. In engaging his contemporary philosophical terrain, then, Spinoza was compelled to work out the demonstrations adding up to something like such a proof, for only in this way would he be in a position to fight his philosophical battle.

Moreover, one can effectively engage a philosophical terrain only to the extent that one addresses the worries that this terrain grammatically allows, however radical one's approach to these worries might be. And the philosophical terrain of Spinoza's day was not propitious to—if it was not altogether sealed against—the fundamentally grammatical question as to the perspective from which a philosophical discourse such as his could be issued. To the Scholastics, for example, theological and philosophical discourse could emerge only from a position that was by definition—and hence unquestionably—subordinate to God's; for them, we humans can conceive of God and His Attributes as the unsurpassable limits in perfection of the most admirable worldly characteristics only because God endowed us with the capacity of reason, and we always form our conceptions from "below." In this view, proofs regarding any aspect of God can be undertaken only by assuming this subordinate position, and hence questions about the locus of enunciation for these proofs are already answered in advance.

In the preceding chapter I discussed Descartes's position in this respect: by placing the solitary human ego at the center of his system, he took an indisputably huge step forward as regards the self-sufficiency of human reason. Given the conditions of elaboration we can perhaps now imagine,[11] this step might have allowed Descartes to open the question presently at issue. But Descartes's haste to subsume human reason under God's benevolent guarantee swiftly closed the grammatical matrix subtending his approach to any development along such lines. It follows that neither the Scholastics nor Descartes could offer Spinoza any grammatical leeway for posing this question with respect to his own philosophical discourse.

Admittedly, this defensive argument can protect Spinoza only historically, not logically. Yet his undeviatingly rigorous reasoning leads us to expect more, so that we should try searching deeper, for perhaps Spinoza had saved himself logically long before this attempt to save him historically.

To that end, note first that Spinoza was particularly sensitive to (what I call) grammatical structure. His reply to Jelles (L 50) emphasizes that nowhere in part I of the *Ethics* does he use the locution "one and only one" in referring to Substance, although this is standard Euclidean terminology that, on the face of it, would have strengthened the rhetorical force of his argument. The reasons he adduces are revealing: "We do not conceive things under the category of numbers unless they are included in a common class. . . . Hence it is clear that a thing cannot be called one or single unless another thing has been conceived which . . . agrees with it. . . . [It] is certain that he who calls God one or single has no true idea of God, or he is speaking of him very improperly." That is, we can assign the number one to some entity only if we can somehow antecedently imagine the existence of some other entity of the same nature without contradicting ourselves. Consequently, by qualifying something as "the one and only one," we implicitly assert that another thing of the same nature might exist in possibility while (since the thing is the only one) denying this in actuality. A locution such as "one and only one" denies in actuality what it asserts as a possibility in order to say what it says. Consequently, its grammatical structure is somewhat analogous to that of Weinberg's self-contradictory definition of an electromagnetic field.

But the two differ in a salient manner. The locution "one and only one thing" is not self-destructive in contexts where the possibility that other things of the same nature might exist is not ruled out logically or grammatically by the very nature of the thing in question. For example, the proposition that there is one and only one right triangle with base *a* and hypotenuse *b* is not self-destructive, for absent the relevant proof, nothing in the items involved precludes us from considering that there might be more such triangles or none at all. But the locution is self-destructive in contexts where it is assigned to a "thing" that cannot be a thing at all because, for example, it includes all things without logical or grammatical exception.[12] For Spinoza, God is precisely this, and hence qualifying Him as "the one and only one" would be as nonsensical as saying that "everything is one and only one."[13] It follows that even if he was in no position to relate the issue to grammar and logic as we understand them today, Spinoza was acutely aware of (what we call) grammaticality. On his part, Wittgenstein agrees emphatically and says exactly the same with respect to logic: the "logical forms are anumerical" and hence "there is no preeminent number" in it (TLP 4.128).

Given that Spinoza was aware of grammatical issues, then, we may continue to investigate whether he might have also been aware that the philosophical content he advances self-destructs. Determining whether a second movement of Spinoza's strategy lies underneath the surface of his text hinges on coming up with such evidence.

Spinoza's Strategy: Second Round

In fact, we need not search far. I already noted a short comment from Spinoza to the effect that the "so-called transcendental terms such as 'entity,' 'thing' 'something' . . . signify ideas confused in the highest degree" (E II p40s). Postponing the examination of Spinoza's conception of signification, and assuming that he took transcendence much as we do today, namely, the opposite of immanence, the terms he considers as most confusing are those issued from a vantage point transcending God/Nature. Nonetheless, Spinoza makes unrestrained use of these terms throughout his treatise, and particularly in part I of the *Ethics*. We seem almost forced to infer that Spinoza was aware that his discourse self-destructs into nonsense.

Della Rocca (1996, 132–34) does not comment on the transcendental—and hence confusing—character of such terms but does proposes a criterion for distinguishing the semantic category to which they belong. He limits his discussion to properties, proposing the designation "neutral" for those characterizing items independent of Attribute and the associated language of description. For example, "having length x" and "being true" are not neutral, for they characterize the implicated items in the language of the Attribute of Extension or of Thought, respectively. By contrast, "is a complex individual" or "bears (some number of) effects" characterizes an item in an Attribute-neutral language. We may generalize Della Rocca's designation to all terms encompassed by an Attribute-neutral language. These include "entity," "thing," "something," "essence," "existence," and so on—that is, Spinoza's transcendental terms—as well as "notions which have been called universal, such as 'man,' 'horse,' 'dog,' etc." (E II p40s), which signify ideas that Spinoza viewed as almost equally confused. Nonetheless, although this neutral language engages Substance and its inhabitants from the outside, its constituent terms are clearly indispensable for composing a metaphysical theory. Spinoza is *forced* to employ terms "signifying ideas confused in the highest degree," despite their confusing character.

Savan (1958), however, argues that if the transcendental and universal terms in question are confusing, as Spinoza admits, then such confusion, together with a whole array of other logical inconsistencies dispersed in his corpus[14]—but unac-knowledged by him, at least directly—lead to the conclusion that Spinoza himself rules out the possibility of linguistically formulating the philosophical content he intended to expound in the *Ethics*. To nevertheless make sense of that work, Savan invokes Spinoza's theory of "entities of reason," sometimes also called "entities of the imagination" (1958, 68, 69), as they are presented at various junctures of the same corpus.[15]

According to Savan (1958, 68), Spinoza considers the entities of reason or imag-ination to have no existence outside the intellect; they are not ideas that can be called true or false. Such entities are of use "only if they function as tools or mental aids and are not treated as if they had some independent status" (ibid.). Yet most philosophers have failed—Savan takes Spinoza as maintaining—to "distinguish the intellect from imagination clearly enough," and by having thus been led to as-sume that "the words they used were names of entities existing outside the intellect" (69), they have been driven to "confusion." In other words—as Savan construes Spi-noza's position—a "characteristic error" of philosophers, "misled by words associ-ated with entities of reason," has been to "hypostasize them and ascribe to them some reality outside the mind" (68). Savan also cites a letter to Peter Balling to the effect that imagination "follows in the wake of the intellect in all things, linking together and interconnecting its images and words just as the intellect does its dem-onstrations"[16] (70), and he concludes by maintaining that the theory of "the entities of reason" in fact "underlies [Spinoza's] method in the *Ethics*" (68). Accordingly, Spinoza's task largely amounts "to show[ing] the philosophers how many of their errors originate in the confusion of entities of reason with entities existing outside the intellect" and thus, "once the limitations of language are recognized," to make them and everybody else "conceive of Substance and its modes through their own *living* ideas" (71, emphasis added).

Savan's essay can be translated into the language of the *Tractatus* with almost no loss. Thus entities of reason or of imagination are essentially what is posited by Wittgenstein's elucidatory "comments" (i.e., all propositions in the *Tractatus* other than the seven head ones), with the intellect interconnecting these comments by simulating the order it follows in demonstrations or, which should be fundamen-tally the same, according to their logical importance (TLP 1n). Since the propo-sitions within which these entities figure bear no truth value, however, they are nonsensical in Wittgenstein's conception of nonsense. In that capacity, they can function only as tools or mental aids for the "clarification of thought" (TLP 4.112) or as judiciously telling nonsense. Many philosophers tend to ascribe reality to what cannot possibly possess it and hence to conflate the different ways words signify (TLP 3.323); these are misleading linguistic practices filling philosophy with "*confu-sions*" (TLP 3.324). It follows that the task of the *Ethics* is to show how the "misun-derstanding of the logic of our language" (TLP Pr ¶2) leads to such confusions and, in achieving this goal, to make its readers conceive Substance (the world) through ideas appropriate to it—or equivalently, to "see the world rightly." The underlying method of Spinoza's treatise, which is not readily apparent on the surface of his text, relies on the entities in question—on judiciously telling nonsense—and hence this text can be considered as deploying what I have been calling a second movement to its strategy. To sum up, Savan's detailed analysis of Spinoza's theory of entities of reason or imagination proves of immense value to those who still wonder what Wittgenstein might have meant by nonsense, why he uses it in the *Tractatus* as he does, and exactly how he uses it.

In reading the *Ethics* in this manner, Savan does not mention Wittgenstein. This is hardly surprising given that in 1958, when Savan's essay was published, practically nobody would have been reading the *Tractatus* in a way highlighting the role of nonsense in it.[17] Wittgenstein's work had probably not even passed through his mind when Savan was reading the *Ethics* as he did. Eleven years later, however, Parkinson (1969, 73) was in a position to reply to Savan considerately and in detail while simultaneously bringing the *Tractatus* into play. The way he does it constitutes an elegant gesture on his part, for he presents his interlocutor with one possible matrix for reading the *Ethics*—a matrix with which Parkinson explicitly disagrees—that could accommodate almost everything that this interlocutor advances. Thus Parkinson (93) writes that the contradictions Savan attributes to Spinoza

> can be avoided by distinguishing, in a Wittgensteinian manner, between showing and stating. The inadequacy in language, it may be argued, is not so much stated by Spinoza as shown—shown by the contradictions in the *Ethics* itself which, it might be suggested, is regarded by Spinoza as the most coherent system that can be constructed with words. This seems to be a valid answer though it will be noted that it implies that the *Ethics* does contain serious contradictions and that this was recognized by Spinoza.

He goes on to argue, against Savan, that Spinoza thinks that "philosophical knowledge can [nevertheless] be expressed in words" (Parkinson 1969, 94), that the theory expounded by the *Ethics* is based on what Spinoza (E II p38 and p38c) calls "common notions" (98), that this basis makes Spinoza's treatise "belong" (95) to the adequate second kind of knowledge, and finally, that "there is no reason to believe that Spinoza thought that [even] intuitive knowledge [or "knowledge of the third kind"] is ineffable" (99). Basically, Parkinson maintains that the *Ethics* constitutes what I have been calling a bona fide philosophical treatise.

The debate between Savan and Parkinson thus directly concerns the question about strategy: is there or is there not a discernible second movement in Spinoza's strategy? And if there is, how does Spinoza mark his awareness of it? As Parkinson casts it, Savan is maintaining that there is only a *second* movement, which is to say that the *Ethics* makes up *just* a ladder of nonsensical rungs.[18] Parkinson, however, maintains that there is only a *first* movement, that the *Ethics* makes up nothing but a bona fide philosophical treatise. In fact, both authors are fundamentally right in their way of describing one movement, while both are wrong in losing sight of the other movement. In other words, the textual evidence adduced by both parties makes my case perfectly: the *Ethics*, just like the *Tractatus*, is both a treatise and a ladder,[19] while the strategy it deploys includes the corresponding two movements unfurling conjointly. Still to be examined, however, are the textual moves Spinoza makes to mark his awareness that something like a second movement is at work within his text, even if he did not recognize that something to be a proper movement leading to the self-destruction of his treatise as a whole.

But first, note that the debate between Savan and Parkinson is centered almost verbatim on the issue fueling the still-raging controversies over the interpretation of the *Tractatus*, namely, the sense in which its nonsense should be taken. Given what has been happening in that Wittgensteinian battle, it is almost safe to presume that, had the controversy between Savan and Parkinson rebounded, it would have become interminable, replicating practically the same battle with respect to Spinoza. Inversely, if the gracious and unruffled Savan-Parkinson debate had been studied by Wittgenstein scholars, it certainly would have given them food for thought, perhaps even cooling down the heavy emotions brought into play (Conant 2007).

Spinoza's awareness that something like a second movement is at work underneath the surface of the *Ethics* is marked first of all by a textual move already noted. Spinoza begins almost all his definitions with "I mean" or "I understand."[20] And to see why this marks the awareness in question, one may start from the obvious: definitions (and axioms) strong enough to bear all the relevant weight make up the foundations of any geometrical edifice. Barring stylistic idiosyncrasies that have no place here, it follows that Spinoza placed the "I" of his self as the first brick of his construction because he considered an impersonal linguistic formulation to be insufficient for securing what he intended to convey. It seems that to understand the *Ethics,* the reader must recognize that it was Spinoza himself who was proposing the definitions in question and, by the same token, that he intended these to be taken precisely as he conceived them.

The following considerations may help elaborate this. For one thing, the rigor with which Spinoza reasons compels us to assume that he did all he could to clarify what he had in mind and therefore that the linguistic formulations of his definitions were as accurate as he could make them. At the same time, this rigor presumably compelled Spinoza to realize that even the most accurate linguistic formulations could not secure those definitions—and thereby the edifice erected on them—from all ambiguity and the attendant misunderstandings. Again, purpose or intention is not a thought and hence cannot be expressed in language with absolutely binding rigor; it can only be displayed in the activity it guides—here, the activity composing the *Ethics*—and by that activity itself. And this implies that no linguistic formulation, no matter how meticulously worked out, can suffice for presenting its intended sense perfectly univocally[21] and thus make its corresponding idea stand out perfectly clearly and distinctly.

Even if his intention could never be perfectly expressed in language, however, Spinoza did think that he could at least indicate that his readers must take it into account. He thus introduces his definitions with a first-person clause to invite readers to acknowledge his intention in formulating them the way he did and, in that sense, come to meet his mind. Without this precautionary textual move, his definitions risked remaining inert pieces of language, susceptible to whatever meaning the reader's imagination passively conferred to them as a matter of course. To be understood correctly, then, these definitions should be considered as *animated* by Spinoza's intentions. Savan's talk about Spinoza's "*live* ideas" can be taken as referring precisely to this.

The distinction between an inert linguistic formulation and what its author intends to convey by it obliges us to examine the signifying relation between words and ideas as Spinoza understood it. To the extent that they are not "common notions," words can express only confused ideas. Knowledge by "words" or by "symbols" is only "knowledge of the first kind" (E II p40s2), which "is the only cause of falsity" (E II p41). Conversely, "all inadequate and confused ideas belong to the first kind of knowledge" (E II p41c). Most language not only "disguises thought" (TLP 4.002) for Spinoza but, much more strongly, confuses it. Savan's argument finds its fundamental grounds here.

Nonetheless, Spinoza also holds that "there are certain ideas . . . common to all men" (E II p38c) that, precisely by being common, "are adequate in the human mind" (E II p39d). Hence humans are indeed endowed with the capacity to think correctly and to attain truth and knowledge; in addition, if someone happens to have "a true idea, [then he] knows at the same time that he has a true idea and cannot doubt its truth" (E II p43). To the extent that the ideas Spinoza himself entertained and intended to formulate linguistically were true, he necessarily knew them to be true without possibility of mistake. Parkinson's argument is for the most part grounded here.

At this point it is natural to assume that, before expressing his ideas in the shared medium of language, Spinoza had thought them through meticulously, distinguished them clearly and distinctly in the solitude of his own mind, and ascertained thereby their truth.[22] (Recall that thinking a strategy is possible for Wittgenstein, too.) But to go on and formulate these ideas linguistically, Spinoza was compelled to employ words that did not constitute common notions, for common notions cannot be formulated independent of their linguistic medium. The *Ethics* could be formulated only in colloquial language, the only available medium for philosophical communication. This is to repeat that no philosophical treatise (indeed, no text whatsoever) can do away with the failings or shortcomings of colloquial language, for the logical (here, the geometrical) order can never appear in unadulterated purity. Given that Spinoza was compelled to engage in philosophical battle, and hence to use weapons appropriate for the purpose, in order to convey what he intended to convey, we may infer that he had to employ not just confusing terms but the "most confusing" of them all. Because his opponents copiously and indiscriminately used the transcendental terms he criticizes, Spinoza sought to counter them by appropriating these terms and turning them resolutely against his opponents. This was, one might say, Spinoza's method for meeting his *opponents halfway*.[23]

The distinction between the inert content of a linguistic expression, on the one hand, and its author's intention in formulating as he or she does, on the other, thus becomes clearer. Everybody, Spinoza included, has to take precautionary measures and provide textual indicators to secure that the relevant intention is conveyed as accurately as possible while keeping in mind that this aim, although mandatory (for no text can do without it), is unattainable.

Given this, Spinoza's problem in the case at hand was how to mark the distinction between the confused ideas that the words in his definitions risked expressing publicly and the indubitably true ideas that he intended the words to express and did express in his mind. On the basis of his understanding of language, the most appropriate means for marking this distinction was to point directly at his own mind, where the true ideas he intended by these words were residing. The simple word "I" preceding the majority of his definitions constitutes the most appropriate such pointer, for it stands in for the bearer of these intentions, namely, Spinoza's mind.

In fact, Spinoza states this almost explicitly. For example, he says that "most errors result solely from the incorrect application of words to things. . . . [Yet] if we *look only to the minds* [of people whom we judge to have erred, we can see] that they are indeed not mistaken; they seem to be wrong [only] because we think they have in mind" something other than what they do have in mind (E II p47s). It follows that "most controversies"—and hence the controversies the *Ethics* would inevitably fuel—"arise [either from the fact that] men do not correctly express what is in *their* minds or [from the fact that] they misunderstand *another's* mind" (ibid., emphasis added). Two points follow. First, if Spinoza worked out his ideas in his mind the best he could, he could not be wrong about them. Second, since he was obliged to express these ideas in colloquial language, he could not ensure that they would be correctly understood even if he expressed them as rigorously as possible. By opening his definitions in the first person, then, he emphasizes that readers who desire to understand his work should try their best to forgo the words and strive instead to "look to his mind."[24]

To make the issue still clearer, note that the meaning intended by the author of a linguistic expression—what makes the expression alive—is taken for granted in most instances of communication. If the meaning is not immediately obvious, however, and the expression appears bizarre or abnormal, anyone who wants to understand the communication will inevitably inquire about the speaker's or author's intention. Wittgenstein says as much, and it is instructive to see why.

Consider the following inversion of a Fregean position: "Frege says: every legitimately constructed proposition must have a sense; and I say: every possible proposition is legitimately constructed, and if it has no sense this can only be because *we* [emphasis added] have given no *meaning* to some of its constituent parts" (TLP 5.4733).

This inversion is telling. In maintaining that every legitimately constructed proposition must have a sense, Frege is subjecting the sense of propositions to (logical) laws of legitimate construction, which he understands to lie outside and thus overarch language; a proposition has sense because it obeys these overarching laws. Wittgenstein has no issue with laws of legitimate construction: if every possible

proposition (any piece of language) is legitimately constructed, there can be no issue of legitimizing laws at all. What makes a piece of language legitimately constructed can thus be only the mere fact that we humans are always already inside
language—as we are always already inside the world—the *mere fact* that we employ
language. Every *piece* of language is legitimately constructed merely because it is a
piece of *language*.

Of course, some pieces of language appear nonsensical—inert, idle, exempt
from the work they should be doing as pieces of language. According to the preceding discussion, however, these pieces of language appear as nonsensical not because they are not legitimately constructed but only because we cannot recognize
the intention that, by conferring meaning to their constitutive parts, would have
made them work as pieces of language and thus animated them. The proposition
"Socrates is identical" is nonsensical only insofar as "we have not made some arbitrary determination" (TLP 5.473) of the way the adjective *identical* is to be taken
here. Once we are told such a determination—that is, once we come to recognize,
though always imperfectly, the intention behind its formulation—the proposition,
and any piece of language, can do the work it should be doing and becomes, in this
sense, animate. It follows that Wittgenstein's "we" in the context at hand and the
"I" preceding Spinoza's definitions indicate exactly the same point, namely, the role
of intention in conferring meaning to linguistic formulations and thus animating
them in a given context.

It is worthwhile noting that Wittgenstein was aware that taking the "we" as
the agency conferring meaning to linguistic expressions risked falling short of the
fundamental principle of Frege's approach, according to which the logical should
be kept separate from the psychological thoroughly and systematically. He thus acknowledged that by invoking intention in the midst of his "method" (TLP 4.1121)—a
move equivalent to "demonstrating" to his generic philosophical opponents that
they "had given no meaning to certain signs of [their] propositions" (TLP 6.53)—
he creates the "danger" that his work[25] will become "entangled in unessential psychological investigations" (TLP 4.1121). In the following chapters I will show how
Wittgenstein tried to pin down the logical dimension of intention so as to mark its
separation from the psychological dimension as clearly as possible.

We can now see in ampler terms why Diamond (2000) was right to hold that
we can come to understand the *Tractatus* only on the condition that we have already understood Wittgenstein's intentions. Yet the preceding discussion is meant
not only to clarify the role of intention with respect to language and to point out
that the *Ethics* vindicates Diamond's reading of the *Tractatus* but also to stress that
Spinoza was aware that a second, self-destructive movement of his strategy lies underneath the text of the *Ethics,* at least to the extent that the historical context allowed him to recognize this.[26]

Epistemology I

It is time to discharge some of the debts I have incurred. I must try to spell out
Spinoza's conception of language as it is intertwined with the kinds of knowledge

he proposes and hence with that particular aspect of his epistemology. Having discussed his third kind of knowledge in chapter 2, here I will address only Spinoza's two other kinds. To shed light on these matters, it is profitable to lay them out more or less in the order that Spinoza considers them in part II of the *Ethics* and hence to start from the unity between the human body and the human mind.

To begin with, then, Spinoza holds that "whatever happens in the object of the idea constituting the human mind is bound to be perceived by the human mind, [while conversely] nothing can happen in that body without its being perceived by the mind" (E II p12). The human mind perceives everything that happens to the human body, and everything that happens to the human body is perceived by the human mind. The indissoluble union of the two is set into place early in part II of the *Ethics,* before the digression on the nature of bodies. Since "the object of the idea constituting the human mind is the body . . . and nothing else" (E II p13), it follows that the ideas forming the human mind (E II p15) are and can only be ideas of the way the human body is affected by other bodies (E II p16). The ideas in question "must involve the nature of the human body together with the nature of the external bodies" (ibid.), and since one basic axiom of the *Ethics* is that "the knowledge of an effect depends on, and involves, the knowledge of the cause" (E I a4), all causal interactions of the human body with other bodies are and must be presented or captured by corresponding ideas, no matter how confused, in the human mind.

To tackle knowledge, Spinoza prepares the ground, so to speak, by defining what he calls "images," which he characterizes as "affections of the human body the ideas of which set forth external bodies as if they were present to us" (E II p17s). That is, images are those affections or modifications of the human body the ideas of which present ("set forth") in the mind the external bodies proximately or remotely responsible for the modifications[27] as if they were present. By specifying that "when the mind regards bodies in this way, we shall say that it 'imagines'" (E II p17s), Spinoza apparently allows the term *images* to refer to the ideas capturing such modifications as well: "The human mind, being limited, is capable of forming in itself only a certain number of distinct images" (E II p40s1). The reason Spinoza takes the liberty of such a double use of the term will appear in a moment.

Spinoza entrusts the imagination with a lot of work in the *Ethics*. Among other things, it plays a crucial role in his conception of memory, which forms a deductive prerequisite for formulating the different kinds of knowledge. Thus memory is "simply a linking of ideas involving the nature of things outside the human body, a linking which occurs in the mind parallel to the order of linking of the affections of the human body" (E II p18s). Spinoza specifies that this sort of linking makes "the mind, from thinking of one thing, pass on straightway to thinking of another thing which has no likeness to the first" (ibid.). This linking is directly related to habit, for in undergoing it, "everyone passes on from one thought to another according as habit in each case has arranged the images in his body" (ibid.).

Habit is the relatively frequent occurrence of bodily modifications in a certain order. These modifications tend to become deeply impressed in the body and persist there as more or less ineradicable traces retaining the imprint of whatever caused them:[28] lingering alterations of the human body left behind after the associated proximate causes have finished their work. Given the previous considerations, ideas "correspond to" these modifications. We may thus presume that to the degree that such bodily modifications persist, the ideas capturing or presenting them persist—remain vivid—to an analogous degree in the mind. The ideas that persist in this manner make up the stock of ideas in memory, with vividness varying according to persistence. Memory thus appears to be the mental correlate of habit. Nevertheless, memory induces us, unaware, to link *subsequent* ideas (ideas corresponding to subsequent modifications of our body) with some idea already in its stock. That is, the idea corresponding to a new modification of the body tends to be spontaneously associated with the idea corresponding to some of the modifications that have already been caused by habit. Moreover, passing spontaneously from one idea to another according to the order habit has given the images in the body is a kind of "free association."[29]

As I will explicate more fully in a moment, Spinoza considers this "associative" linking of ideas following the order of habit to constitute the first kind of knowledge, which he also calls "opinion" or "imagination" (E II p40s2). The ideas implicated here are "inadequate," like "conclusions without premises" (E II p28d), while the knowledge they constitute makes up only "fragmentary and confused knowledge" (E II p29c). By contrast, Spinoza has already distinguished (in E II p18s) the associative linking in question, which can constitute only fragmentary and confused knowledge, from the "linking of ideas in accordance with the order of the intellect whereby the mind perceives things through their first causes, and which is the same in all men" (ibid.). It is this linking that can lead to adequate knowledge of the second kind.

Spinoza's distinction can be spelled out as follows. First of all, the human body is continuously undergoing modifications caused by external bodies. These modifications occur one after another and are thus linked by their order of occurrence. At the same time, each of these modifications is presented in the mind by its corresponding idea, which in some way captures both the modification in question and the body that caused it (E II p16). Now the distinction at issue implies that the mind has two ways of taking stock of the order linking these modifications. One way proceeds "in accordance with the order of the intellect" (E II p18s). Since the laws of divine nature (E I p29) necessitate the order of occurrence for these modifications, the mind can conceive this order *actively*, by doing the necessary work; that is, it can identify the chain of causes in operation and connect it, as far as human finitude allows, to the laws in question. This amounts to comprehending the order of the modifications of our body as necessary (E I p29 and p33). The other way of assessing the same order proceeds in accordance with associative linking. In this process, the mind perceives *passively*, merely recording without doing any real work, the occurrence of the modifications in question and orders these ideas through the spontaneous workings of memory. This way can lead only to the fragmentary and confused knowledge of imagination and to apprehending the order of bodily modifications as due to imaginary causes, or as contingent (E I p29).

Note that since "the linking of ideas in accordance of the order of the intellect is the same for all men" (E II p18s), they are all in a position to attain exactly the same knowledge once they comprehend the necessary order of the modifications of their bodies. From the fact that "all bodies agree in certain respects" (E II p38c), it follows that the second kind of knowledge, as knowledge of what is thus common to all things, is objective: it is knowledge comprehending the necessary connections among the relevant causes as expressing the laws of divine nature (E I p29). Since the order of the intellect is the same for all, however, this knowledge is intersubjective: it is the same for all and in principle attainable by all.[30] Because it is objective, this knowledge is adequate, and because it is intersubjective, it is equally adequate for all. Moreover, in maintaining that people are endowed with this capacity for adequate knowledge because "all bodies [including their own] agree in certain respects" (E II p38c), Spinoza is maintaining that they have this capacity because they are themselves a subordinate part of the order of Nature, sharing things in common with all its other parts, each a particular part on a par with all the others despite the all too obvious variety and multifarious wealth of our surroundings.

In addition, since "those things that are common to all things and are equally in the part as in the whole can be conceived only adequately" (E II p38), it follows that people are endowed with the capacity of adequately knowing all things in what they share in common, or in other words, of knowing all there is to know up to the limit of human finitude. Moreover, once one knows, one cannot doubt that one knows, for "he who has a true idea knows at the same time that he has a true idea, and cannot doubt its truth" (E II p43). Hence "nobody who has a true idea is unaware that a true idea involves certainty" (E II p43s).[31] By thus attributing our capacity for adequate knowledge (which in addition self-authenticates as certain) to our being part of Nature and hence to our being internal to It, Spinoza explicitly presents the perspective of radical immanence from the *epistemological angle*. By the same token, the traditional question as to how knowledge can be possible is swept away.

Allow me to summarize. The "ideas common to all men" are common to all precisely because "all bodies"—including human bodies—"agree in certain respects" (E II p38c). The mind is capable of taking account of this agreement, making it "the basis of its reasoning processes" (E II p40s1), and working to attain "knowledge of the second kind," or "reason" (E II p40s2). This knowledge consists in the adequate ideas (E II p38) presenting or capturing the shared respects in question, as well as properties of things that can be inferred rigorously from these ideas (E II p40s2). The ideas in this precise sense common to all people "are called common notions" (E II p40s1).

Spinoza's use of the passive voice here ("are called") almost certainly refers to Euclid. In fact, immediately after presenting the twenty-three definitions and the five postulates of Book 1 of the *Elements,* Euclid formulates five additional propositions, comprising six claims, that he places under the heading of "Common Notions." These too are axioms (or postulates), but they do not pertain to geometry in particular. They concern equality (and inequality) in general and are thus of universal validity. To wit (Euclid 1956): Things that are equal to the same thing are equal to one another (1); if equals be added to (or subtracted from) equals, the wholes (or remainders) are equals (2 and 3); things that coincide with one another are equal to one another (5); and the whole is greater than the part (6). The common notions in question not only cover all things without exception; as universal, they are also presupposed by geometry and are essential for the effective flow of the geometrical order itself. In setting them apart from the purely geometrical postulates, Euclid circumscribes the particular subject matter of geometry while keeping it united to the universal order of things. The formulation of the common notions in the rigorous geometrical form infuses the rigor of geometry to all things covered by these notions outside that discipline, that is, to everything.

These characteristics must have appealed to Spinoza, for they perfectly suit the job he required them to perform. Not restricted to Euclid's five synonymous propositions and wisely remaining unspecified,[32] Spinoza's common notions are the ideas that present what all things share in common. The aspects of things shared in this way and the ideas presenting them guarantee, at both the ontological and epistemological levels and with the rigor of Euclid's system, the unity of Nature while leaving all particulars (Spinoza's "singular essences") free to do their work (deploy God's power) on their own. Spinoza's common notions mentally present and order the grand metaphysical picture of things in a way that makes them all cohere while allowing each to pursue its own course.

Language

Given Spinoza's rigor, we should not consider his characterization of notions as the adequate "*ideas* common to all men" to be either fortuitous or an unreflective loan from Euclid's terminology. In fact, at one point Spinoza uses the terms "common notions," "transcendental terms," "notions called universal" (E II 40s1), "words," and "symbols" (E II p40s2) in a single context; this part of the *Ethics* presents the core of his conception of language. We are therefore justified in inferring that Spinoza refers to notions as adequate ideas because he understands them as not only mental but also *linguistic* items. The prominent position given to common notions justifies what Parkinson maintains: given that Spinoza was compelled to employ "confusing" universal and transcendental terms in expressing his views, common notions do express in language the adequate ideas bearing the same name in his philosophical vocabulary. Since language is altogether confusing, however, we must try to understand how common notions manage to escape this confusion.

First, note that Spinoza takes linguistic entities to be *bodily* affections: "it is essential to distinguish between ideas and the words we use to signify things, . . . for the essence of words and images is constituted *solely by corporeal motions* far removed from the concept of thought" (E II p49s, emphasis added). Spinoza clarifies his meaning by explaining how the word *apple* acquires its signification: "For example from hearing the word 'pomum' [apple,] a Roman will straightway fall to thinking of the fruit, which has no likeness to that articulated sound or anything in common with it other than that the man's body has often been affected by them both; that is, the man has often heard the word 'pomum' while seeing the fruit. So everyone will pass from one thought to another according as habit in each case has arranged the image in his body" (E II p18s).

Spinoza's view may be spelled out, with a Mediterranean bent, as follows. What the word *olive* expresses for us is the confused idea correlative to the many instances of the bodily modifications we have undergone by having come into "casual" (E II p40s2) contact with various kinds of olives (Corfu olives, Cretan olives, Kalamata olives, Lucca olives, industrially processed olives, and so forth) together with hearing (by means of our ears, a bodily organ) the sound of the word or seeing (by means of our eyes) its written sign. "The human body is limited" (E II p40s1) and hence cannot separately and sharply host all the modifications it undergoes by coming into contact with different olives: it "can form only a certain number of distinct images" (ibid.). Thus, the significant differences between olives get wiped out in the body, leaving behind only a smeared-out bodily impression coming from the superimposition of all these olive-modifications.

At the same time, all olive-modifications have become unified in the body because it has always undergone one particular modification in all these occasions: *all* modifications brought about by bodily contact with olives have been accompanied by an experience of the same aural or visual expression of the word *olive*. Our bodily "olive habit" is constituted in precisely this way. By the same token, all the ideas correlative to the bodily modifications constituting such an olive habit are brought blurrily together in the form of olive memories.[33] In that guise, they make up a fragmentary and confused idea of what an olive is or what the word *olive* expresses for us. The mind, to the extent that it does not expend the necessary effort to sort out things, can perceive these ideas only unclearly and indistinctly, only as "opaque and blurred" (TLP 4.112). The word *olive* constitutes a confusing "universal term" in that the idea it expresses is necessarily confused. Spinoza gives exactly the same account for the more pronounced case of transcendental terms, where "all images" (E II p40s1) become murkily juxtaposed in such a manner. This account can be extended straightforwardly—by some causal theory of reference—to cases where we relate to words without having come into direct bodily contact with any of the things these words express for other people.

Presumably this is why Spinoza juxtaposes the terms *word* and *symbol* as synonymous (E II p40s2). A word is a symbol because it *stands in for* something different from it: the modification of my body consisting in my bodily uttering the sound of the word *olive* (or bodily writing its sign) stands in for my bodily olive habit. More generally, a *bodily modification* of the right sort stands in for another bodily modification of the relevant sort, or rather, for the smeared-out total of bodily modifications constituting the corresponding habit. And since my olive habit will typically differ from any interlocutors' olive habits, misunderstanding in communication is the rule rather than the exception: my bodily acts of uttering (or writing) the word *olive* express my olive memory and stands in for my olive habit, while my interlocutors' bodily acts of hearing (or reading) the word attach the corresponding sound (or sight) to their olive memories and stand in for their olive habits. And their olive habits and olive memories usually differ from mine. "Most controversies" arise from such "misunderstanding" (E II p47s), and "therefore it is not surprising that so many controversies have arisen among philosophers" (E II p40s2).

Note that the preceding picture of things easily accounts for meaning change. Thus the next occurrence of my hearing (or reading) the word *olive* might take place in a context that does not allow the associated bodily modification to fit well with—to become typically superimposed on—my already constituted and anyway murkily demarcated olive habit. In such a case, if appropriate overall conditions are satisfied, this new occurrence will tend to enlarge, restrict, or even drastically change my constituted olive habit. In such cases, my mind will concomitantly realize that my old idea of an olive should be abandoned to the profit of the novel idea, and my olive memory will adjust correspondingly. The possibility of working thus on the confused ideas constituting our memories is tantamount to the capacity for coming up with corresponding adequate ideas in our minds.

Further, in expressing some fraction of memory and in standing in for the correlative habit, the word *olive* appears to connect the realm of ideas with the realm external to the mind and thus to mediate between Spinoza's two Attributes. In itself, this or any other word possesses two inseparable facets (as with a coin),[34] one bodily and one mental, so that it appears to participate in these two realms simultaneously. In my view, Spinoza uses the word *image* for both bodily and mental items because he acknowledges and respects this fundamental characteristic of linguistic entities: it is *in the nature of language*—in that of the signifying relation, with its two facets of "standing in for" within the body and of "expressing" within the mind—that all linguistic items should appear to play such a mediating role and thus seem to partake of both Attributes at once.

If this were not the case, if such an apparent mediator did not exist, there would be no means for expressing, receiving, or communicating ideas (items of the mind), for such expression, reception, and communication can take place only by means of the body. Absent this apparent mediator, moreover, it would be impossible to deal with mental and extramental items in a unified mode and thus engage in philosophical discourse.

Nonetheless, for deep epistemological reasons yet to be examined, Spinoza's two Attributes should be absolutely separate, with no connection, no relation, and no channel of communication between them. This uncompromising separation, imposed by God's nature, destroys the possibility of any kind of link between the two apart from the kind of correspondence already mentioned. It follows that all mediators presenting themselves as innocent bridges are confusing in some degree. Language—the signifying relation—which appears in the guise of such a mediator, is altogether confusing precisely for this reason.

Yet the situation as described implies that some linguistic items manage to escape this plight, namely, common notions. Common notions, as *notions,* are linguistic items. But the way Spinoza understands them makes them the only linguistic items in the position to display the previously mentioned correspondence between bodily modifications and ideas without violating the absolute separation of the two realms. And they perform this feat not because they manage to do the job of the impossible mediator but because they retain the absolute separation while simultaneously displaying the correspondence; the ability to do this amounts to the *constitutive* capacity of ideas to capture or to present things belonging to an absolutely separate realm. This absolute separation together with that constitutive capacity is linguistically borne by one key word of Spinoza's conception of the perspective of radical immanence and hence of his epistemology: the small and apparently insignificant word *common*.

As we saw, adequate knowledge is possible because all bodies, including human ones, share things in common. This is what makes them parts of unified Nature, and it is on this basis that they all follow Nature's necessary order. Concurrently, ideas, too, are parts of unified Nature and so must follow its necessary order. But one can attain adequate knowledge of bodies different from one's own only on the basis of what it shares in common with the bodies acting on it causally. Hence adequate knowledge of the necessary order of Nature presupposes that all bodies are connected causally. At the same time, such knowledge is adequate only if the ideas composing it are logically (geometrically) connected with one another. All ideas are logically (geometrically) interconnected to other ideas; no idea can ever be isolated from other ideas, making interconnection a property common to all ideas. These apparently different uses of the word *common* are brought together by a fundamental thesis of Spinoza's epistemology: "The order and the connection of ideas is the same as the order and the connection of things" (E II p7), a thesis I will discuss in the following chapters.

As linguistic entities, then, common notions separately and sharply *materialize* in the bodily realm what they distinctly and clearly *express* in the realm of ideas, namely, some well-delimited ingredient of the shared (common) order and connection at issue. A given common notion is thus a bodily modification that stands in for—acts as symbol of—some other bodily modification, one that, in being separately and sharply marked, is not subordinate to habit; simultaneously, the same common notion expresses in the mind the idea correlative to that modification, the idea that, in being adequate, is not subordinate to memory. Thus the ideational facets of common notions can be taken as the rigorous metaphysical concepts that are necessary and sufficient for laying out adequate knowledge, though they are sufficiently few, general, and void of content to leave the scientific endeavor free to go on with its business. Spinoza's metaphysics is minimal with respect to this epistemological aspect as well.

Going back to the *Tractatus* after concentrating on the *Ethics* so long, it suffices to note one thing. Once we acknowledge that Spinoza does not appear to distinguish words from propositions and ordinary worldly things or bodily modifications from facts, and once we take into account that Wittgenstein was not unduly concerned with the characteristics of concepts, metaphysical or otherwise, the extent to which the one's conception of language matches that of the other again proves arresting.

To wit: While for Spinoza "the essence of words and images is constituted solely by corporeal motions of the human body" (E II p49s), for Wittgenstein "language is part of the human organism and not less complicated than it" (TLP 4.002). While Spinoza holds that a word is a bodily modifications that stands in for another bodily modification, Wittgenstein counters that "a proposition is a picture of reality" (TLP 4.021), and since "the picture is a fact" (TLP 2.141), a proposition is a fact that stands in for another fact (TLP 5.542). While Spinoza maintains that a word is a bodily item expressing some idea (however confused), Wittgenstein says that "in the proposition a thought is expressed perceptibly through the senses" (TLP 3.1). More specifically, Wittgenstein's distinction between "the part of the proposition which characterizes its sense" (TLP 3.31) and the part that is "perceptible by the senses" (TLP 3.32) almost exactly matches Spinoza's distinction between the bodily and the mental facet of words as I have explicated it.

In addition, we have seen that "language disguises the thought" (TLP 4.002) for Wittgenstein and confuses it for Spinoza, while both see philosophical controversies as arising for fundamentally the same reasons. Moreover, with respect to knowledge in particular, recall from chapter 2 that Spinoza's knowledge of the third kind, or intuition, impressively matches Wittgenstein's notion of contemplation and that both treatises lead to philosophical silence after each has gone through all the fundamental aspects of philosophy.

Of course, the match is not perfect. For example, Spinoza focuses on words while Wittgenstein focuses on propositions, and the former does not seem to talk about facts at all while the latter talks about them all the time. In addition, one refers to thoughts whereas the other speaks of ideas. And the list could go on. The reason for and the significance of these discrepancies will occupy the next chapter, but to clear the ground for that chapter, recall that an overall discrepancy inevitably remains: since the *Tractatus* is not a textbook, the workings of language in its relation to Spinoza's first kind of knowledge do not interest Wittgenstein for their own sake, and hence they cannot be put up for matching. At the same time, Wittgenstein's overall understanding of imagination seems to allow it to work in accordance with the way it works for Spinoza, that is, through a kind of free association and as a tool or mental aid (in the guise of telling nonsense, among other things) for helping us out.

So again, Spinoza's third kind of knowledge nicely correlates with Wittgenstein's contemplation, but to what extent, if at all, does the logical requirements that the *Tractatus* lays down with respect to thought match Spinoza's second kind of knowledge? Before answering that, however, we must look at Wittgenstein's conception of substance.

Substance II

For Wittgenstein, "substance is what exists independently of what is the case" (TLP 2.024), and since "the world is *everything* that is the case" (TLP 1, emphasis added), substance must be what makes or allows anything to be the case. This is the "what" shared by everything that is the case, the "what" grounding the first occurrence of "is" in TLP 1, the absolutely unqualified "what" referred to by the grammatical subject of this verb.

Wittgenstein spells out this minimalist conception of substance by specifying that what "we need [in order] to understand logic is not that such and such is the case but that something *is*" (TLP 5.552). Substance is thus *what is* in this absolutely unqualified sense,[35] the "content" (TLP 2.025) securing that "there can be a fixed form of the world" (TLP 2.026) and that "reality" can have "a form," the form Wittgenstein calls "logical form" (TLP 2.18). In other words, substance is the absolutely unqualified content that bears logic, the minimal substantive support logic requires if it is to exist (TLP 5.5521). "Logic," while "*preceding* every experience that something is so," and thus coming "before the How," nevertheless cannot come "before the What"[36] at issue, namely, substance (TLP 5.552, emphasis added). In the metaphysical order at issue here, substance is thus the absolutely unqualified "what" that necessarily has precedence even over logic. But since Spinoza's "God or Nature" simply *is,* we may infer that Wittgenstein's substance matches Spinoza's Substance as well as we could wish it to do, at least on this general level.

In discussing substance, however, Wittgenstein goes one step further by maintaining that the "substance of the world is formed by objects" (TLP 2.021). These are substance in being "the fixed" and "the existent" (TLP 2.027), though there is nothing substantive inside or beyond them, for they are "simple" (TLP 2.02). The "combinations of these objects form atomic facts" (TLP 2.01), and each such combination stands on its own without requiring any kind of glue, any additional substantive item, to keep the objects together: these are such as to combine like "links in a chain" (TLP 2.03). Wittgenstein holds further that "signs," which are equally "simple" (TLP 3.201), "name" the objects in question (TLP 3.202), and hence they "mean" them (TLP 3.203) in the context (TLP 4.23) of a "completely analyzed proposition" (TLP 3.201), which by the same token involves only "elementary propositions" (TLP 4.21), each of which "asserts the existence of [the corresponding] atomic fact" (ibid.) while itself being a "concatenation of [the corresponding] names" (TLP 4.22). In addition, since "in propositions thoughts can be so expressed that to the objects of the thought correspond the elements of the propositional sign" (TLP 3.2), *completely* analyzed propositions correspond to thoughts that are crystal clear and perfectly distinct (the positive opposite of the "opaque" and "blurred" thoughts of TLP 4.112) in the sense that the thoughts in question cannot be clarified further.

In case one wonders why Wittgenstein proceeds in equating the mere "what" of substance with the objects in question, recall that, unlike Spinoza, Wittgenstein focuses on the "logic of our language" (TLP Pr ¶2). Given this focus, he is compelled to advance some fundamental postulate allowing him to go on, though given the overall character of his approach, this postulate should not contradict what we attest all the time in our everyday linguistic dealings: "all propositions of our colloquial language are *actually,* just as they are, logically completely in order" (TLP 5.5563, emphasis added). Thus the postulate Wittgenstein advances is as metaphysically innocuous as it could be, for it is none other than "the postulate of the determinateness of sense" (TLP 3.23). Wittgenstein's logical rigor obliges him to acknowledge that he is advancing a postulate here; however, in his typical overly concise manner, he hastens to equate it immediately with the "postulate of the possibility of simple signs" (ibid.). It should be obvious that by meaning the objects we are discussing (TLP 3.203), the simple signs in question presuppose these objects in the metaphysical order. We might infer that Wittgenstein starts his treatise from objects, which he does not openly postulate, rather than the determinateness of sense, which he openly postulates, on the grounds explicated in the preceding chapter.

Metaphysical order dictates that what concerns the world should precede what concerns language, and hence objects should come before the determinateness of sense (and the implied simple signs) in the exposition of philosophical content. This is indeed what happens in the *Tractatus*. Nevertheless, in a discourse that logically requires postulates and reasons from them in a logically rigorous way, different logically equivalent modes of presenting philosophical content are generally possible. The most typical is the axiomatic mode, according to which all postulates grounding the philosophical approach are set forth at the beginning of the exposition. This is, of course, Spinoza's mode.

Yet the overall aim of the *Tractatus* (doing away with philosophical worries and hence with philosophical postulates and philosophical content altogether) dictates that, among logically equivalent modes, the one to be chosen should render the required postulates in the least metaphysically worrisome terms. The postulate of the determinateness of sense is precisely this. Given that, as I will show, this logically implies simple signs and thus objects, Wittgenstein presumably chose to forgo an axiomatic presentation, even if the objects he discusses risk appearing too "metaphysical," before the *Tractatus* has managed to retroactively defuse their metaphysical charge and do away with the philosophical content they put forth. By the same token, an axiomatic presentation would have obscured Wittgenstein's final goal, for it could not have avoided presenting a postulate that would have concerned objects as a particular—and particularly controversial—philosophical thesis, and as we have seen, Wittgenstein had nothing to do with philosophical theses even if the first movement of his strategy compelled him to appear as advancing them. Perhaps Wittgensteinian objects have given rise to conflicting interpretations because, among other things, no one has ever taken these considerations into proper account.

To understand how determinateness of sense implies simple signs, and thus their correlative Wittgensteinian objects, note first that, on the postulate stipulating the determinateness of sense, any proposition leaving its sense indeterminate (in some context not up for examination) should have its sense fully determined by a "complete analysis of this proposition" (TLP 3.25). To claim that the determinateness of sense implies simple signs—"simple names" that "cannot be analyzed further" (TLP 3.26)—is then simply to claim that the analysis in question has a definite terminus. If it lacked one, then "whether a proposition had sense would depend on whether another proposition was true" (TLP 2.0211), thus leading to infinite regress. Put as simply as possible, to the extent that, for example, the sense of the proposition "my rat sleeps on my cat" requires determination in some given context not at issue, achieving that determination would require ascertaining whether my rat is the kind of thing of which it is possible that it sleeps on my cat, whether it is possible for a rat to do the kind of thing in which sleeping on my cat consists, whether this rat is the kind of thing for which it is possible that the rat is mine, and so on. If there were no terminus to all these judgments, nothing would stop the analysis from going on indefinitely, thus keeping the sense of the proposition indefinitely indeterminate.

Wittgenstein holds that in using this complete analysis, "we must come to elementary propositions which consist of names in immediate combination" (TLP 4.221), or "concatenation" (TLP 4.22), and that there is "one and only one" (TLP 3.25) such analysis for a proposition. This short argument explicates why the postulate of the determinateness of sense entails that a complete analysis exists; its uniqueness follows from a simple argument: a second complete analysis of a given proposition would be possible only if the sense of every elementary proposition appearing in the terminus of the first analysis but not in that of the second (and there must be some, for if all the terminal elementary propositions were identical, nothing would effectively distinguish the two analyses) could be cashed out in terms of some combination of elementary propositions appearing in the terminus of the second analysis, for the sense of the initial proposition could be preserved only in this way. But if its sense could be cashed out in such terms, the first elementary proposition would not be elementary, and Wittgenstein forbids us to consider any other option, for "from an elementary proposition no other [elementary proposition] can be inferred" (TLP 5.134). Thus, the fact that a proposition has one and only one complete analysis makes Wittgenstein's postulate of the determinateness of sense entail the postulate of the possibility of simple signs.

Now in being signs, the signs in question must be the signs *of* something, and their status as simple implies that the something of which they are the signs must be simple as well—otherwise we would be requiring simpler signs to mean or name the corresponding simpler constituents—so that the objects that the simple signs mean or name, the "objects forming substance" (TLP 2.021), are correspondingly simple. Thus if "the world had no substance" formed by objects that are thus simple, then whether a proposition has sense would "depend on whether another proposition was true" (TLP 2.0211) up to infinite regress. If the world had no such substance, however, "it would be impossible to form a picture of [it] (true or false)" (TLP 2.0212), for the infinite regress could never allow us to fix that picture precisely. If that were the case, language, which "expresses thought" (TLP 3.1), would be left hanging in the air, and since thought is "a picture of the facts" (TLP 3), a failure to provide for substantive objects would make thought impossible as well. All this shows why Wittgenstein's account requires substantive objects that are simple and how the postulate of the determinateness of sense implies—and is implied by—such objects.[37]

Further, note that, according to Wittgenstein, "logic treats of every possibility and all possibilities are its facts" (TLP 2.0121). Thus, if his objects are the substantive "what" bearing logic in the way suggested, then this "what" bears these possibilities. More precisely put, each object is constitutively associated with its characteristic *manifold of logical possibilities,* which makes up and is identical to the "form" of that object. Wittgenstein states this concisely: "Substance is form and content" (TLP 2.025); content is the substantive "facet" of an object; and "form is the possibility of [the object's] occurrence in atomic facts" (TLP 2.0141). Every object thus lies within a "space of possible atomic facts" (TLP 2.013). This space is made available by the logical manifolds associated with other objects (their attending form), which can combine with the first in a chainlike fashion to form a particular atomic fact. Thus "a spatial object must lie in infinite space," while "a speck in the visual field has, so to speak, a color space around it," a "tone," a "pitch," space, "the object of the sense of touch," a space of "hardness, etc." (TLP 2.0131).

Since a Wittgensteinian object cannot be thought of without the space, or manifold, of its associated logical possibilities (TLP 2.013), this object is merely the content that bears or grounds this manifold. In other words, objects are merely the "what" (the content) that bears manifolds of logical possibilities, and they are there (postulated) only to bear these manifolds. Conversely, these logical manifolds are merely the "hooks" (the form) that allow objects to combine with one another in an atomic fact, and they are there only to secure the possibility of this combination, thus guaranteeing the determinateness of sense. Consequently, content and form are here two inseparably united facets,[38] and the unit they constitute is metaphysically minimal enough to be compatible with the contentless nature of logic, whose propositions "treat of nothing" (TLP 6.124).

Since objects combine in the way described, the manifolds in question can be superposed: for example, many everyday things—from the facts involving which (and the propositions picturing these facts) a Wittgensteinian analysis would start—are at one and the same time located, colored, audible, and hard. It follows that the aim of the unique complete analysis is *not* meant to provide some kind of metaphysical foundation for everyday objects and their sensible properties by means of Wittgensteinian objects and their associated manifolds; to carry out such an analysis is simply to set forth the logically—if not practically—attainable limit where simple signs or names join together in the form of the corresponding elementary propositions. This is the aim, for the elementary propositions attained by such an analysis would come to fully *display* (or "show," as I used the term in chapter 4) all the logical possibilities at issue.

Wittgensteinian analysis is not ontologically oriented, aiming at the "ultimate constituents" of the world, whatever these might or might not prove to be, and at the properties allowing them to combine with one another. The analysis is logical in that its terminus is merely the space of logical possibilities (my rat is the kind of thing for which *it is possible to . . .*) ultimately displaying or showing exactly how objects combine to form atomic facts. By the same token, since any elementary proposition of those terminating the analysis is simply a "concatenation of names" (TLP 4.22) that need no additional linguistic item to keep them together, it is self-sufficient. Since "every possibility must lie in the nature of the object and a new possibility cannot subsequently be found" (TLP 2.0123), such an elementary proposition is final: nothing can ever be added to it, and nothing can ever challenge or change it (other than its happening to be true or false). The analysis terminating in such elementary propositions can be deemed *logico*-philosophical for this reason.

It is nevertheless worthwhile noting that Wittgenstein marks off the logical manifold of space—the manifold of logical possibilities attending extension—by beginning the relevant proposition (TLP 2.0131) with a separate paragraph devoted to it. On the one hand, we apparently may infer that this discrimination is a token of generality: that something has color, hardness, or pitch, thus engaging our sense organs in the relevant respect, *presupposes* that it is extended. (Something colored or hard is located somewhere, while a particular space configuration of sound waves is associated with any given sound.) On the other hand, if language is "part of the human organism" (TLP 4.002) and hence a bodily function, extension will apply not just to anything engaging our sense organs but also to everything linguistic, so that all these involve the logical manifold of space even if only in the "stand in for" relation characterizing language. (Every proposition has a "part perceptible by the senses" [TLP 3.32] and so a part that is extended.) In contrast to Spinoza, however, Wittgenstein is not disposed to investigate whether, how, and to what extent all the logical possibilities he envisages might be brought down to the logical manifold of space.

To repeat this in stronger terms, Wittgenstein's objects and their associated logical manifolds do not presuppose, imply, chart, or put forth any kind of ontological account of the furniture of the world. Simple as they are, these objects in no way point at or vindicate atomism in nature—or for that matter, in language, logic, or anywhere else. The "ultimate constituents of matter" might be atomistic or not; they might be discrete, continuous, or simultaneously both; they may turn out to be particles and antiparticles, quarks, superstrings, and fundamental fields or anything else scientific research might find by digging deeper and going further in the directions of the microscopic, the macroscopic, the faster, the more complex, the chaotic, and so on. Wittgensteinian objects are "ontological" only in being empty enough to be compatible with logic, constituting the mere substantive "what"[39] that is logically necessary for bearing logical possibility in general. In their absence, logic itself would hang in the air. Wittgenstein's objects thus form the ontologically minimal—and in that sense logical—substantive item securing the foothold that logical relations, in being relations, must have. They imply and are implied by the determinateness of sense because both thoughts and propositions are pictures *of* something, and these objects make up the *ultimate* something, the mere "what" allowing pictures to be pictures, that is, to be "*of.*" If they were endowed with a richer ontological content, they would go beyond the logical to enter the province of the conceptual and thus interfere with the autonomy of science. Wittgenstein's metaphysics is minimal in that it does not overstep the limits of the logical and hence of the conceptually empty.

Returning to Spinoza, it would perhaps appear too rash to ask whether the

Ethics can match Wittgenstein's objects. But Spinoza's formal rigor may surprise us once more. To see this, however, we must consider a digression Spinoza felt forced to make.

Bodies

In the "brief preface concerning the nature of bodies" (E II p13s) that Spinoza inserts immediately after E II p13, bodies in general are considered in a manner prefacing the treatment of the human body in the rest of part II. This fraction of the *Ethics* presents a rather curious structure. It starts with two general axioms to the effect that "all bodies are either in motion or in rest" (a1) and "each single body can move at varying speeds" (a2) and then goes on to three lemmas, their proofs, and corollary (offering a version of the law of inertia). Next come three more axioms that concern the ways bodies interact, with a definition of the "individual thing" standing between the second and the third axiom of this second class, whose elements are numbered as a group. Another four lemmas, numbered in sequence with the preceding ones (so, lemmas 4 to 7), appear next. Each is followed by its proof, and a final scholium closes the discussion of bodies in general. Immediately afterward Spinoza offers six numbered postulates concerning the human body but provides no argument to justify them. The exposition resumes the thread interrupted in this way through the formulation of E II p14, which relates the human body to the human mind. This, then is the digression to which I referred.

Spinoza himself explains the rationale dictating this structure: "If my intention had been to write a treatise on body, I should have to expand my explications and demonstrations. But I have already declared a different intention, and the only reason for my dealing with this subject is that I may readily deduce therefrom what I have set out to prove" (E II p13sl7s). Spinoza thus expresses what I have been repeating: theses on motion, rest, and speed (the first two axioms) and on the basics of bodily interactions (the remaining three axioms) are the only ingredients he needs to borrow from the science of his day—transforming them to fit his thoughts on the matter—to go on with his general philosophical task. Similarly, one might argue, the discursive separation of the axioms into two categories, the insertion of lemmas numbered separately, and the culmination of the "brief preface" in six unsupported postulates on the human body amount to Spinoza's acknowledgment that the theses he lays out in this way leave all kinds of gaps and thus do not add up to a comprehensive construction.

In the course of this digression, Spinoza distinguishes "the simplest bodies" from "composite" ones (E II p13sca2). All bodies, "the simplest" as well as the "composite," are "distinguished from one another in respect to motion and rest, quickness or slowness, and not in respect to substance" (ibid.). As I will show, bodies cannot be distinguished with respect to Substance because as bodies, they all equally share in being Substance under the Attribute of Extension. Nonetheless, the distinction of bodies with respect to motion is of fundamental importance, for it provides the principle allowing for all the diversity in Nature.

Thus, barring "the simplest bodies," complex individuals are composed as follows: when some bodies, which are always either in motion or at rest (E II p13sa1 and 2), enter for a time into an "unvarying relation of movement" (E II p13sca2def) among themselves—presumably an intricate relation of relative velocities and speeds—they make up a "composite body" or an "individual thing" (ibid.) that lasts for that time, namely, for as long as the relation in question remains unvarying. This unvarying relation is a "mutual relation of motion-and-rest" (E II p13sl5) among the "components" (E II p12sl4d) or the "parts" (E II p13sl5) of that composite individual. This unvarying relation is literally essential, for it constitutes no less than the individual's "nature," or identity, which is preserved for as long as the relation in question persists unvaryingly.

Spinoza clearly shows that the nature, or identity, of the complex individual is preserved for all the time during which "changes in the component bodies" (E II p13s14d) do not touch that unvarying relation. Thus, irrespective of whether parts of the complex individual become "separated" from it and replaced by others of the "same nature" (E II p13sl4), irrespective of whether some other of its parts become "proportionately greater or smaller" (E II p13sl5), irrespective of whether some other of its parts change the "existing direction of their motion" (E II p13sl6), and more generally, irrespective of all "the many ways" in which the complex individual may "be affected" (E II p13sl7s), its identity is preserved insofar as the relation defining its nature remains invariant. This unvarying relation constitutes the individual's form (E II p13sl4d) for Spinoza.

Since bodies do not differ in substance, a complex individual can change only in form, never in substance. But since the form in question constitutes the individual's nature, a change in form is fatal: once the parts composing it change their "mutual relation of motion-and-rest," the individual as such perishes, leaving its various component parts to go their own ways.

It is of paramount importance to note that Spinoza carefully avoids saying that bodies are *made up* from the simplest bodies. On the contrary, whenever he refers to complex individuals, he always takes them to be composed of parts that are identifiable complex individuals in their own right, characterized by their own particular unvarying relation or form. Thus the form of the overall individual is taken to be some kind of synthesis of the unvarying relations—the forms—of the complex individuals composing it. Moreover, if Spinoza speaks only tangentially about the lower limit of complexity, located in the simplest bodies, he does not hesitate to invoke the upper limit, which is Nature as a whole. Thus the "continuation" of the process of composition can go up "to infinity," where "the whole of Nature" (E II p13sl7s) becomes a complex individual in its own right. For Spinoza, Nature can thus be "readily conceived as one individual whose parts vary in infinite ways without any change in the individual as a whole" (ibid.). This is, then, an individual that retains its identity forever and therefore cannot perish.

Note first that Spinoza avoids spelling out the nature of these simplest bodies. Again, he entrusts the examination of all substantive ontological issues to science, excepting what he needs to borrow from it to further his own purposes. And of course, the scientific terrain was rapidly changing in his time, for the Newtonian synthesis that would eventually settle it down was still some time off. Many of the ontological (and conceptual) ingredients of what would become later classical mechanics were thus not yet set into place and hence not available for helping Spinoza more sharply demarcate exactly what his loan from that quarter should be.[40] Moreover, since all available evidence indicates that he must have been acutely aware of that unsettled state, his reluctance to delve more deeply into the characteristics of this loan becomes a mark of philosophical wisdom. Be that as it may, what seems clear is that Spinoza could not have been an atomist in the more or less standard sense of the term, for the simplest bodies cannot be equated to the kind of discrete and indivisible building blocks that atomism requires.

To see why the simplest bodies are restricted in this way, we must return to an important proposition mentioned earlier. In his scholium to E I p15 ("Whatever is, is in God, and nothing can be or be conceived without God"), Spinoza discusses infinity and infinite divisibility in contrast to finitude and discreteness while displaying at the same time the paradoxes (the "absurdities") involved in trying to conceive of infinity as composed from finite parts or of finite things as composed from infinitesimal parts. He admits that the issue "is very difficult"[41] and remains strikingly reserved in what he advances, speaking of "all those alleged absurdities (if indeed they are absurdities, which is not now under discussion)" (E I p15s).

The move by which Spinoza tries to reconcile infinity and infinite divisibility with finitude and discreteness at the ontological level follows the pattern of the move by which he tries to reconcile infinity and finitude at the causal level. Thus he says that

> we conceive quantity in two ways. . . . [We do so either] abstractly and superficially as represented in the imagination or as substance, which we do only through the intellect. [By the first way] we find [quantity] finite, divisible and made out of parts. . . . [By the second way we find it] infinite, one, and indivisible. . . . [Yet] matter is everywhere the same and there are no distinct parts in it except insofar as we conceive [it] as modified in various ways. Then its parts are distinct *not really but modally*. (E I p15s, emphasis added)

Since, as the next chapter will reveal, modes are for the most part finite, while reality, in being referred to Substance and explicitly equated to perfection ("By reality and perfection I mean the same thing" [E IId6]), is infinite, the distinction between really and modally here amounts to the distinction between God's infinity and the finitude of everyday things at the ontological level. And Spinoza's way of distinguishing the two correlates exactly to his way of distinguishing the infinite character of God's causal power from the finite causes involved in our everyday dealings, a distinction at the causal level.

In the discussion of finitude and infinitude Spinoza furthermore states starkly that "there is no vacuum in Nature" and mentions having discussed the issue elsewhere (E I p15s).[42] It follows that the simplest bodies cannot be conceived as atoms even modally: to be conceived in this way implies that they are discrete and indivisible entities, capable of moving more or less freely on their own so as to enter at

various moments into "mutual relations of motion-and-rest," which would compose the corresponding complex individuals. But even independent of the issue of
indivisibility—defining the very idea of an atom—which Spinoza has already characterized as difficult, if not absurd, conceiving the motion and rest of the simplest
bodies construed as such atoms obviously presupposes the existence of the vacuum:
if these bodies were conceived as moving in some material medium, however im-
ponderable, the question of the atoms composing *that* medium would inevitably
arise. Hence "the simplest bodies" would no longer be the simplest.[43]

To see how Spinoza conceives his simplest bodies and the role he reserves for
them, consider the distinction between substance and form but explicitly restrict it
to bodies that are *not* the simplest. Recall that all bodies share in being Substance
under the Attribute of Extension—in fact, bodies are *modes* of Extension—and
hence are in this respect the same. What distinguishes them from one another is
their form, which, precisely because it distinguishes them, is identical to their "nature." This nature, or form, is always a relation—namely, the "mutual relation of
motion-and-rest," which *in the last analysis and only in the last analysis* attaches to
and involves the simplest bodies.

That is, to examine the composition of bodies, which is what interests him at
this stage, Spinoza first distinguishes the substantive from the relational or formal
facet of each composite body. He then underplays the substantive facet, since it is
the same for all bodies, while he highlights the relational or formal facet, for only
this facet makes the composition of bodies possible. Moreover, the identification
of the nature of a composite body with its form implies that it is this form, and it
alone, that holds together the body's parts, thus constituting the body's identity.
Nothing more, no additional substantive item, no other glue, is required to make
it the body it is. In this way Spinoza's simplest bodies seem to retreat into the background and begin to resemble Wittgenstein's objects.

Afterlife

Before examining the grounds and the character of this resemblance, it is
profitable to try answering a question left open in chapter 2, namely, why Spinoza's
claim that a part of the human mind is eternal does not imply an eternal afterlife
for human individuals. Spinoza's understanding of bodies provides us the means
for doing this.

Recall that the eternal part of somebody's mind (let's call the person "Karl")
is the sum of the adequate ideas Karl entertained when living, the ideas having
corresponded to some of the modifications of his body brought about by its causal
interactions with external bodies. Since Karl retains his identity as long as the various parts of his body are held together by an unvarying relation among them, he
remains alive as long as the modifications of any such part do not alter this relation;
his death breaks up this relation. Karl's body disintegrates, and its various parts,
having become "liberated" from that relation, go their own ways as complex individuals in their own right. However, since "knowledge [and hence the idea] of the
effect depends on and involves the knowledge [and hence the idea] of the cause" (E

I a4), the adequate ideas Karl's mind had entertained continue to be *adequately* involved in the ideas that capture these posthumous interactions in the mind of God. This makes these ideas eternal truths, providing the sense in which they form the eternal part of Karl's mind.

Now since "the mind can exercise neither imagination nor memory save while the body endures" (E V p21), what perishes when Karl's body perishes is his imagination and memory, that is, all inadequate ideas in his mind as *inadequate* ideas. By corresponding in the mind of God to posthumous interactions, the ideas in question become disentangled and clarified and by the same token are reordered along the eternal order of Nature. As previously discussed, however, the unvarying relations among Karl's bodily parts had constituted the form of his living body and hence his bodily identity. But during his lifetime, one idea prominent among those in his mind had to have been the idea corresponding to that nature itself, namely, *the idea of that bodily identity*.[44] We may thus infer that this had been the idea expressing Karl's self-awareness or self-consciousness, the idea of his inalienable *personal* identity, the idea legitimizing the use of the first-person singular—in short, the idea consisting in his "I." Because it corresponded to Karl's nature, this idea should have been *necessarily* prominent. Spinoza seems to support this line of thought in distinguishing "the idea of Peter which is in another man" from the "idea *of Peter* which *constitutes* [the Latin is *constituit*] Peter's mind" (E II p17s, emphasis added). The latter is presumably the idea Peter entertains with respect to himself, that is, the idea of Peter's "I" as entertained by Peter himself. Since this is the idea *constituting* Peter's mind, it must be prominent in that mind.

In chapter 2 I tried to explain why the "I" in question is necessarily prominent: it overarches all ideas in Karl's live mind, adequate as well as inadequate, by virtue of its being the *sole* idea that functions only in and by taking itself as the organization and control center of *all* other ideas in his mind since and insofar as it is his mind that has them. But in assuming by its mere existence this overarching position, the "I" tends to take itself as somehow determining all other ideas in Karl's mind and hence—since the ideas in question are ideas of his body as it relates causally to other parts of Nature—as relatively autonomous from those other parts, as some kind of "kingdom within a kingdom." For this precise reason, the "I" can only be an "imaginary" or inadequate idea.[45] Spinoza's disagreement with Descartes's use of the "I" finds perhaps here its deeper roots.

It would, moreover, have been impossible for Karl to work on his own self in a way that could fully transform his "I" into an *adequate* idea, no matter how hard he might have striven for a self-conception as a part of Nature, on a par and in harmony with everything in It. To be capable of having achieved this, Karl would have had to be capable of adequately knowing the unvarying relation constituting his nature or, equivalently, of adequately knowing his entire body in all its "parts, . . . [which] are very composite individual things" (E II p24d). This is obviously beyond human capacities, for "the human mind does not involve an adequate knowledge of the component parts of the human body" (E II p24). The idea of his being a relatively autonomous self could never have been completely done away with while Karl was alive, for the very idea of being a self is a fundamental part of what makes up human finitude.

In being thus incurably inadequate, the idea consisting in the overarching "I" belongs to the perishable part of Karl's mind and hence perishes with his death. No matter how large the eternal part of his mind might have become through his having striven for that effect, the eternal part of his mind could never have come to include his "I," that is, his idea of being a person. His idea of being a person, *as Karl had entertained it when living,* his idea of the identity of his live body, cannot survive his death.

Looking at this from yet another angle may help clarify the matter. Recall from chapter 2 that the moments when we "feel and experience that we are eternal" (E V p23s) are precisely the moments when the "I" in question retreats and tends to disappear. We cannot feel and experience that we are eternal while clinging to the overarching "I": the "I" of personal identity cannot enjoy eternal afterlife, for it is precisely this idea that is compelled to retreat and disappear at the very moments we attain the feeling and experience of eternity. Eternity and personal afterlife thus necessarily exclude each other; we cannot personally enjoy an eternal afterlife precisely because we are persons.

Nonetheless, since "there is necessarily in God an idea which expresses the essence of this or that human body under the form of eternity" (E V p22), the fact that while alive, Karl's body did or underwent something while simultaneously entertaining the corresponding ideas, adequate or inadequate, *is* an eternal truth. But this eternal truth, the eternal truth expressing one's essence in the "form of eternity," is inaccessible to one's self. The form of eternity is separated from the form of duration by the barrier consisting in one's ineradicable sense of being oneself.

Substance III

Returning once more to Substance, recall that Wittgenstein postulates objects that are simple in the sense that there is nothing inside or beyond them. Using the terms of Spinoza's vocabulary, one might say that these objects cannot be decomposed further and are in this sense "the simplest." Further, Wittgenstein draws a distinction between the substantive content and the form of his objects, and I tried to show why and how his objects are there only to provide the minimal substantive foothold necessary for their own combination in atomic facts. To again use the terms of Spinoza's vocabulary, one might reasonably say that these objects are there to *compose* an atomic fact. I have already discussed the sense in which the form of these objects could be termed relational in Spinoza's vocabulary: this form is no more than a logical manifold, the manifold of the logical possibilities allowing the corresponding object to combine with others in, to compose, an atomic fact. No additional substantive item, no other glue, is required to keep objects together in atomic facts. Lastly, Wittgenstein's objects are ultimately displayed at the terminus of the logical (and philosophical) analysis he seeks: the simple signs to be finally attained by such an analysis are names concatenated in a chainlike fashion so as to permit the strict correspondence of each simple sign with the object it names.

Spinoza, however, though he never discusses analysis, unmistakably implies that the analysis of a composite body would find its ultimate terminus in his simplest bodies, which as was previously demonstrated *cannot be atoms*.

Given this striking similarity between the two authors' conceptions of the simple or "simplest," the role Spinoza reserves to his simplest bodies may become clearer: independent of ontological status, independent of what science might say on the constituents of matter, simplest bodies must necessarily exist so as ultimately to *substantively ground* the form and hence the identity of composite bodies. As we saw in the case of Wittgenstein, such a substantive grounding is, *mutatis mutandis,* logically necessary, for in its absence, bodies would be merely relational bundles relating nothing, mere relations freely floating in the big nowhere and hence no bodies at all. There would be nothing to bear the relational identity of a composite body and by the same token nothing to bear the "laws of God's nature," which in Spinoza's times had already begun to appear as relational. Spinoza's simplest bodies are, we might claim, the substantive and absolutely minimal "what" that logically secures the material existence (the existence in the Attribute of Extension) of any material thing and hence the material existence even of particles and fields, of quarks and superstrings, or of whatever science might discover or concoct in searching for the constitution of matter.

It follows that the existence of simplest bodies of absolutely minimal ontological thinness is logically necessary for the discipline of physics—as it is, as it has been, and as it must continue to be carried out—if it is to tackle its tasks, even though physicists might not suspect that a logical necessity grounds their work. We might claim therefore that Spinoza refrains from endowing his simplest bodies with any particular ontological content not only because of the historical reasons already canvassed and not only because he sharply and wisely distinguishes science from philosophy but also, and most important, because doing so would allow his simplest bodies to play their required logical role, one that in the last analysis has nothing to do with the debates concerning atomism and Spinoza's stand with respect to it.

Be that as it may, differences between the approaches of the two authors remain in realms beyond merely philosophical vocabulary. For example, Spinoza's simplest bodies at least appear to be basically ontological items, while Wittgenstein's objects are purely and expressly logical items. That they might involve extension is of little concern to him, if not fully irrelevant to his project. In fact, it would be absurd to try fixing the number of his objects by means of logic (TLP 5.553–5.554); in that respect, he considers only that there might be infinitely many (TLP 4.2211). Spinoza, however, is concerned with only one Wittgensteinian object, namely, that which substantively grounds the logical manifold of space. Thus, speaking from the logical—rather than ontological—point of view, we may say that Spinoza's simplest bodies actualize the substantive foothold of the possibilities making up the logical manifold of Extension. At the same time, this might help explain why Spinoza does not distinguish the logical function of his simplest bodies from their ontological appearance: if there were only one Wittgensteinian object, its logical role would be barely distinguishable from some sort of ontological status. We might thus surmise that one aspect of the historical distance separating Spinoza from Wittgenstein is expressed by the difficulty Spinoza finds in pinning down such a difference between logical role and ontological status. As I will discuss in the next chapter, however, this is not his last word on the matter. Spinoza compensates for this paucity in logical objects by multiplying his Attributes to infinity.

Putting this in the two vocabularies conjointly, we may say that both authors understand Substance to be *what* offers itself to the logical possibility of being extended. By thus amounting to the logical possibility of something's occupying a spatial location, Extension is nothing but that logical possibility itself. In that capacity, Extension *qualifies* or *modifies* Substance in the corresponding respect, making Extension an Attribute of Substance. Inversely, Extension is the way by which Substance offers itself to being "*perceived*" (E I def4) as spacelike, the way by which it *expresses itself* as spacelike (Deleuze 1968). Obviously, the infinity characterizing Spinoza's Attribute of Extension is identical to the infinity characterizing Wittgenstein's logical manifold of space.

A further line of thought may make this clearer. On the one hand, the mere "what" that constitutes Substance—Wittgensteinian objects—bears all logical manifolds and hence logic as such. On the other hand, since "logic comes before the How," it "*precedes* every experience that something is *so*" (TLP 5.552). In thus forming the precondition of all experience, logic forms the precondition of thought and of language. If we adopt for a moment yet another philosophical vocabulary, this would be equivalent to saying that logic forms the condition of possibility for thought and language. For this reason, it may indeed be called "transcendental" (TLP 6.13).[46] Obviously, this condition of possibility is epistemological if anything. Recall from chapter 2 that the ethical is inseparable from human purpose or intention, that it precedes our engaging in activities and hence our bringing things about as well as our undergoing things through those activities. In that sense, it constitutes a precondition—a condition of possibility—for our hooking onto the world as human agents. Accordingly, "ethics is transcendental" (TLP 6.421), too, and for analogous reasons. Obviously, this condition of possibility concerns what we do and what happens to us, that is, it is the condition for the possibility of human agency. Note that Wittgenstein reserves the qualification *transcendental* for logic and for ethics only.

Of course, Kant's terminology was not available to Spinoza. If it had been, he presumably would have readily endorsed the term in this specific context, for it is perfectly innocuous with respect to his thoroughly immanentist position. Calling logic and ethics "transcendental" adds nothing to and subtracts nothing from the relation that Spinoza's Substance entertains with Its Attributes. Thus to call logic "transcendental" is only to point to the epistemological condition of possibility that allows our minds to ideate our bodies (and thence other extended bodies) and to express in language, however confusedly, the corresponding ideas. To call ethics "transcendental" is only to point to the condition of possibility that allows our bodies to causally share in Extension and thus to act on other bodies and to be acted on by them. To hold that only logic and ethics are transcendental is simply to hold that the conditions of possibility allowing us to share in Substance involve and can only involve our minds and our bodies, for we are bodies and minds and nothing else. In strictly Spinozistic terminology, this is to say that the modes we are can be expressed in the Attribute of Thought and in the Attribute of Extension and in these two Attributes only.

CHAPTER SEVEN

Matching Content

> [Intellectual objectivity is] the ability *to control* one's Pro and Con and to dispose of them, so that one knows how to employ a *variety* of perspectives . . . in the service of knowledge.
>
> —Friedrich Nietzsche, *Toward a Genealogy of Morals,* III, §12

THE MINIMALIST METAPHYSICS we have been examining envisions us as inseparably bonded bodies and minds thrown into a world that consists of other bodies and other minds and, in the last analysis, nothing else. It remains to see how Spinoza and Wittgenstein take account of these absolutely fundamental facts, as well as whether and to what extent their approaches match each other in the corresponding respects. Having landed on Spinoza's Attributes, it is from here we have to start.

Spinoza's Attributes

Spinoza defines what he means by Attribute early in the *Ethics*: "By attribute I mean what the intellect perceives of substance as constituting its essence" (E I def4). We saw that he defines God as "consist[ing] of infinite attributes, each of which expresses eternal and infinite essence" (E I def6) and that Substance is identical to God. From His side, God *expresses* His essence through each of His infinitely many Attributes, while from the other side, as it were, the intellect *perceives* that essence through the same Attributes. Spinoza distinguishes perception from conception: "conception seems to expresses an activity of the mind" whereas "perception seems to indicate that the mind is passive to its object" (E II def3ex).

Now since Substance—and hence its essence—"can be conceived only through itself" (E I def3), an Attribute that expresses this essence "must [also] be conceived only through itself" (E I p10). It follows that an Attribute is *epistemically isolated*: whatever it might include can be conceived—and hence known, understood, and explained—only in terms of items belonging to that Attribute alone and in the language pertaining to it exclusively.[1] At the same time, "all [the infinitely many Attributes expressing God's essence] have always been [in God] simultaneously," and hence "no attribute could be produced by another" (E I p10s). It follows that Attributes do not generate others or interact causally with one another; each Attribute is *causally isolated* as well. In short, there is no way Attributes might combine or mingle. Each is perfectly self-sufficient along both the epistemic and the causal dimensions, expressing in its own way the perfect self-sufficiency of God.

Seen from one side, then, Attributes are the independent channels through which God expresses Himself and deploys His power; from the "opposite" side, they are the channels through which God is perceived by the intellect and His power is borne by the products of that power. This means that the Attributes are not objective ingredients that add up to compose God or Substance: the same all-inclusive Substance expresses itself in its entirety and deploys all its undivided power through each of them in exactly the same manner and to exactly the same extent. Nevertheless, the Attributes are not subjective appearances presenting Substance as the perceiving intellect makes it out to be but leaving it inaccessible in itself. The perceiving intellect is a product of the all-powerful Substance, and hence its perception is objective in exactly the same manner and to exactly the same extent that Substance, which produces it, is objective. Substance does not hide its "in itself" from its products, and there is no loss in objectivity by the intellect's perception of Substance under one Attribute or another. As Spinoza puts it succinctly, "the human mind has an adequate knowledge of the eternal and infinite essence of God" (E II p47). We may thus say that the Attributes are the infinitely many *objective perspectives* Substance determines, the objective perspectives along which Substance deploys its power while simultaneously allowing the intellect to perceive it objectively, as it is (Conant 2005, 2006).

A necessarily inadequate metaphor might be helpful. Because we are objectively confined to the surface of the Earth, we can perceive only one face of the Moon—or equivalently, the Moon expresses itself with respect to us through only one face—although we (assuming or having demonstrated that it is a perfect sphere) know that the Moon possesses infinitely many such faces. We know, in addition, that each of these faces is objectively different from the others in that it is offered to perceivers situated at objectively different places; that to the extent that the Moon is a perfect sphere, such vantage points are epistemically equivalent; and finally, that these vantage points cannot communicate in the sense that we cannot occupy two at the same time. It is in the Moon's nature, we may say, to offer itself to us and similar perceivers only through one of its faces each time. In that sense, each face of the Moon is epistemically isolated from the others.

Now imagine further that the Moon is the sole causal power that determines all the vantage points under which it can be objectively perceived; that it has produced us (by the "laws of its own nature") as perceivers in all our capacities for objective knowledge; and that it has confined us to a single unchangeable vantage point. If these conditions are satisfied, then it follows that our objectively determined vantage point obliges us to recognize that the *whole and undivided* causal power in question acts on us through the face of the Moon we are in a position to perceive and that all perceivers necessarily do the same with respect to the face each perceives. Since it is the Moon's whole and undivided causal power that is thus felt to affect us through only one of its faces each time, it makes no sense to say that the faces causally interfere or mingle in any way with one another. All faces of the Moon are causally isolated and fully self-standing on all counts. Each face is self-sufficient along both the epistemic and the causal dimensions, expressing without loss the corresponding perfect self-sufficiency of the Moon as such.

But it would make little sense to say either that the Moon is composed of its different faces or that any of its faces is subjective in the sense of merely appearing as the corresponding perceiver makes it out: perceivers are objectively situated where they are, they know this fact together with all the previously mentioned facts, and they take account of all this knowledge in the very act of perceiving.[2] Our ability to perceive only one face of the perfectly spherical Moon entails no loss of objectivity, for all perceivers endowed with the same capacities would come up with exactly the same knowledge of the Moon as such. This, I claim, is a rough picture of an objective perspective and of Attributes as Spinoza understands them.[3]

Spinoza defines "mode" as "the affections [or modifications] of Substance, that which is in [resides in and is causally dependent on] something else and is conceived in [i.e., depends epistemically on] something else" (E I def5). He further specifies that "all things that are, are in themselves or in something else" (E I a1) and that "if a thing can be conceived as not existing, its essence does not involve existence" (E I a7). It follows that a thing that can be conceived as not existing is necessarily conceived "in something else" and hence that it is a mode. Taking into account that "the essence of man does not involve necessary existence" (E II a1), it follows that man is a mode (E II p10c). In addition, "since from the order of Nature it is equally possible that a certain man exists or does not exist" (E II a1), the order of Nature can have any person come to be or to perish. As was discussed in the previous chapter, humans constitute *finite modes*.

But not all the infinitely many Attributes are accessible to this finite mode. A person "thinks" (E II a2) and "feel[s] bodies" (E II a4) and "feel[s] and perceive[s] *only* bodies and modes of thinking" (E II a5, emphasis added). It follows that only two Attributes of God's infinitely many are accessible to humankind: the "Attribute of Thought" (E II p1) and the "Attribute of Extension" (E II p2). These are the only two channels through which God expresses Himself to humans, the only two objective perspectives that make us relate to Him and understand Him to the limit of our finitude. To the extent that bodies are the sole inhabitants of the Attribute of Extension, ideas are the sole inhabitants of the Attribute of Thought: "an idea is a conception of the Mind which the Mind forms because it is a thinking thing" (E II def3). By the same token, bodies are modes in the Attribute of Extension, and ideas are modes in the Attribute of Thought. The preceding chapter showed that bodies connect causally while ideas connect logically (according to the geometrical order). Borrowing a different terminology (Sellars 1991; McDowell 1996), we might thus say that the Attribute of Extension constitutes the "realm of causes" while the Attribute of Thought constitutes the "realm of reasons."[4] Still, both Attributes (realms) are epistemically and causally isolated, while each is perfectly self-sufficient along both those dimensions.

So how does Spinoza understands the infinitely many Attributes required by the infinitude of God but inaccessible to humankind? This question cannot be disregarded, for it crucially involves the way Spinoza understands the Attribute of Thought and thus his whole epistemology. The issue may be formulated as follows.[5] Given that any Attribute amounts to that which "the intellect perceives of substance as constituting its essence" (E I def4), while the intellect can be deployed within the Attribute of Thought alone, there seem to be only two ways to explain how the Attribute of Thought connects with the inaccessible Attributes so as to provide the corresponding knowledge in God's mind (and only in God's mind), and both are unsatisfactory. On the one hand, if the Attribute of Thought is reserved for perceiving the "essence" of Substance as expressed in the Attribute of Extension (and in itself), then the essence of Substance as expressed in each inaccessible Attribute X could be "perceived"[6]—by God's infinite mind alone, since by definition Attribute X is inaccessible to humankind—only by another Attribute of Thought, the one particularly reserved for X. In this case, there would be infinitely many distinct Attributes of Thought expressing God's mind, thus making it difficult to hold the necessary unity of that mind.

On the other hand, if God's mind conceives the essence of Substance as expressed in all the infinitely many Attributes through a unique Attribute of Thought, then this Attribute would be unduly privileged, not remaining on a par with the others, as it appears the case should be. Moreover, the Attribute of Thought would be internally compartmentalized: an insuperable barrier would separate the part conceiving the essence of Substance under the Attribute of Extension—the part including the finite intellect of humankind—from all its other parts, and additional insuperable barriers would perhaps separate all those other parts from one another. Yet there seems to be no principle in Spinoza's philosophy capable of both accounting for such compartmentalization and explaining, as it should, why only one such compartment can involve the human intellect.

The issue has been a problem for others, too. It was formulated, almost in the terms I employ, by Shuler (L 63), speaking for Tschirnhaus, as well as by Tschirnhaus himself some time later (L 65) and was addressed directly to Spinoza. The author's answer, however (L 64 and 66), evokes only E I p10, according to which "each attribute . . . must be conceived through itself," and simply refers to the passage in the *Ethics* where Spinoza says that "thinking substance and extended substance are one and the same substance, comprehended now under this attribute now under that [while] a mode of Extension and the idea of that mode are one and the same thing, expressed in two different ways" (E II p7s).

Psychoanalysis

To address this conundrum, first note that we can hardly consider an inaccessible Attribute X as containing items that are essentially mysterious in that they *necessarily* cannot be conceived by man's finite intellect. Since these items can be conceived by God's infinite mind and since human intellect, even if only finite, is of the same nature as God's mind—recall that "we feel and we experience that we are eternal" (E V p23s) in a godlike fashion and that "the human mind has an adequate knowledge of the eternal and infinite essence of God" (E II p47)—it seems that we should take these items to be only *contingently* "unperceivable," that is, as inconceivable by science (and by philosophy) in Spinoza's time but perhaps not in other times. This would make Attribute X only contingently inaccessible.

To say that Attribute X constitutes an Attribute is to say that it should be on a par with the two familiar ones: the items it contains should be self-consistent and cohere perfectly with one another, following a rigorous order and connection within it. Such items should possess all the characteristics allowing God's essence to be expressed by Attribute X as perfectly as it is expressed by the other two. But saying that Attribute X is only contingently inaccessible and that its items are only contingently inconceivable cannot undercut the fundamental facts of Spinoza's metaphysics pinpointed earlier, namely, that we are nothing more than bodies and minds thrown into the world. Accordingly, the additional Attributes must implicate bodies and mental items (ideas) in any event, for at the level of those fundamental facts, there is nothing else to implicate. To say that the items residing in any additional Attribute are contingently inconceivable—that is, were inconceivable in Spinoza's time—is therefore to say that the bodies and the ideas implicated in that Attribute would have appeared to Spinoza as extraordinary mixtures of extended and ideational items that would have amounted to incomprehensibly delimited bits of extended matter if understood through the Attribute of Extension and as fully confused ideas if understood through the Attribute of Thought.

Given this, one might tentatively advance an account of a contingently inaccessible Attribute.

To that effect, recall the discussion of expert knowledge in chapter 2: in possessing expert knowledge, the expert acts as an indivisible body-mind, while the overarching "I" condensing the idea of the acting body in a necessarily inadequate ("imaginary") manner retreats and tends to disappear. To this I now add what I have been intimating in various places, namely, that psychoanalysis is a cognitive (and therapeutic, i.e., causally effective) endeavor *constitutively considering a human subject as an indivisible body-mind* while taking the overarching "I" (the "ego") as being determined by the unconscious. Psychoanalysis thus mixes bodies and minds in ways Spinoza would have found hard to conceive, let alone accept. Now if we are allowed to consider the discipline of psychoanalysis as an objective perspective, irreducible to the Attributes of Extension and of Thought, through which Substance (the mere "what" constituting it) expresses itself while allowing the human intellect to perceive it, then under yet to be specified conditions, psychoanalysis might constitute a Spinozistic Attribute P that—on a first round of approach—can be added alongside the Attributes of Extension and Thought.

To be a candidate for such a role, psychoanalysis must satisfy the following conditions:

1. Psychoanalysis should be all-inclusive. It should be in a position to countenance *everything* accessible to the "perceiving intellect" of humankind (and not just human body-minds) under the objective perspective it constitutes; hence it should be infinite in the way the other Attributes are, encompassing *all* Substance. But since only bodies and ideas are accessible to the human perceiving intellect, psychoanalysis need only be in a position to countenance all bodies and all ideas in its own terms for it to be all-inclusive in the relevant sense. Each body and each idea should have a correlative item proper to psychoanalysis (call it a *P*-item); conversely, each *P*-item should have a correlative body or idea. On Spinoza's understanding of the connection—or rather nonconnection—between the familiar two Attributes, the bodies (Extended items, or *E*-items) and the ideas (*I*-items) at issue should be correlative to each other. Given, then, that any body and any idea can be *emotionally invested,* thus affecting the body-mind of a human subject in the ways psychoanalysis lays out, the discipline would perhaps not be too hard-pressed to subscribe to this condition, exorbitant as it might appear at first sight.

2. Psychoanalysis should be epistemically and causally isolated and self-sufficient, conditions it seems to satisfy.[7] First, it provides a theoretical account for the relations holding among *P*-items solely by its own means. Second, it can lay out, again solely by its own means, all psychoanalytic causes (or rather causes-reasons, since the human subject is a body-mind) that generate *P*-items and make them act on a human body-mind. It thus satisfies the epistemic condition. The causal condition is satisfied in the same way, for psychoanalytic theory accounts for the ways *P*-items (emotionally invested bodies and ideas) are generated by and act on the body-mind of a human subject through what Lacan (2006f) calls "*la causalité psychique.*"[8]

3. Psychoanalysis should be expressed in a language proper only to it. Barring the inevitable neutral terms, such as those previously discussed, the vocabulary of psychoanalysis should lay out the field's epistemic and causal self-sufficiency and hence should be systematic and self-sufficient in its own right. For present purposes, we may consider that the vocabulary of psychoanalysis as presented, for example, in the dictionary of Laplanche and Pontalis (1973) is both systematic and self-sufficient in the required ways. This is demonstrated by, among other things, the fact that each psychoanalytic term refers to other psychoanalytic terms and is not grounded on anything outside psychoanalysis other than referents of the relevant neutral terms. This makes the psychoanalytic vocabulary a pure *P*-vocabulary as far as this can go.

To the extent that these conditions are satisfied, psychoanalysis can be identified as an Attribute, independent of the Attributes of Extension and of Thought, on a par with them, and lying alongside them in a way Spinoza would perhaps have accepted. To wit: any *E*-item together with its corresponding *I*-item (e.g., this stick or that purse together with their corresponding ideas) might be emotionally invested so as to psychodynamically affect the body-mind of a human subject thus constituting the correlative *P*-item. Conversely, any *P*-item (something that affects some human body-mind psychodynamically) has an *E*-counterpart (the corresponding *bodily symptom*) as well as an *I*-counterpart (the unconscious idea attached to that bodily symptom).

It is important, however, to stress a particularity. Although psychoanalytic theory orders and connects *P*-items in the same way with respect to all human subjects, the "psychical causality" at work functions differently for different human subjects. *P*-items and their corresponding emotional investments are particular to a given human subject, and hence their order and connection is different for each. This makes the process of therapy correspondingly unique, though the existence and the scope of these differences should be amenable to psychoanalytic theory.[9] These characteristics make the order and connection of *P*-items differ from the common "order and connection" of bodies and ideas (E II p7). This need not be a great concern here, however, for Spinoza nowhere maintains that order and connection should be exactly the same for all the infinitely many inaccessible Attributes he feels compelled to introduce.

This concludes the first round of my effort to suggest a Spinozistic Attribute that might be added to those of Extension and of Thought and to understand psychoanalysis in Spinoza's terms. Now I will hazard a second round. Under the very stringent conditions I will specify, this second round might result in integrating psychoanalysis—or rather something resembling psychoanalysis, for these conditions are harsh indeed—more fully into Spinoza's system, but only at the cost of denying psychoanalysis the status of a third independent Attribute, thus nullifying my previous efforts to find one.

The following three paragraphs enumerate the conditions in question:

First, we must assume that (1) psychoanalytic therapy may be definitive in the sense that it ultimately makes patients become psychoanalytically transparent to themselves; that (2) to be transparent in this way is to emotionally disinvest all bodies and all ideas that had been affecting one's body-mind, with such divestiture implying that one has finally come to see things rightly, as they are; and that (3) such definitive therapy can be attained by appropriating the *Ethics*—while doing the necessary work on one's self, with or without help from another—which should thus be taken as a textbook whose main objective is the emotional divestiture at issue for each of its readers.

Second, we must assume that (4) Spinoza's imagination might be taken as corresponding to the unconscious, whereby the imaginary "I" becomes a function of the unconscious, as psychoanalysis has it (Lacan 2006d), and that (5) the "common notions" at work in the last three parts of the *Ethics* might be taken as corresponding to the conceptual system of psychoanalysis, while those at work in the first two parts correspond to philosophical grounds of that conceptual system in tandem with the philosophical grounds of the conceptual system of natural science, with all implicated concepts being rigorously interconnected.

Third, we must accept some of the proposals already advanced, namely, that (6) in claiming "the object of the idea constituting the human mind" to be "the body"

(E II p13), Spinoza explicitly takes the human subject as a body-mind, which entails that the overarching "I" condensing that idea is necessarily an imaginary construct; that (7) bodily modifications and the inadequate ideas thereof, as Spinoza takes them, differ very little from the bodily affections or bodily symptoms and the conscious cum unconscious ideas thereof as psychoanalysis construes them; and that (8) concomitantly, the order of the imagination by which a human subject acquires inadequate ideas while undergoing the corresponding bodily modifications, as Spinoza has it, does not differ much from the order by which a human subject becomes mentally *and* bodily affected, as psychoanalysis has it.

If all these conditions are satisfied, therapy would amount to reordering and reconnecting the ideas corresponding to the bodily modifications of the patient, a process that would make these ideas correspond distinctly and clearly to the causal interactions between the affecting and the affected bodies according to the order and connection of things in Nature. The outcome of therapy would thus be a human subject who enjoys perfect psychodynamic health and has come to fully accept his or her place in Nature and whose overarching emotion has become the "intellectual love of God." Under these conditions, something resembling psychoanalysis becomes fully integrated in Spinoza's system—but they also deny Attribute P an independent status. If these conditions are satisfied, P becomes a mere replicate of the Attribute of Thought (insofar as emotions are taken as pertaining to the mind) or the Attribute of Extension (insofar as emotions are taken as pertaining to the body), and the two are identical on the basis of Spinoza's thesis that "the order and connection of ideas is the same as the order and connection of things" (E II p7).

Historical Materialism

To rescue the issue from this undecided state, we might equally tentatively suggest that Althusser's conception of what goes by the misnomer "historical materialism" (HM) could furnish the third Attribute. For Althusser (1972), HM constitutes the "scientific continent"[10] of the social and the historical in that it encompasses within a unique consistent system all the concepts required to account for all the "scientific regions" making up the different social and historical disciplines. Centered on concepts such as "mode of production," "forces and relations of production," "social formation," "social classes," "political and ideological relations," "state," "ideological formations," "ideological state apparatuses," and so forth (Harnecker 1974; Baltas 1991), this system, once appropriately elaborated, provides the scientific knowledge—in the sense proper to HM—of everything constituting the social and the historical.

One might argue on this basis that the "scientific continent" of HM forms an objective perspective in the previously defined sense: it is all-inclusive (since everything human possesses sociohistorical aspects); it is epistemically and causally isolated and self-sufficient (for by definition, so to speak, the conceptual system of HM presents itself as self-supporting and in a position to provide a theoretical account for every sociohistorical fact and event as well as for the causal chains linking such facts and events); and finally, it is linguistically self-sufficient, excepting again inevitable neutral terms. For these reasons, it does pose as a candidate for an additional Spinozistic Attribute.

To play that role, HM need only be able to cash out everything "the intellect perceives of substance as constituting its essence" (E I def4) in its own terms. In other words, HM need only be able to "perceive" everything accessible to the human intellect as some *HM*-item that should simultaneously possesses an *E*-correlate, an *I*-correlate, and perhaps even a *P*-correlate; conversely, each *E*-item, *I*-item, and *P*-item should have an *HM*-correlate. To that effect, we may argue—sketchily and briefly—that the *E*-correlates of *HM*-items are the spatial aspects of sociohistorical facts and events studied by contemporary social, political, and cultural geography,[11] that the *I*-correlates are the ideological formations by which social classes take stock of sociohistorical facts and events, and that the *P*-correlates are something related to what Freud calls the "discontents of civilization." Conversely, the *HM*-correlate of an *E*-item or an *I*-item would be the sociohistorical aspect of that item (since all bodies are caught in a sociohistorical web in the wider sense and ideas always possesses a sociohistorical aspect), while the *HM*-correlate of a *P*-item would be the *P*-item's own sociohistorical aspect, for no item of the unconscious can avoid being caught in the sociohistorical web in the wider sense. At the same time, it seems obvious that the order and connection linking *HM*-items cannot be identified either with the order and connection shared by *E*-items and *I*-items or with that linking *P*-items: even if causal entailment could be identified with deductive entailment, as Spinoza holds to be the case, it cannot be identified with either psychical or sociohistorical causation.

It follows that the objective perspective of HM might be identified with an additional Spinozistic Attribute only at the cost of keeping its own order and connection, for in contrast to the *P*-Attribute, a second round turning the perspective of HM into a replica of either the Attribute of Extension or that of Thought seems to be barred. This difference between the *P*-Attribute and the *HM*-Attribute stems from the way Spinoza conceives the social and the historical. As his political writings attest,[12] he conceives a given society as a composite body made up of parts that seem finally to reduce to individuals. But HM is not "humanism" (Althusser 1996b) in that it does not take sociohistorical facts and events to result from the actions of individuals even in the last analysis. The concepts of HM involve only collectivities ("social classes," "class fractions," and so forth) theoretically irreducible to the individuals who, from Spinoza's angle, might be thought to compose them; by the same token, the conceptual system of HM takes individual human subjects as agents of social practices and as bearers of social relations and as always situated, and hence determined, sociohistorically. In other words, human subjecthood does not ground the conceptual system of HM; the occupant of the corresponding role—a sociohistorical subject—is derived from sociohistorical concepts. Since psychoanalysis is directly concerned with human subjects in that it fundamentally abstracts away their sociohistorical determinations, something resembling it could replicate the Attribute of Extension or of Thought and thus become integrated in the core of Spinoza's system; in contrast, HM might constitute an additional Spinozistic Attribute only by keeping the order and connection proper to it different from that shared by the Attributes of Extension and Thought.

This largely speculative discussion might be summed up as follows. If we could somehow establish that a science (a major scientific discipline) can be profitably described as an objective, all-inclusive perspective[13] on the mere "what" of Substance, one that is epistemically, causally, and linguistically self-sufficient in the previously discussed senses,[14] then to the extent that psychoanalysis and the Althusserian version of HM pass the test, they should both be counted as sciences in their own right, irreducible to physics or to natural science generally.[15] And once these sciences are equated to Spinozistic Attributes, we are allowed to advance, always very tentatively, that any Spinozistic Attribute may be interpreted today as a *scientific perspective,* as a science or a major scientific discipline—a "scientific continent"— that is already in place or might come about. Since there is nothing to prevent history from coming up with indefinitely many such perspectives, Spinoza's assertion that the Attributes of Substance are infinitely many acquires some plausibility, and his additional Attributes can be interpreted without inflicting too much violence on his overall approach.[16]

A point made at the beginning of the present discussion buttresses this interpretation: Spinoza was historically in no position to envisage objective, all-inclusive, and multiply self-sufficient perspectives implicating bodies and ideas but not basing themselves directly, ontologically and conceptually, on bodies or ideas as such. It was almost impossible for him to conceive that bodies and ideas could become jumbled in ways that could constitute nevertheless objective perspectives presenting all the virtues his Attributes had to show for themselves. It is interesting to note, however, that Curley (1969 152–53) adduces evidence to the effect that Spinoza might have been aware that this limitation was indeed historically contingent: "After the preceding considerations of Nature, we have *so far* been able to find in it only two attributes which belong to this all-perfect being. And these attributes *give us nothing to satisfy us that they are the only two ones* which constitute the perfect being" (ST I, i, emphasis added).[17]

Scientific Perspectives versus Logical Manifolds

It turns out, then, that the proposed candidates for items that might instantiate a third Attribute, *P*-items and *HM*-items, cannot do away with bodies and ideas but implicate them in ways exceeding the absolute separation of the Attributes of Extension and Thought as Spinoza conceived it. But a situation where bodies and ideas become thus inextricably—and inevitably—entangled in an additional Attribute raises a problem with respect to the two familiar Attributes that a present-day Spinoza would have to take in hand.

This problem may be formulated as follows. The historical Spinoza's inability to conceive more than the two Attributes he discusses amounts to his ability to countenance *only two scientific perspectives* on the "what" of Substance, namely, the scientific perspective of physics, or of natural science generally, which he took as coextensive with the Attribute of Extension, and the scientific perspective of psychology (or perhaps we might say psychoanalysis) which he took as coextensive with the Attribute of Thought. But if additional scientific perspectives of equal rank and epistemological dignity can be added to those two, if each such perspective cannot do away with bodies and ideas but has to implicate both, and if it can implicate both only inextricably, in a way incapable of coherently separating what belongs to Extension from what belongs to Thought, then the two Attributes of the historical Spinoza require further elaboration. Such elaboration would have to start from the realization that each of the two familiar Attributes carries two different functions without proceeding to the necessary disambiguation: on the one hand, each Attribute sets forth the grounding conceptual layout of the corresponding scientific perspective (recall the role Spinoza's "brief preface" played in grounding the conceptual layout of physics or natural science; I will later try to identify the grounding conceptual layout for psychology); on the other hand, each Attribute addresses from its own angle the relevant fundamental facts, namely, that we are nothing but indivisible body-minds thrown into the world as just that.

To disambiguate between the two functions and to resolve the attendant tension consistently, a present-day Spinoza should begin by carefully distinguishing the conceptual from the logical in either attribute, acknowledging that it is the ultimate (logical) level that addresses the absolutely fundamental facts in question and, accordingly, that this is the level that displays the possibilities the conceptual level can exploit. He should then recognize that even the most general—or most empty—*conceptual* layout of the scientific perspective associated with either Attribute must be historically contingent and hence fallible, so that it cannot directly concern the logical level as such. Moreover, he should construe the logical level as comprising what is minimally (logically) required for relations (natural laws) to obtain with respect to the Attribute of Extension and to what is minimally (logically) required for ideas to be ideas with respect to the Attribute of Thought. He should, finally, assess the ways these two sets of requirements intimately interconnect, taking his cue from—but also trying to clarify—the thesis that the order and connection of items in one Attribute parallels that of items in the other.

This program presumes that the Attributes of Extension and Thought cannot be identified fully with and reduced to the scientific perspectives they encompass. To say that something purely logical remains in both after all conceptual items have been abstracted away and consigned to the conceptual level of the relevant scientific perspective is to say that both Extension and Thought include something more fundamental than a scientific perspective, something not amenable to science, for science necessarily presupposes it. This logical remainder is the privilege of Extension and of Thought and of no other Spinozistic Attribute because of the aforementioned fundamental facts: we are nothing but body-minds always already thrown into the world. We can do whatever we can do and we can think whatever we can think—including setting up scientific perspectives of unlimited conceptual variety—only by "possessing" such an indivisible body-mind unity and only because we have been thrown into the world as such body-mind unities.

The logical "must" of this situation lies at the roots of any scientific perspective, "beneath" the conceptual level. Any scientific perspective on the world is compelled to implicate both bodies and minds at the level of *logical necessity* expressed by the fact of this situation. The logical part remaining after the present-day Spinoza has assigned all concepts to the relevant scientific perspectives is precisely what makes up this logical necessity. And since the bodies we "possess"—the bodies we *are*—always already take part in the world of Extension, the logical necessity at issue concerns not three items in their interconnections (i.e., the world, our bodies, and our minds) but only two and whatever they may entail, namely, the items of the Attribute of Extension and those of the Attribute of Thought.

The present-day Spinoza who would proceed to such a separation of the conceptual (or grammatical) from the logical while relegating the conceptual to science has the figure of the young Wittgenstein. Again, Wittgenstein is not concerned with conceptual issues and hence with scientific perspectives, but he nonetheless remains sensitive to the fact that scientific perspectives must rely on manifolds of logical possibilities that cannot be pinpointed or enumerated by scientific investigation. We saw that Wittgenstein provides for the possibility of indefinitely many scientific perspectives—and by the same token of radical paradigm shifts within any scientific perspective—by multiplying his logical objects and their associated manifolds indefinitely.[18] We may say, therefore, that the logical manifolds in question supply these perspectives and those shifts with their necessary logical grounding.

Wittgenstein distinguishes the realm of facts from the realm of thoughts (and of propositions) while making the two correspond strictly to each other. This is his way of tackling the logical necessities implicated in the absolutely fundamental facts at issue, necessities inseparable from the logical possibilities allowed us by the same token. As we saw, Spinoza keeps the Attribute of Extension absolutely separate from that of Thought while instituting a strict correspondence (often called "psychophysical parallelism")[19] between extended modes and their correlative ideas.

Ideas versus Thoughts; Extended Modes versus Facts

At this point we should examine whether and to what extent Spinoza's Attribute of Extension matches Wittgenstein's world of facts (TLP 1.1) and Spinoza's Attribute of Thought matches the realm of what Wittgenstein calls "thought"—that is, whether and to what extent Wittgenstein's facts match Spinoza's extended modes and Wittgenstein's thoughts match Spinoza's ideas. It is convenient to start from Spinoza's Attribute of Thought and the ideas it includes. Since Spinoza's ideas encompass lots of things we now tend to view as separate and more or less independent,[20] we should begin by trying to sort out what is involved.

First, note that a mind, the individualized item in the Attribute of Thought correlative to an individual body in the Attribute of Extension, "forms ideas because mind is a thinking thing" (E II d3). Thinking is the activity proper to the mind, what the mind as such does. Individual minds constitute modes, but thinking, too, possesses modes or presents itself in modes; these divide fundamentally between the intellect and the emotions. Thus, "by intellect . . . we understand a

definite mode of thinking which differs from other modes such as desire, love, etc."
(E I p31d). As we saw, minds have the capacity of ordering and connecting the ideas
they form—and hence include—either actively according to the intellect or passively according to the temporal sequence of occurrences of their correlative bodily
modifications.

An idea is "not some dumb thing like a picture on a tablet but . . . a mode of
thinking [that constitutes] the very act of understanding" (E II p43s). By the same
token, ideas "involve affirmation and negation" (E II p49). Yet ideas always have an
"*ideatum*," or object; ideas are always ideas *of* their objects (E II a4 and a4ex). Even
the ideas constituting the emotions involve the *objects* of the emotions: "modes of
thinking such as love, desire[, and so on] do not occur unless there is in the same
individual the idea of the thing loved, desired, etc. But the idea [of that thing] can
be without any other mode of thinking" (E II a3), and thus such an idea can inhabit
the corresponding mind dispassionately. The general conceptual layout of psychology I sought is organized fundamentally in this manner.

Since ideas constitute the "very act of understanding" and "involve affirmation
and negation," they are judgments; it follows that their objects in the Attribute of
Extension—their corresponding extended modes—should be viewed not as being
individual extended things but as constituting *facts,* though of course ones that in
some way implicate individual extended things.[21] Through this indirect route, at
least, Wittgenstein's world of facts (TLP 1.1) appears to match Spinoza's Attribute
of Extension.

Spinoza sometimes appears to contradict this in the *Ethics.* For example, he
distinguishes the "idea of Peter which constitutes Peter's mind" from the "the idea
of Peter which is in another man" (E II p17s), seeming to imply that ideas might
refer to individual extended things (in this case, the thing, or man, named Peter)
and not to facts. Yet appearances are misleading here. As the passage in question
makes perfectly clear, the idea of Peter includes the ideas correlative to facts such
as that Peter is a particular height, Peter weighs a particular amount, Peter wears
eyeglasses, Peter's action on some one occasion had such and such an effect, and so
forth. The ideas correlative to all those facts are brought together under the idea of
Peter, even as this becomes a complex idea confusedly summarizing the ideas correlative to the many facts. The idea expressed in language by a word naming a thing
is in most cases a complex idea that is similarly always correlative to an equally
complex set of facts. And the idea of Peter constituting Peter's mind will inevitably
differ from the idea of Peter that any other person might have, for that idea will be
correlative to the bodily modifications that person underwent when encountering
Peter's body on various occasions.

Since an individual body is almost always composite in the previously discussed manner, the idea correlative to that body, no matter which mind it inhabits,
will also be correspondingly complex, involving the ideas correlative to the facts
that in extension lay out the complexity of that composition. This means that Spinoza is not speaking infelicitously in using words that to us sound like names. As
the example of the word *olive* demonstrated, a word expresses the usually confused

synopsis of all the ideas correlative to all the facts involving the modifications that one's body has undergone in coming to contact with the relevant external bodies.

Moreover, the world of Wittgensteinian facts does not include only ordinary worldly items, such as cats sleeping on mats, rats sleeping on cats, and the like. Recall that for Wittgenstein "a proposition is a picture of reality" (TLP 4.021), and since such a "picture is a fact" (TLP 2.141), propositions too are facts (TLP 3.14), inhabiting the same worldly realm as do "ordinary" facts. Recall also that propositions possess this status because they cannot fail to include a part "perceptible by the senses" (TLP 3.32). For his part, Spinoza holds that linguistic expressions of ideas always share in the Attribute of Extension, for "the essence of words and images is constituted solely by corporeal motions of the human body" (E II p49s). It follows that the match between Spinoza's Attribute of Extension and Wittgenstein's world of facts is preserved throughout, item by item.

Again, Spinoza takes the emotions to be "modes of thinking," and hence ideas, always involving the objects of the emotions. If so, then ideas will usually bear an emotional charge. As was discussed in chapter 2, however, Wittgenstein relegates emotions to the domain of empirical psychology and takes thoughts to be totally dispassionate pictures of facts. Nonetheless, Wittgenstein considers something like the linguistic expression of emotions by examining the corresponding propositional attitudes, that is, "A thinks p, A thinks that p is the case," and the like (TLP 5.542), and I have rehearsed what I take to be the reasons making him differ from Spinoza in this respect.

Of course, Wittgenstein treats propositional attitudes from the angle of his logico(-philosophical) analysis. Thus the proposition "A thinks p" reduces to the form "'p' says p" (TLP 5.542), whereby "we have no coordination of a fact and an object [namely, A, as TLP 5.541 makes clear], but a coordination of facts by means of a coordination of their objects" (TLP 5.542). That is, the fact consisting of that p (that p is or is not the case) is coordinated with the fact consisting in the proposition "p" (since propositions are facts) by means of the coordination of their objects, namely by A's enunciating (voicing or writing) "p" and, say, B's (or even A's) hearing or reading the utterance. That the sense of "p" is p secures—or rather is identical to—the coordination.[22] In this way, A's thinking, which "contemporary superficial psychology" takes to be carried out in A's "composite soul" (TLP 5.5421), is reduced away as irrelevant to the analysis. What remains of A, of A's beliefs and "composite soul," is merely the *logically pure position* from which "p" has been enunciated. The following chapter addresses the significance of this.

Note that such an absolutely minimal position of enunciation and the concomitant logically pure and absolutely thin subject of enunciation, with no "inner world" to clutter it in any way, is *logically necessary*: the existence of language presupposes a subject carrying out linguistic acts. From the point of view of logic, however, "there is no thinking, presenting subject" (TLP 5.631) but only a logically pure position of enunciation. This is why Wittgenstein called the psychology of his day "superficial," for it did not pose this logical necessity at its starting point, confusing it with the thought processes making up its empirical object of investigation. Similarly, "philosophers, who hold that thought processes are essential to the philosophy of logic, got entangled for the most part in [analogous] unessential psychological investigations" (TLP 4.1121), the result being some baggy "composite soul" that "would not be a soul any longer" (TLP 5.5421), for the logical necessity grounding it will be lost from sight.

Although Spinoza is crucially interested in empirical psychology, his differentiation from Wittgenstein in the present respect can be circumscribed accurately, allowing us to largely ignore it in what follows: again, working appropriately on the emotional charge of ideas would divest their objects of this charge and thus make the objects stand out dispassionately, as they really are. After such work is done, Spinoza's ideas can compare to Wittgenstein's thoughts, which are dispassionate, so to speak, by definition. Accordingly, we may limit ourselves to asking whether and to what extent such dispassionate ideas match such dispassionate thoughts.

Epistemology II

For Spinoza, ideas can be adequate or inadequate. Adequate ideas are "clear and distinct," while inadequate ideas are either "confused" (E II p28) or both "confused and fragmentary" (E II p35). For his part, Wittgenstein understands philosophy as aiming to make "clear and sharply delimited thoughts which otherwise are, as it were, opaque and blurred" (TLP 4.112). This goal is attainable, for "anything that can be thought at all can be thought clearly" (TLP 4.116). Since opaque thoughts are unclear thoughts and blurred thoughts are indistinct thoughts, if we take opaque and blurred thoughts to be thoughts held confusedly in one's mind, we may infer that the characterization by which Spinoza distinguishes his two kinds of ideas is almost identical to the characterization by which Wittgenstein distinguishes his two kinds of thoughts.

Spinoza uses his characterization to pinpoint what we might call the "internal" mark of truth: "By an adequate idea I mean an idea which, insofar as it is considered in itself without relation to its object, has all the properties—that is, the intrinsic characteristics—of a true idea" (E II def4). In addition, he carefully specifies that he says "intrinsic so as to exclude the extrinsic characteristics—to wit, the agreement of the idea with that of which it is an idea" (E II def3ex). For Wittgenstein, however, "thought is the logical picture of the facts" (TLP 3), making thoughts that are clear and sharply delimited equivalent to logical pictures that presumably are themselves clear and sharply delimited. Ignoring for the moment any worries about the qualification of "the logical" here, we would perhaps not inflict much violence on Wittgenstein's understanding of the matter at hand were we to suppose that possessing a clear and sharply delimited logical picture allows one to read off from the world the *truth value* of the "significant proposition" expressing the corresponding thought (TLP 4). It follows that for Wittgenstein, too, clarity and sharp delimitation (distinctness) constitute the mark of truth internal to the logical picture constituting thought. He further asserts, however, that the task of philosophy rightly conceived is to make thoughts carry the internal mark of truth on their faces. And that is the task he sets himself in the *Tractatus*.

I have now arrived at a position allowing me to honor the debt I incurred when claiming that the ladder constituted by the *Tractatus* could also assume the figure of a scaffold.

The claim that "philosophy is not a theory but an activity" (TLP 4.112) implies that the business of the *Tractatus* as a logico-*philosophical* treatise is neither to produce thoughts—particular "logical pictures of facts"—that would constitute knowledge of the world nor to present thoughts that might have been produced. These tasks may concern science but not philosophy. One might thus say that no fact whatsoever is logically pictured in the *Tractatus* and that the *Tractatus* does not contain one single thought in that sense. As a *logico*-philosophical treatise, in contrast, the *Tractatus* addresses the logical status of thoughts—what makes thoughts be thoughts in the first place—and hence its task is to show or display that thoughts are logical pictures of facts. From this vantage point, thoughts may be considered as always already given, as always already standing imperturbably by themselves outside the *Tractatus*, as (to use Spinoza's idiom) eternal truths in the mind of God.

To show that thought enjoys this logical status is at the same time to show the logical connections thought entertains both with the world and with language or, equivalently, to display the logical necessities we are "compelled" to heed together with the manifolds of logical possibilities "allowed" us. Since traditional philosophy has been quasi-systematically hiding, entangling, and mystifying these logical connections, we require a scaffold from which to remove the associated grime. The figure of the scaffold thus allows us to portray the *Tractatus* as the provisional construction erected to help clear away, once and for all, the layers of confusion built up on the "edifice" of these logical necessities and possibilities and thus to reveal that edifice in its pristine purity.

The scaffold consists of elucidatory "planks" of telling nonsense. It follows that the "thoughts [that] are expressed" (TLP Pr ¶7) in the *Tractatus* are *not* logical pictures of facts. They are only the means and the only means that philosophical activity (an activity of *thinking*) makes available for clearing away confusion and thus clarifying and sharply delimiting thoughts generally (i.e., logical pictures of facts). Coming to perceive thoughts as logical pictures of facts is thus the *outcome* of the philosophical activity composing the *Tractatus*, and one can express thoughts as such logical pictures only after having been liberated from confusion by that activity. Since the elucidations at issue are the only means that philosophical thinking offers for carrying out its task, since in this sense they consist only of thinking, they may be justifiably called *performatively effective thoughts*, even though in being that, they cannot be identified with logical pictures of the facts. This is also to say, as we saw in chapter 4, that the realm of thoughts as that of the logical pictures of all facts is *not* coextensive with the realm of colloquial language, for the latter also includes nonsense, both telling or performatively effective nonsense and idle or pointless nonsense. Composing the *Tractatus* is possible because the activity of thinking can judiciously employ this additional leeway as an instrument—employ nonsense as telling—so as to clear up confusion and make thoughts stand out clearly and delimit them sharply.

As logical pictures of facts, thoughts do not have the out-of-this-world status that the preceding comments might appear to imply. At least from the point of view of eternity, which Wittgenstein espouses, thoughts are neither discovered nor invented. As the "significant propositions" that they are (TLP 4), thoughts as such, together with the logical necessities and possibilities involving them, animate colloquial language, for a propositional sign functions as a significant proposition if and only if it is "thought out" (TLP 3.5), while "all the propositions of colloquial language are actually, just as they are, logically completely in order" (TLP 5.5563). Again, the problem is that "colloquial language disguises the thought" (TLP 4.002). Hence philosophy's task is to strip off the "external clothes" making up this disguise so as to display the "form of the body" (ibid.) clearly and distinctly. The *Tractatus* may therefore be further portrayed as a scaffold from which to apply the logico-linguistic analysis that, once carried out, would strip off this clothing and make the body of thought visible in all the glory of its nudity.

Returning to the pictorial character of thought, note that Wittgenstein begins by stating that "we make to ourselves pictures of facts" (TLP 2.1) and specifies that such pictures may be straightforwardly "spatial" (TLP 2.182), that is, pictures in the more or less standard sense of the term, or as varied as "gramophone record[s], [musical] score[s], waves of sound" (TLP 4.014), and so on. "Every [such] picture is *also* a logical picture" (TLP 2.182), whereas "thought is *the* logical picture of the facts" (TLP 3, emphasis added). It follows that thought is implicated in, or rather underlies, all varieties of pictures that are not purely logical in the way thought is. *The* logical picture "agrees with reality or not," thus being "true or false" (TLP 2. 21), while *any* picture, like thought, "*includes the pictorial relationship* which makes it into a picture" (TLP 2.1513, emphasis added).

Disregarding differences in terminology, Spinoza considers these matters in a nearly identical manner. Given that ideas are the items "the mind forms because it is a thinking thing" (E II d3) and that "a true idea must agree with that of which it is the idea" (E I a6), it is literally axiomatic for him that "man thinks" (E II a2), that "each mode of thinking" cannot be without its object (E II a3), and hence that, much as they do for Wittgenstein, ideas include the "relationship" that makes them ideas *of* their objects.

Thus both Wittgensteinian thoughts and Spinozistic ideas include in themselves the "arrow" pointing at their objects, the "hook" attaching them to their object, the "element" making each the thought or the idea *of* its object. Without this arrow or this hook, without being of their objects, thoughts or ideas could not be thoughts or ideas in the first place, or to put it differently, thoughts or ideas are thoughts or ideas only in virtue of having objects and of being thoughts or ideas of those objects. Consequently, Spinoza's axiom that we think and Wittgenstein's claim that we make for ourselves pictures of facts are *logically inseparable* from the "fact" that all our ideas or all our thoughts hook onto their objects and are *of* their objects.[23]

To relate this connection to the previous discussion of our nature as body-minds, we may say that the thesis that "man thinks" or, equivalently, that "we make to ourselves pictures of facts" states a *logical necessity*: we are logically compelled, as it were, to think or to make to ourselves pictures of facts by the logical necessities associated with our being part of the world and having been thrown into it as unified body-minds. In other words, thinking or making to ourselves pictures of facts is logically connected to "possessing" human bodies and human minds inseparably wedded to each other, to being human.[24] Our thoughts or ideas are congenitally *of* the world, hook congenitally *onto* the world, because our having always already been thrown into the world in this way makes it logically necessary that they are.

As a result, there is no logical or metaphysical issue concerning how ideas or thoughts can possibly be of something radically different from the mind, how the mind can come to know something radically external to itself, or how the truth of ideas for Spinoza or of propositions expressing thoughts for Wittgenstein can be secured. If the thoughts and ideas in our minds are of something radically different from our minds by logical necessity, there can be no question as to the way this link is established. And if this link is secured by logical necessity, there can be no issue as to the possibility of knowledge, even as all kinds of psychological questions regarding our thought processes may remain open to the appropriate investigations. Furthermore, if the adequacy (clarity and distinctness) of an idea is the internal mark of its truth for Spinoza, while clarity and sharp delimitation constitute the internal mark of truth or of falsity for Wittgenstein, no logical or metaphysical issue about truth in general can arise, and hence undertaking to formulate a "theory of truth" that would take the term *theory* seriously and try to go beyond elucidations would necessarily be circular and congenitally confused.

To further clarify what is at issue, it is important to examine the difference just noted, namely, that for Spinoza, truth is predicated of ideas and hence of mental items, whereas for Wittgenstein it is predicated of propositions and hence of linguistic items, while the internal mark of truth for the one seems to amount to the internal mark for *either truth or falsity* for the other.

Thus Wittgenstein predicates truth and falsity of propositions and not of thoughts but does so in a way that intimately involves thoughts. As we saw, "in the proposition the thought is expressed perceptibly through the senses" (TLP 3.1) by a "propositional sign" (TLP 3.12) that is nothing other than the "proposition [itself] *in its projective relation to the world*" (TLP 3.12, emphasis added). Linguistic expressions amount to "using the sensibly perceptible sign" (TLP 3.11) to link thought to the world, while the hook consists in that "projection" itself. Again, this projection—this hook—is already included in the proposition, and Wittgenstein specifies that it is included in virtue of the proposition's being a "picture of reality" (TLP 4.021). Such hooks constitute the acts of thinking that animate propositions: "the method of projection is the *thinking* of the sense of the proposition" (TLP 3.11, emphasis added).

Furthermore, a proposition is the "description of a fact" (TLP 4.023), which can be "true or false" in virtue of the proposition's being "a picture of reality" (TLP 4.06),[25] and "one needs only to say 'Yes' or 'No' to make the proposition agree with reality" (TLP 4.023). The proposition is true if the "fact it describes" (TLP 4.023) obtains, and it is false if the same fact does not obtain. This presupposes, of course, that the fact in question is at the outset "thinkable" (TLP 3.001), and since "what is thinkable is also possible" (TLP 3.02), "the fact the proposition describes" is logically possible. The presupposed logical possibility that the described fact might obtain allows a mere judgment of affirmation or denial to make the corresponding proposition agree with reality.

At this point it might be worthwhile to recall something discussed in a slightly different context in the preceding chapter, namely that "thinking the sense of the proposition" (TLP 3.11) is identical to "applying" and thinking through—"thinking out"—"the propositional sign" at issue (TLP 3.5). On the one hand, this act of thinking through endows the proposition with its sense; on the other, the act renders the thought that the proposition expresses clear and sharply delimited.

Nonetheless, Spinoza mistrusts language more than Wittgenstein does. Barring the common notions, which are both adequate ideas and linguistic items managing to express these ideas accurately in language (at least as far as this can go, for additional confusing linguistic items cannot fail to appear), language is the source of confusion, and hence linguistic items are not good candidates for truth predication. Thus, as was noted in a different context, "most errors result solely from the incorrect application of words to things. . . . [Yet if we] look only to the minds [of the people concerned, we can see] that they are indeed not mistaken; they seem to be wrong [only] because we think they have [actually] in mind" (E II p46s) something we consider to be an error. In consequence, "most controversies arise from this, that men do not correctly express what is in their mind or they misunderstand another's mind" (ibid.). It follows that truth and falsity can be predicated only of mental modes: adequate ideas make up true knowledge, while "falsity consists in the privation of knowledge which inadequate ideas, that is fragmentary and confused ideas, involve" (E II p35).[26]

Furthermore, although Spinoza maintains that "truth is the standard both of itself and of falsity" (E II p43s), although he holds that an idea in itself, that is, "insofar as it is an idea, involves affirmation and negation" (E II p49), he does not fully understand the symmetry between truth and falsity as Wittgenstein does. Specifically, he sees no internal mark of falsity. If one "look[s] to the meaning and not to words alone" and thus thinks this apparent symmetry through, one will realize that truth and falsity "are related to one another as being to nonbeing" (E II p49s). Now, since being and nonbeing refer to the realm of causes, and since "from a cause there necessarily follows an effect" (E I a3), while "nothing exists from whose nature an effect does not follow" (E I p36), there is no possibility of *pure uncaused* nonbeing, that is, nonbeing taken absolutely and not as referred to a finite mode that has come to perish by the action of some cause. There is no place available for its correlate within the "order and connection of ideas" (E II p7), which is identical to the causal order and connection of things. But this is also to say that there is no internal mark of falsity: "there is nothing in ideas that constitutes the form of falsity" (E II p49s).

Since Wittgenstein holds that "it cannot be discovered from the picture alone whether it is true or false" (TLP 2.224), he presumably would readily agree with this last formulation by Spinoza. As he would agree, moreover, with Spinoza's position that an idea in itself (a thought for him) "involves affirmation and negation" (E II p49), the distance between the two on the issue of truth (and falsity) might not be that great after all. The main difference seems to arise because Spinoza does not sever truth and falsity from knowledge as it resides in the mind and hence cannot predicate truth or falsity of items possessing an extramental (extended) component, such as propositions or linguistic expressions in general. He thus appears to be pursuing a direction opposite to that of the tradition inaugurated by Frege, whose central tenets include the drastic separation of the logical (and thus conceptual) from the mental (psychological), together with the concomitant requirement that truth (and falsity) be predicated of extramental items exclusively. Wittgenstein takes his distance from that tradition on this point: what endows a proposition with sense—ensuring by the same token the link between language and world—is a *mental* act, that of thinking through the propositional sign (TLP 3.5). By identifying the thinkable, or what the mind *can* think, with what is logically possible (TLP 3.02), he restricts the mental contribution to both sense and truth to just what is logically necessary for thought and language to be possible. Thus even with respect to points where the two authors appear to diverge hopelessly, the distance effectively separating them proves not only amenable to explication but also minimal.

Nonetheless, this is not the end of the matter. To assess the present difference more accurately, we must examine the issues to which it relates and how the two authors treat them. Prominent among these is the way Spinoza conceives the possible, namely, finite modes that do not exist but that the laws of God's nature nevertheless allow to exist, as well as how the possible relates to the necessary both for Spinoza and for Wittgenstein.

CHAPTER EIGHT

Matching Form

> But one will have gathered what I am driving at, namely, that
> there always remains a metaphysical faith upon which our faith
> in science rests.
>
> —Friedrich Nietzsche, *The Gay Science,* §344

W E JUST SAW HOW Wittgensteinian facts and their pictures match Spinozistic extended modes and their ideas taken one by one. But Spinoza holds that both extended modes and ideas are "ordered and connected," with the order and connection identical in the corresponding two Attributes. The task at this juncture is thus to clarify this relationship and to examine whether and to what extent Wittgenstein's approach includes an analogue. This concern might be broken down into several issues. Is there a Wittgensteinian analogue of order and connection linking facts? Is there one linking thoughts? If such analogues exist, are they identical to each other, as their counterparts are for Spinoza? Finally, if the answer to all these questions is yes, to what extent does Wittgenstein's resulting sole order and connection match Spinoza's?

Possible Facts versus Possible Extended Modes

To carry out these tasks, we must first examine how Wittgenstein and Spinoza understand possibility. We have to ask what the former says about possible facts and the thoughts picturing them, what the latter says about possible extended modes and the ideas presenting them, and finally, whether and to what extent the overall sphere including all possible facts and their correlative thoughts for the former matches the overall sphere including all possible extended modes and their correlative ideas for the latter.

Starting with Wittgenstein, note that "a fact" that obtains is "the existence of atomic facts" (TLP 2), and recall that an atomic fact is a chainlike concatenation of "objects," which are simple (TLP 2.03). Recall as well that a proposition both "describes a fact" (TLP 4.023) and "pictures reality" (TLP 4.021). The proposition that expresses the corresponding thought in language can be analyzed into the corresponding "simplest propositions, the elementary propositions, [which] assert the existence of [the corresponding] atomic facts" (TLP 4.21). Each such elementary proposition is a chainlike "concatenation" (TLP 4.22) of "simple" signs that "name" (TLP 3.202) the objects involved.

But from Wittgenstein's *logico*-philosophical perspective, the issue from the outset is not actuality (which facts exist) but possibility. Thus "the facts of logic are [just] possibilities" (2.0121), and from the vantage point of logic, there can be no distinction between actual and "merely" possible facts. As we saw in the previous chapter, there is no internal mark of falsity—"it cannot be discovered from the picture alone whether it is true or false" (TLP 2.224)—and hence the proposition does not reveal on its face whether it is true or false, or to put it differently, whether or not the fact "it describes" (TLP 4.023) obtains. The corresponding "picture represents what it represents independently of its truth or falsehood" (TLP 2.22), that is, "one can understand the proposition without knowing whether it is true or not" (TLP 4.024), and hence if the proposition is to "agree with reality," one must also "say Yes or No" (TLP 4.023). In short, the proposition in itself can picture only a *possible* fact. It is not up to logic—or to philosophy—to voice what is the case, for the question as to which facts exist concerns the "how of the world," which comes "after logic"; answering this question is the affair only of science and of "experience" generally (TLP 5.552).

It follows that Wittgenstein's world of facts, consisting of the totality of facts that obtain, with "these being *all* the facts" (TLP 1.1), should not be considered separately from the wider realm of logically possible facts. Wittgenstein marks off this wider realm by giving it a particular name: "the existence and nonexistence of atomic facts is *the reality*" (TLP 2.06, emphasis added). Now given that the "atomic facts are independent from one another" (TLP 2.061), which implies that "from the existence or nonexistence of an atomic fact we cannot infer the existence or nonexistence of another" (TLP 2.062), and given that nothing inherent to a fact makes it actual rather than possible—for a "thinkable [fact] is also [a] possible [fact]" (TLP 3.02) while the corresponding proposition does not announce whether the fact it pictures is actual—to go from "the reality" of all possible facts to the "world" of actual facts (and to all of them), we need to add something to the reality consisting of all possible facts.

What we need to add is the "metafact" (which cannot be properly called a fact) stating in all required generality that, taken together, the actual facts, that is, the "world" (TLP 1.1 and 1.11), constitute a *subset* of possible facts and nothing else. Adding this metafact to the reality of all possible facts turns it into "the total reality" (TLP 2.063). By the same token, the total reality is sealed off from anything external to possible facts; because it is total, nothing could be added to or interfere with it. But adding this particular metafact to the reality of all possible facts is adding something perfectly innocuous or perfectly empty. The metafact in question is nothing other than the principle picking all actual facts from all possible ones.

This is the principle that identifies the world (all the existing facts), and for this reason Wittgenstein may say, "The total reality is the world" (ibid.). At the same time, Wittgenstein notes that no other principle distinguishing the actual from the possible could ever emerge from outside the realm of possible facts, that is, from outside logic. Logical possibility—logically possible facts—is all that a logico-philosophical treatise may address, for "outside logic all is accident" (TLP 6.3): which facts obtain is not a matter of logic or of philosophy.

Nonetheless, since the truth value of a proposition cannot be read off from its face, the proposition may be picturing either an actual or a merely possible fact. And since the thought a proposition expresses can be the thought of either an actual or a merely possible fact (again, what is possible is thinkable), it follows that both language and thought picture both actual and merely possible facts, doing so equally and equally well. From the point of view of logic, these two sorts of facts and their pictures in either thought or language are perfectly on a par with each other.

From the angle of Spinoza's naturalism, the principal issue should be the existent extended modes, not the merely possible ones. Moreover, because Spinoza works at the level of concepts and not of logic, he takes the possible to be what "the laws of God's nature" allow, not everything logically possible, as Wittgenstein has it. Within these restrictions, Spinoza does accept possible extended modes, and he characterizes them in a way that makes the relation between possible but not actual extended modes and their correlative ideas similar to the relation between Wittgenstein's possible facts and their "pictures" in thought.

First, Spinoza's claim that "the essence of things produced by God does not involve existence" (E I p24) implies the possibility of nonexistent extended modes, and the fact that "God is the efficient cause not only of the existence of things but also of their essence" (E I p25) implies the restriction just noted: the essence of things must conform to the laws of God's nature, and hence both the actual and merely possible extended modes will include only those allowed by these laws.

Moreover, Spinoza holds that "the ideas of individual things or modes which do not exist must be comprehended in the infinite idea of God in the same way as the formal essences of individual things or modes are contained in the attributes of God" (E II p8), which implies that the existence of an extended mode is not necessary for God to entertain its corresponding idea. Yet to understand the further implications of this proposition, we have to get clear on the term "formal essence."

Basically, formal essence is to be distinguished from "objective essence." Formal essence is that which makes a thing or mode exist in whichever Attribute it is expressed—the thing or mode has form (the previously discussed unvarying relation among its parts)—while objective essence is that which offers a thing or mode to thought, that which makes it an object of thought (Shirley 1982, 26–27). If the idea of a thing or mode that does not exist must be comprehended as a formal essence, then this idea must be comprehended as form is comprehended, that is, as an existing thing or mode is comprehended. It follows that the idea of a nonexistent thing or mode has precisely the same status as do the ideas of existing things or modes;

moreover—excluding God and His infinite Attributes, whose essence involves existence—nothing internal to such an idea can imply that the thing or mode of which it is the idea exists or does not exist. Just as was the case for Wittgenstein, then, idea of an extended mode taken in itself does not in itself let us read off whether the extended mode exists, for the idea does not include in itself any information about existence. In current terminology, existence is not a predicate.[1]

It follows that the sphere of the possible is wider than that of the actual in more or less the same manner for both authors: the sort of relations that Spinoza's nonexistent but possible modes bear to their correlative ideas is the same as the one that Wittgenstein's possible facts (those inhabiting "the reality") bear to the thoughts picturing them (as well as to the propositions expressing those thoughts). But then a hard question emerges, particularly for Spinoza: how can possible (i.e., allowed by the "laws of God's nature") but not actual extended modes fit the order and connection that E II p7 says should be identical for both the Attributes of Thought and Extension? Since the ideas of possible but not actual extended modes exist in God's mind, how can their correlatives in the Attribute of Extension not exist if the order and connection is the same in both? I will try to answer this question after clarifying what order and connection mean for Wittgenstein.

Natural Space versus Logical Space

Whereas Spinoza's extended modes reside in natural (Euclidean) space, Wittgenstein's facts reside in logical space:[2] "the facts in logical space are the world" (TLP 1.13). For Spinoza, moreover, because the extended modes and their correlative thoughts are radically heterogeneous, only the former resides in natural space, but for Wittgenstein, both facts and their correlative pictures inhabit logical space: "The picture presents in logical space the facts, the existence and nonexistence of atomic facts" (TLP 2.11). A proposition, which "describes" a fact (TLP 4.021), "determines a place" in logical space, while "the whole logical space [is] given by it: [the] proposition reaches throughout the whole of logical space" (TLP 3.42).

To understand why Wittgenstein takes any proposition to reach throughout logical space, note that the picture a proposition constitutes carries "logical scaffolding around it" (TLP 3.42). This logical scaffolding consists in the proposition's inherent possibilities for combining with other propositions through "denial, logical sum, logical product, etc." (ibid.), that is, through the logical constants. To say that these possibilities are inherent in the proposition is only to say that it is the "place the proposition determines in logical space" (TLP 3.4)—the "coordinates" defining this place (TLP 3.41)—alone that constitutes the logical scaffolding in question. If these possibilities were not thus inherent in the proposition, if the place the proposition determines did not offer itself spontaneously, as it were, to the possibility of combining with other propositions, then the logical constants would have to be "additional elements" (TLP 3.42) ensuring precisely such a coordination.

It becomes thus clear why a proposition does not simply picture the fact it describes but constitutes a "picture of reality" (TLP 4.021) generally: since the proposition reaches out to the propositions describing all possible facts, the proposition becomes not just the picture of the particular fact it describes but the picture of all possible facts that, by residing in the same logical space, are inherently coordinated with the particular place the proposition determines. According to TLP 2.06, this is precisely a picture of "the reality." In other words, a proposition always entails a picture of all the facts that can be related to that proposition by logical constants.

The issue here might become clearer if we note an important distinction. Wittgenstein holds that "the logic of the facts cannot be represented"—or equivalently, that "the logical constants do not represent" (TLP 4.0312);[3] at the same time, he asserts that "the possibility of propositions is based upon the principle of the representation of objects by signs" (ibid.). The point here seems to be that if the very possibility of propositions is based on the principle of representation, there must be something identical in the representing proposition and the represented fact "in order that the one can represent the other at all" (TLP 2.161). Thus, if logic could, *per impossibile,* represent or be represented, it would not be logic that makes a proposition to represent its corresponding fact. Securing the necessary correspondence between the realm of facts and the realm of propositions would then be the role of some kind of metalogic. But the same issue would arise for the metalogic, and so on into an infinite regress. Consequently, facts and their pictures must, as a matter of logical necessity, share logical space.

Now logical space is the realm of logic, the realm Wittgenstein sees as pervading the world, thought, and language: "Logic fills the world: the limits of the world are also its limits. We cannot therefore say in logic: the world has this in it and this but not that. . . . We cannot think what we cannot think; so what we cannot think we cannot *say* either" (TLP 5.61). Logic fills the world because no fact that exists (or is logically possible), no thought, and no proposition can escape logic. Nothing can exceed logic, for any fact that exists or can exist is conceivable in thought (TLP 3.02), while each thought is expressible in language, for "man possesses the capacity of constructing languages in which *every sense* can be expressed" (TLP 4.002, emphasis added).

Conversely, thought cannot exceed the limits of the world, for "however different from the real one an imagined world may be, it must have something in common—a form—with the real world" (TLP 2.022). (I will examine this "form" a bit later.) Similarly, since "the propositions are everything which follows from the totality of the elementary propositions (and of course also from the fact that it is the totality of them all)" (TLP 4.52), and since propositions describe their corresponding facts (TLP 4.03), no proposition having sense—that is, agreeing or disagreeing with possibilities of the existence of atomic facts (TLP 4.2)—can exceed the same limit: "On the other side of the limit lies simply nonsense" (TLP Pr ¶4).

All logical constants reduce to "simultaneous negation" following from the Sheffer stroke (and its successive applications), which implies that there is "one and only [one] general primitive sign in logic" (TLP 5.472). The adequate symbolism should take in such scantiness, and all "superfluous" (TLP 5.132) signs should be accordingly eliminated. At the level of elementary propositions, however, logical constants have no role, for again, an elementary proposition consists in a chain-like concatenation of simple names with nothing else holding the names together. Thus one can construct any proposition out of elementary propositions by means of simultaneous negation, and each elementary proposition can be either true or false. Proposing the logical possibility of such a construction—the inverse procedure constituting "the one and only complete analysis of a proposition" (TLP 3.25)—is a way of formulating the logical necessities discussed earlier: there is the world of facts, and we, who are simultaneously bodies and minds, are thrown into that world; there are thoughts, made by us, *of* the facts of the world; there are proposi-tions we use to express thought in language; and there is logic—a unique logical space—shared by the world, thought, and language and thus securing the logically necessary correspondences, while the truth value of a proposition remains undeter-mined because a proposition cannot by itself indicate whether the fact it describes obtains. Given this, room necessarily remains for science and for experience: only they can determine which facts obtain.

Since logical space is what facts share with thought and language and what secures the necessary correspondences, we might be tempted to take it as the Witt-gensteinian analogue of Spinoza's order and connection, which, because both Spi-nozistic Attributes share this, secures the equivalent correspondences. Moreover, like Spinoza, Wittgenstein denies any communication between world and thought: thoughts are just pictures of facts, and hence the relation between the picture and the pictured can be readily construed as matching the relation between Spinozistic ideas and the extended modes that form their objects. Terminological quibbles may remain; for example, Spinoza expressly disallows taking ideas as "dumb pictures" (E II p43s), while Wittgenstein considers propositions as *animated* by thought, whereby it follows that both ultimately hold the same position. But setting such quibbles aside, and given that an idea is solely what presents its object to thought,[4] to assert such a match is to say that an idea might well be conceived as a "picture" of its object in the Attribute of Thought.

Further, given that the order and connection within the Attribute of Thought is that of deductive entailment, and deductive entailment is obviously related to the logical connections at work in logical space, identifying the match in such a manner might not be too wide of the mark. It is certainly unsatisfactory, however, because Spinoza holds that both Attributes exhibit the same order and connection primarily to establish the identity between deductive entailment and causal entail-ment. Causation and the productive power it implicates are all-important for Spi-noza, a significance further manifested by his refusal to consider ideas as "dumb pictures," presenting them instead as the "very *acts* of understanding" (E II p43s, emphasis added). Given this fundamental feature of Spinoza's approach, and be-cause there is nothing involving causation in the account of Wittgenstein's logical space, we must search deeper for the proposed match. To find the Wittgensteinian analogue of order and connection, then, we must look at the level of objects and their associated manifolds of possibilities, that is, at the level of Substance.

Order and Connection versus Form and Structure

Wittgenstein's logical space is shared by propositions, thoughts, and facts because logical space connects propositions, and thereby the thoughts that propositions express, in the same way it connects the facts pictured by these propositions. Remaining within Wittgenstein's framework (and thus excluding causation altogether), while for the moment downplaying propositions, we might stretch the notion a bit and say that logical space is just the pure order and connection that facts share with thoughts. Given that logical space runs throughout the realms of the world and thought, Wittgenstein would presumably find no great difficulty in accepting that "the order and connection of thoughts is the same as the order and connection of facts": "order and connection" is taken here as synonymous with "logical space."

Concomitantly, Spinoza, too, would find no great difficulty in accepting this drastically truncated version of E II p7 within the relevant confines: Wittgenstein's thoughts—"clear and sharply delimited"—match his adequate ideas; his misgivings about language have become pacified, for the thoughts (or ideas) in question can be expressed linguistically by his "common notions"; and finally, once causation and its power are excluded from the discussion, he can accept that the only remaining relation between facts and thoughts is that of causally inert depiction, while the one-to-one correspondence between the two ensures that the order and connection among thoughts and the order and connection among facts must be the same. Thus Spinoza should be willing to accept a Wittgensteinian translation of "order and connection" as "logical space," for the latter would simply be a way of naming the logical grounds of deduction as Spinoza practices it. But how is depiction cashed out, and how does causation enter Wittgenstein's game?

In the preceding chapter I discussed Wittgensteinian pictures but did not stress that the picture and the pictured share something fundamental: "The picture has the *logical form of representation* in common with what it pictures" (TLP 2.2, emphasis added). Now "the form is the possibility of the structure" (TLP 2.033), while structure is "the way in which objects hang together in an atomic fact" (TLP 2.032) as well as what "the connection of the elements in the [corresponding] picture" is "called" (TLP 2.15). In sharing logical form, atomic facts and their pictures (the elementary propositions) share structure, too; by the same token, the structure they share is what allows facts and their pictures to hold together in themselves, that is, their structure is the order and connection in question as it appears at the "bottom" level of atomic facts and of elementary propositions.

At the same time, the form of representation at issue is "the logical form of representation" (TLP 2.2), while this logical form, the possibility of structure, is itself "the form of reality" (TLP 2.18), that is, the possibility of structuring all facts that can exist (TLP 2.06). Again, for Wittgenstein "the thought is the logical picture of reality" (TLP 3), which implies that the thought, insofar as it is a picture, "presents the facts in logical space" (TLP 2.11), and insofar as it is "the *logical* picture," it cannot include anything beyond logic; it presents the facts in logical space in purely logical terms, as the mere logical possibilities structuring them.

Finally, recall that a "picture is a fact" (TLP 2.141). If it is to be capable of representing the fact of which it is the picture, therefore, the two must share the same logical form of representation. But since, once again, the form is the possibility of the structure, the form of representation shared by both picture and pictured is just the *possibility* that the corresponding items—objects and simple names—inherently possess for connecting in chainlike concatenations. If such corresponding concatenations are to exist (i.e., if the elementary proposition is to be the picture of the corresponding atomic fact), the concatenation in itself should be identical in both fact and proposition, and hence the manifolds of possibilities allowing this concatenation should be identical, too. It follows that at this "bottom level" *there can be no real distinction* between objects and simple names: if a Wittgensteinian object is merely the substantive "what" that bears a manifold of logical possibilities, its corresponding name, since it bears exactly the same manifold of logical possibilities, is *logically indistinguishable* from it. With respect to logical possibility, the object and the name of the object are strictly identical. Consequently, at the bottom level of substance, language, with all the "disguises" it carries, disappears completely, leaving behind it only simple names, which are logically indistinguishable from the objects they name.

This is presumably why Wittgenstein says that "logic is not a field in which we express what *we* wish with the help of signs" (TLP 6.124): logic is not a means of expression at all, so that it is not up to us to express anything in it or by it. Logic is, rather, a "field in which the nature of the absolutely necessary signs speaks for itself" (ibid.). It follows that "logic is not a body of doctrine but a mirror-image reflection of the world" (TLP 6.13), with names perfectly reflecting their corresponding objects, which is why "logic must take care of itself" (TLP 5.473) as well as why "it is always possible to construe logic in such a way that every proposition is its own proof" (TLP 6.1265). The idea that "logic is transcendental" (TLP 6.13) can be approached from this angle also.

Therefore, what matches order and connection at the bottom level of substance—always excluding causality—is not just logical space taken amorphously. It is the *pure concatenation as such* of Wittgensteinian objects; it is the way these objects hang together without any glue securing the connection; it is, in a word, what Wittgenstein's calls "structure." And if form is indeed just the possibility of structure, then we may say conversely that structure is actualized "form," the way objects, which are the minimal substantive bearers of form, *actually* hang together and thus constitute existing atomic facts, the facts ultimately making up the world of everything that is the case.

In sum, Wittgenstein's assertion that at the bottom level of mere logical possibilities there is no language matches Spinoza's relative lack of emphasis on language: the level of mere logical possibilities is the level of Substance as such, the level focusing Spinoza's concerns from the very beginning. But because he remains at the level of concepts, Spinoza does not envision logical possibilities in Wittgenstein's way. He is exclusively interested in the actual world, in the world that is the case. This is, of course, the world wherein both Spinoza and Wittgenstein write their treatises, the world into which each of us has already been thrown as a body-mind, and thus the world that always already includes thought and language. Succinctly put, this is the world wherein "*we* make *to ourselves* pictures of facts" (TLP 2.1, emphasis added).

The Metaphysical Subject

I invoke TLP 2.1 at this juncture to stress that the world should be considered as always already including the *agency* that performs the fundamental act of "making pictures of facts," or equivalently, the agency that has thought (Haugeland 2000), bears language, and carries out linguistic acts. This agency—human agency—inhabits the world alongside logic, that is, "before the How" putting down which particular facts exist and what our "experience" (TLP 5.552) of those facts might be. To say that "*ethics* is transcendental" (TLP 6.421) in the sense previously outlined is therefore to say that the very possibility of thought and language implies the human agent who can have thought, bear language, and carry out linguistic acts, the agent who by the same token necessarily possesses the purpose or intention that must precede and preside over every activity the agent undertakes. If the very possibility of thought and language necessarily depends on such an agent, a particular place must be reserved for this agent at the heart of the network of logical necessities and possibilities at issue.

The claim that human agency occupies this place marks the direction along which our examination should continue, for agency is intimately related to causation. To answer whether and to what extent Spinoza's order and connection can find a Wittgensteinian analogue with respect to causation as well, we have to examine human agency, the human subject, at the purely logical level, the level at which Wittgenstein considers it.

To locate the logical position reserved to the human subject, Wittgenstein reverts to the first-person singular: "*The limits of my language* mean the limits of my world" (TLP 5.6). Such a reversion here is perfectly appropriate for a number of reasons, most of which I have canvassed. First of all, as was previously discussed, any human who has thought and bears language must have already entered the human order—always concurrently linguistic, social, and gendered—which is precisely the order marked by the capacity to use the first person. Moreover, we saw that the "I" overarches all ideas contained in Spinoza's mind, so that Wittgenstein's use of the "I" as the means for determining the logical position of human agency makes his approach on this point match Spinoza's perfectly.

Recall, furthermore, that the use of the "I" effectively marks one's assumption of ethical responsibility over all the activities structuring one's life. To the extent that the ethical precedes and presides over all actions, "ethics is transcendental" (TLP 6.421), and if the attendant responsibility can be only one's own, then only the "I" of one's self can bear it. The "I" in question, the "I" associated with the "transcendental" character of the ethical, must then be purely logical, the "I" of the "metaphysical subject" (TLP 5.641). Hence the position occupied by this "I" is a purely logical position, a position absolutely thin and completely unadorned.

Moreover, if the proposition "A believes *p*" (TLP 5.541) may be reduced to the proposition "'*p*' says *p*" (TLP 5.542), then by invoking Kant, one is warranted in adding the clause "I think . . . ," "I believe . . . ," and so forth to any thought one has and to any proposition one enunciates with no loss to the objective status of the thought or proposition. This invariance goes together with an absolutely thin "soul" (TLP 5.5421) having no "inner world" to clutter it. This "soul" can be only the absolutely thin soul of the metaphysical subject, the one that accompanies a logically pure position of enunciation.

None of this makes anyone's language different from language as I have been discussing it. My language is not private, not an idiolect; it is and must be language *as mine,* language as I am using it from the vantage point of the "metaphysical subject" I presently occupy. "The language which I understand" (TLP 5.62) is language (the language) as I use it and understand it. This is presumably the sense of language in which "the limits of my language mean [*bedeuten*] the limits of my world" (TLP 5.6)

Since language (the language) can express "every sense" (TLP 4.002) and so every thought, *my* language, the language of the "metaphysical subject," can express any of my thoughts. Furthermore, since language, which consists in "the totality of propositions" (TLP 4.001), can "represent the whole reality" (TLP 4.12), my language can represent the whole of *my* reality, that is, of "my world"—"the totality of [my] reality" (TLP 2.063)—as my language delimits it. From my vantage point, that of my language and of my thought, the world as I see it, my world, is perfectly coextensive with my thought and with the "significant propositions" I can formulate. But *my* world in this sense is my (the) *whole* world.

Why this is the case is not difficult to understand. Since "logic fills the world" (TLP 5.6), it fills my world, for occupying the vantage point of the metaphysical subject does not render me illogical. In addition, since my world is perfectly coextensive with what I can think and with what I can express with significant propositions, there can be no outside to my thought or to my language: "we cannot think what we cannot think; so what we cannot think we cannot *say* either" (TLP 5.61). Wittgenstein equates the limits of one's language with the limits of one's world for precisely this reason. It follows that everything in my world is an object of my thought and of my language, *my* object, for it is I who thinks them. Consequently, everything in my world can be pictured in my thought, and in that sense everything belongs to my mind; from my vantage point, my mind pictures absolutely everything in my world. Since this world can have no outside, this is *the whole* world for me. "The truth in solipsism" (TLP 5.62) is precisely this. By the same token, since I have thus become identical to the world and since my language is perfectly coextensive with it, there is no position *in* the world from which this truth in solipsism can be "said"; nonetheless, "it shows itself" (ibid.).

But where is the metaphysical subject located? From what perspective can the true discourse of solipsism be issued? Wittgenstein's answer follows immediately. Since my world is the whole world for me, since everything in my world is pictured in my thought and expressed in my language, my thought *exhausts* the world. My world has no additional corner I can occupy when issuing the discourse of solipsism. If I wrote a book truly describing "the world as I found it" (TLP 5.631), I could include in the description absolutely everything in my world, even a full "report of my body, saying which members obey my will and which not and so forth" (TLP 5.631). But it would be logically impossible to say anything about the "I" who has authored my book: "of the subject [who has written the book] alone no mention could be made" (ibid.). I as the author of my book cannot come to occupy a place in my world. I, as the author of my book, *cannot exist* in my world.[5]

If I cannot exist in my world, then nothing in my world can characterize me. I who issue the true discourse of solipsism cannot be endowed with anything belonging to my world, for I have already allocated my body to it. I who have authored the book fully describing the world, including my body, am not constituted through the impact of my world on me; this "I" cannot have any experience whatsoever. For, on the one hand, "no part of our experience is at the same time a priori. Whatever we see could be other than it is. Whatever we can describe at all could be other than it is. There is no a priori order of things" (TLP 5.634); on the other hand, in issuing the true discourse of solipsism, I am located outside the world, and I am therefore the logically pure "I" and hence the "I" a priori.[6]

Yet none of this undermines the logical possibility of writing the book in question. Even if "the subject does not belong to the world" (TLP 5.632), its existence is logically possible, or more accurately, given what I have been saying about the conditions of possibility regarding thought and language, *logically necessary*. Hence it should be "something," and Wittgenstein specifies that it is indeed "something": "it is a limit of the world" (ibid.) whose "field of sight," however, obviously cannot encompass itself (TLP 5.6331) or any other limit. This is the purely logical position of utterance. In Wittgenstein's words, "The philosophical self, . . . the metaphysical subject, . . . is the limit of the world—not part of it" (TLP 5.641), for the "I" in question "shrinks to a point without extension and there remains the reality coordinated with it" (TLP 5.64). It "shrinks" to a *purely logical position*.

But the reality in question is the reality that remains after the metaphysical subject—the subject issuing the truth in solipsism—has been abstracted away as its limit. Hence this is the total reality as such, the total reality *independent of the subject,* the total reality as it is. And this is exactly the reality that realism intends. Therefore, it can indeed "be seen here that solipsism, when [its implications are] followed out strictly [down to the bottom level of logic] coincides with pure realism" (TLP 5.64). At the level of logic, even the most extreme philosophical oppositions are wiped out.

Fractals

"The total reality" (or the world) involved in the identification of solipsism with realism allows Spinoza to resume his conversation with Wittgenstein. What interests Spinoza in the present context is not the subject's role in solipsism but the opposite pole of the identification, that of reality, which Wittgenstein leaves here more or less in the dark. We may say that Spinoza is even keen to resume the conversation at this juncture, for Wittgenstein has expressly relegated the body of his solipsist (or his own body) to the world of facts, the realm matching Spinoza's Attribute of Extension. And Spinoza has relegated his own body to this realm from the very beginning.

At issue here is the total reality or world of everything that is the case (TLP 2.06). And thus, if I am the subject issuing the true discourse of solipsism and writing the book on "the world as I found it," the world includes my body with all its organs, properties, manifestations, and modifications and in all its extremely varied states. Since my solipsistic mind includes everything, all these items must appear in my book, doing so as pictures of the facts laying out the corresponding interactions in the world, namely, those involving various worldly things as they impinge on my body and determine its states. My body and its states can be presented in my book only as outcomes of the *world's natural history*.

Wittgenstein, who does not appear to be very interested in his body, nevertheless implies (e.g., in TLP 5.552) that we acquire experiences through our sense organs, and these experiences always result from the impact of worldly things on the various parts of our bodies and so make up various facts. Since his solipsistic mind includes everything, there must be thoughts picturing these facts, even if they would not present themselves—at least initially, before elucidation has finished with its job—as clear and sharply delimited in the way he pursues. Thus Wittgenstein could fairly easily accept, at least in part, a version of Spinoza's thesis on the relation between mind and body: the facts laying out what Wittgenstein's body experiences are being pictured in his mind as correlative thoughts or, to switch vocabularies, as mostly inadequate (certainly for Spinoza but not necessarily for Wittgenstein) ideas whose objects are extended modes as they relate to his body.

To be more accurate, Wittgenstein would not be too hard-pressed in accepting that "whatever happens in the object of the idea constituting the human mind [i.e., the corresponding human body] is bound to be perceived by the human mind; i.e., the idea of that thing will necessarily be in the human mind. That is to say, if the object of the idea constituting the human mind is a body [its corresponding body], nothing can happen in that body without its being perceived by the mind" (E II p12).

But recall that Spinoza also holds the converse of E II p12, namely, that the "the object of the idea constituting the human mind," taken as a whole, "is the [corresponding] body," also taken as a whole, and "nothing else" (E II p13). In less succinct terms, there can be no idea in the human mind unless this idea has an object that in some way involves the corresponding human body as it undergoes modification by the impact of other bodies on it.

Strange as it might appear at first sight, Wittgenstein would have no difficulty in subscribing to the core of this converse thesis, too: as we saw, "colloquial language" is for him "a part of the human organism," and "man has the capacity of constructing languages in which every sense can be expressed" (TLP 4.002). It follows that every "significant proposition," every proposition expressing sense, has some correlate in the human organism and hence a bodily correlate. And since a "significant proposition is the [corresponding] thought" (TLP 4), the thought that any significant proposition expresses has a bodily correlate as well. Even if such a Wittgensteinian thought, a picture of some remote fact of the world, need not be identical to the picture of *its own* bodily correlate—as the idea matching that thought would be for Spinoza—Wittgenstein's solipsistic mind can have no thought whatsoever without the thought's correlate in his body.

Before going any further, we should explore what this discrepancy involves. I just said that Wittgenstein's solipsistic mind necessarily includes thoughts, or pictures, both of the remote fact R of the world and of the modification M of Wittgenstein's body that is the bodily correlate (because language is a "part of the human organism") of the thought of R. Wittgenstein does not take a position as to whether these two thoughts/pictures coincide, but we might say the following on his behalf: if the two pictures do not coincide, then Wittgenstein's solipsistic mind would comprise two distinct regions, one that straightforwardly contains pictures of facts, which should include both the picture of R and the picture of M (let us call it the proper region), and one that contains the pictures of the bodily modifications correlative to the first pictures (let us call it the parasitic region). Since the picture of M is obviously included in both the proper and the parasitic regions, the two regions overlap, for the picture of M is included twice, first as straightforwardly the picture of M and second as the picture correlative to the bodily modification correlative to the picture of R. It is therefore much simpler to take Wittgenstein as assuming that the two regions coincide: the thought/picture of R is identical with the thought/picture of M. In other words, the picture of the bodily correlate of the picture of R is just the picture of R. But to say this is simply to use Wittgenstein's idiom to translate Spinoza's double thesis to the effect that the human mind is the idea of the human body and of that body alone.

It may be helpful to liberate ourselves briefly from both the Spinozistic and the Wittgensteinian frameworks and formulate the issue in the vernacular. Having the thought/picture of the remote fact R in my mind "involves"[7] undergoing some modification M of my body, a modification placing it, say, in state S. This appears naturalistic and straightforward enough.[8] In addition, however, we might further posit that in this situation, I also have some sort of mental picture or, perhaps more accurately, mental sense of state S, whether consciously or not.[9] If this is so, then one would expect that the thought of R and the mental sense of S (or of M) would be somehow related to each other and that this mental sense would be somehow connected to its correlative bodily state S. For example, my thought of Hopper's painting hanging on my wall should have something to do with the distinct apprehension I feel in contemplating it, while this apprehension should be somehow related to my correlative bodily state S (or the modification M having brought S about).

Now if these proposed relations are not mere appearances, if we are forced for some reason to exclude body-mind causation or any other interaction between bodies and minds, and if we stick to causation regarding things in the world (extended modes), then there seems to be only one way to account for those relations, namely, some version of Spinoza's psychophysical parallelism. Thus, naturally enough with respect both to the overall framework of his approach and to the logical implacability with which he confronts philosophical issues, Spinoza decisively cuts the Gordian knot by postulating that the remote fact R, which is an extended mode, and the bodily modification M (or the bodily state S) are themselves causally linked, with nothing else involved. And that is that.

Having thus taken Wittgenstein as acquiescing to both horns of Spinoza's thesis, we may allow the authors to resume their conversation. Since Wittgenstein's solipsistic mind includes thoughts/pictures of everything that is the case, his mind may be said to form a picture or map of the entire world, one that is accurate insofar as the thoughts in question are (or have become through elucidation) clear and delimited sharply. Concurrently, since each such thought has one and only one correlate in Wittgenstein's body, these bodily correlates constitute a map of the entire world on Wittgenstein's body, thus making it a body-map of the world at its microscopic (or rather "microcosmic" [TLP 5.63]) scale. If these bodily correlates can become unequivocally and distinctly delineated so as to correspond exactly to their correlative thoughts, this map will be as accurate as its corresponding picture. And since this map accurately replicates the world at its microcosmic scale, Wittgenstein's body may be considered, naturally enough if thoroughly unexpectedly, to be a kind of fractal.[10]

Now everything that is the case constitutes a particular state of the world, specifically, the state obtaining at a particular moment. Accordingly, the map consisting of either the pictures in Wittgenstein's mind or of the correlates of those pictures in his body involves a state of both that mind and that body at the same time. But Wittgenstein maintains that "the world and life are one" (TLP 5.621), and since life always involves a dynamic or evolving aspect, the map constituted by either his mind or his body must evolve from one of its states to the next just as the world evolves from one of its states to the next. This confirms, therefore, though from a novel angle, that Wittgenstein's mind and Wittgenstein's body concurrently constitute dynamic maps of the world. Wittgenstein's body is thus an *evolving fractal* that maintains its strict correspondence with the world from state to state.

Wittgenstein, however, would likely retort at this juncture that he is focusing his work at the level of logic. His mind is solipsistic in that he has come to occupy the position of the metaphysical subject who is not a "thinking, presenting subject" (TLP 5.631) but "a point without extension" (TLP 5.64), a "limit of the world" (TLP 5.632), a logically pure position of enunciation. At the level of logic, his body, relegated as it is to the world of facts, can be the body only of the metaphysical subject and as such should be identified with the *world as such*, the world that realism intends. The body of the metaphysical subject, the body "coordinated" (TLP 5.64) with the solipsistic mind of the metaphysical subject, is in this sense the world as body. If Wittgenstein's body is a fractal, it is only a "limit" fractal coinciding perfectly with that of which it is the fractal.

And Wittgenstein might then go on to remark that it is rather Spinoza's body that should be viewed as a proper fractal at its finite, properly microcosmic scale. Since for Spinoza the mind contains the ideas of what happens in the corresponding body and only those ideas, he goes in the opposite direction of forsaking the world outside the body: the only thing his philosophical theory allows the mind to conceive with respect to the workings of the world is that corresponding body alone. Wittgenstein would no doubt concede that Spinoza might be able to adduce compelling reasons for the thesis that only the body can provide adequate knowledge of the world at large, but solely to point out that this necessarily makes Spinoza's body or any other equally complex worldly body replicate at its scale the structure of the world, thus making *it* a proper fractal. In addition, Wittgenstein could continue, because Spinoza focuses on the level of concepts, he cannot countenance the metaphysical subject and thus cannot directly acknowledge the truth in solipsism. But he likely would nevertheless concede that there is no metaphysical subject for Spinoza simply because that role is reserved for God, who is the absolute solipsist: His Mind contains absolutely everything, and His Body is perfectly coextensive with—strictly identical to—the entire world His Absolute Power produces.

The difference here between Spinoza and Wittgenstein might be put more succinctly. Both authors place human bodies within the realm of facts (extended modes), which is radically different from the realm of thoughts (ideas) and thus should remain separate from it, but they differ in their concern with the world as such. Wittgenstein accommodates the homogeneity of bodies in general by identifying the body of the solipsist with the entire world taken as a body. Spinoza, however, explains the "common nature" of all bodies, of their homogeneity, by going in the opposite direction: after having established the objectivity and accuracy of the mapping involved—after having established that his finite body is a dynamic fractal—he forsakes the world as such and takes his own body as the exclusive object of investigation.

Spinoza would no doubt accept this and simply remark that, although he fully respects the restrictions that Wittgenstein's logical interests impose on his treatise, his focus on logical possibility makes Wittgenstein overly downplay the world he too inhabits. In evidence of this, he could point out that Wittgenstein says practically nothing about our finitude and very little on the workings of the world apart from asserting that "no part of our experience is a priori" and that "there is no a priori order of things" (TLP 5.634). Spinoza, however, takes his body as the sole object of his mind and of all ideas residing therein, which allows it to undergo the causal interactions modifying it: by taking his own finite body as a proper fractal, he leaves room for causation. Pursuing this line of thought, he would press Wittgenstein to clarify his position, underlining that he himself, by invoking God's productive power and the necessity of the "laws of God's own nature" (E I p17), hoists the matter of causation to prime importance, just as any sober naturalist ought to do.

The difference between the two here appears stark, for on the one hand, Spinoza expressly holds that "nothing in nature is contingent, but all things are from the necessity of the divine nature determined to exist and to act in a definite way" (E I p29) and that "things could not have been produced by God in any other way or in any other order than is the case" (E I p33). On the other hand, Wittgenstein expressly holds that "outside logic everything is accidental" (TLP 6.3) and that "a necessity for one thing to happen because another has happened does not exist. There is only logical necessity" (TLP 6.37). But we have seen that such appearances of disagreement frequently prove misleading. It might well be the case, therefore, that this difference is similarly less momentous that it appeared to be at first sight, simply manifesting the strategic difference in focus for the two men.

Natural History

Starting with Spinoza, then, recall that he portrays things as being connected through causal chains involving an infinite number of links. If we restrict the scope to just two "things," however—say, my body and another finite extended mode R, however remote—we might initially simplify matters by supposing these two things to be connected via a chain possessing a *finite* number of links, no matter how large that number. If we assume, first, that R and my body are linked in this fashion, making the causal chain connecting them, *per impossibile,* isolated from the rest of Nature, and second, that the chain starting at R ends at P, which is the proximate cause of the modification M of my body, we come to an interesting question whose answer leads to remarkable consequences.

Note first that if, as E I p13 says, my mind can contain *only* ideas of my body and its modifications—presumably brought about by proximate causes—and if my mind possesses the idea of R, however inadequately, then my body must necessarily have causally interacted with R. But then how can my mind possess the idea of R if the proximate cause of M is P? Or to put it differently, how can R remain present, as it were, throughout the causal chain from R to M? Spinoza initially seems to say that it does so only indirectly, for it passes through the Attribute of Thought: "The knowledge of an effect depends on, and involves, the knowledge of the cause" (E I a4), from which it follows that my mind's possessing the idea of M "depends on and involves" the idea of P and thereby the idea of the mode that caused P, and so on, link by link, to the idea of R. Spinoza also stipulates that "things which have nothing in common cannot be understood through each other; that is, the conception of the one does not involve the conception of the other" (E I a5). As a result, all the links of the causal chain from R to M must have "something in common." But what can this shared something be?

Spinoza's answer relies on characteristics of the human body and, by extension, to analogous characteristics of other bodies acting causally on the human body. Thus "the human body can undergo many changes and nevertheless *retain impressions* or *traces* of objects (see II post 5) and consequently the same image of things; for the definition of which see ii p17s" (E III post 2, emphasis added). The postulate to which Spinoza refers us here is one of those closing "the brief preface concerning the nature of bodies": "when a liquid part of the human body is determined by an external body to impinge frequently on another part which is soft, it changes the surface of that part and *impresses on it certain traces* of the external body acting upon it" (E II post 5, emphasis added). His other reference takes us back to the definition of the term *image* discussed earlier in relation to habit and to memory.

This provides a basis for understanding what happens in the causal chain going from R to M. In any interaction within that causal chain, the causing body modifies the affected body so that some trace of the causing body remains *permanently impressed* on the affected body. And this trace remains impressed in a way allowing it to be transmitted to the other bodies implicated in the causal chain leading to P. Thus P both retains the traces of all those bodies and transmits these traces to my body. The associated modification of my body M thus includes all those traces, which means that my idea of M includes not just the idea of P but also the ideas of all the bodies implicated in the causal chain, which evidently are the ideas of the corresponding traces. It is assumed that all bodies implicated are characterized by the complexity permitting such plasticity.[11] Thus, insofar as such plasticity is common to all finite but sufficiently complex things in Nature, the causal efficacy of bodily interactions becomes specified, and our knowledge (adequate or not) of things in Nature and their causal connections is ensured by our knowledge (adequate or not) just of the modifications of our own bodies. At the same time, this form of causal efficacy and interconnectedness makes any sufficiently complex extended mode, including my body, the faithful repository of the traces of all bodies with which it has interacted. Its present state is thus the outcome of all the transformations it has correspondingly undergone, and in this sense it is the causal outcome of the *world's natural history.*

If we give up the simplification and return to Spinoza's thesis that these causal chains comprise an infinite number of links (no cause "can exist and be determined to act" unless another cause brings it about and "determines it [too] to act, and so on ad infinitum" [E I p28]), the way in which Spinoza's body makes up a proper fractal becomes clear. For one, the causal chain we have been talking about does not stop at R. The mode R has been brought about and "determined to act" by a causal chain connecting it to R_1; R_1, by a causal chain connecting it to R_2; and so on, ad infinitum. In addition, if R is complex enough, each of its constituent parts will be a proper individual in its own right. Each such individual has been brought about by some other infinitely long causal chain that reaches up to the body that undergoes modification M. Thus, to say that my body is modified by the proximate cause P that connects causally to R is to say that it is causally modified by *all* the links of *all* those causal chains in that it retains the traces of them all. The modifications my body has undergone up to any given moment of its history are nothing but the traces these links have left on it.

Thus, naturally enough, my body is causally modified by the air I breathe, the food I eat, the sounds I hear, the resistance that my limbs encounter in coming into contact with other bodies, and so on, as well as by all the causes that produced and causally affected that particular volume of air, that particular chunk of food, those particular sounds, and so on. Concurrently, each sufficiently complex individual part of my body, however large or how small (my blood, my stomach, my liver, my tissues, each of my cells), is causally modified by particular things (finite extended modes) acting on it at its own proper scale, as well as by the causes that produced all those particular things and determined their actions. Ultimately, there is no single thing in Nature that cannot—and does not, up to the limit of my body's finitude—connect causally, by near or by far, with my body and its parts via some causal chain.[12] My body is the outcome of the world's natural history, for it shares in the causal fabric of Nature. Nonetheless, my body's capacity to causally interact with and be modified by any thing in Nature is strictly concomitant to my mind's ability to conceive that same thing. This makes everything in Nature in principle knowable.

At the same time, my body is a fractal, for the modifications it undergoes in the course of its history are the traces that things in Nature leave on it, while those traces themselves are *copies* of those things at the "microcosmic" scale of my body. My body is a fractal because the modifications it has undergone are replicas at its scale of all of those things in Nature. This also helps explain why God is the immanent cause of finite things: His *infinitude* is intrinsically involved in the causal production as well as in the causal modifications of any finite extended mode whatsoever.

To clarify still further why my body is a fractal, note that the foregoing considerations imply that, if my body is offered sufficient time, it will run though all its possible states, with the succession of those states being the succession of the causal interactions my body undergoes with the various things in Nature. These causal interactions are what "connect" my body to those things, making it part of the causal fabric of Nature according to the "necessity" that "determines" all bodies to act (E I p29) as they do. My body's connection to the other things in Nature follows the "necessity of divine nature" (ibid.) and nothing else. At the same time, since "things could not have been produced by God in any other . . . order than is the case" (E I p33), the order in which these interactions follow one another, and hence the order of succession of my bodily states, must be none other than the immutable order expressing the same divine necessity. Thus my body's connection to other things in Nature and the order through which its successive states run is the unique order and connection expressing that unique necessity. My body is a fractal in precisely this sense and for precisely this reason. And while my body is running causally through all its possible states, my mind (or rather, its eternal part) will be simultaneously running deductively through the ideas of those states. The limit here is, once again, the limit of our finitude: a human body will perish before it runs through all its possible states.

Note that by referring to "natural history," I mean to imply that Spinoza understands causal interconnectedness in a way significantly different from the way contemporary physics characterizes natural processes. This is not only to say that Spinoza could not have had foreknowledge of the way the "science of motion" would develop into classical mechanics but also to underline that this development proved to be grossly at odds with anything that Spinoza would have expected to issue from a science of Extension. Not only did he lack the means for construing natural processes according to the strictures that classical mechanics would eventually set forth, but more important, he lacked the inclination to do so.[13]

The strictures in question fundamentally involve laws in the form of mathematical relations, expressed as differential equations, that interconnect concepts (acceleration, mass, force, momentum, etc.), laws that various phenomena—say, free fall—are taken to obey, while time is taken as a free parameter underlying all phenomena.[14] These fundamental strictures imply[15] that classical mechanics—and *mutatis mutandis*, physics in general—carves out the phenomena it engages in a way that wipes out all the individuating features of the bodies implicated except those pertaining to their motion. Thus, the discipline is *constitutively* incapable of distinguishing, say, a freely falling apple from a freely falling brick if the two have the same mass and conform to identical initial and boundary conditions. Phenomena are *generic* in this particular sense; physics is capable, again *constitutively,* of individuation only to the extent that the relevant initial and boundary conditions can provide this.

To be thus carved out, phenomena must be totally isolated from their environment (which is the function of a laboratory), so that the widely acclaimed predictive power of physical theories may not be so wondrous after all, at least in terms of philosophical considerations, for it is a direct consequence of the isolation of phenomena in conjunction with the undifferentiated nature of time. (Since all time instants are strictly equivalent, the past differs from the future purely conventionally, barring the increased entropy dictated by the second law of thermodynamics, which would require a complex discussion in its own right.) It can be shown (Baltas 2007b) that the predictive power of physics, as well as the repeatability of physical experiments, rests on the discipline's limited power to individuate in the sense at issue here.

By contrast, Spinoza takes all bodies and all extended modes as fully singular individuals, each of which possesses its proper and inalienable identity constituted by the previously discussed unvarying relation. A causal relation is thus what a given singular individual does to another singular individual, while the "laws of God's nature," although relational, can concern only what such individuals and their causal interactions share in common; such laws would thus not undercut the singularity at issue. By the same token, and as the immanence of God in everything requires, the laws of God's nature are not overarching strictures that individuals must *obey* but descriptions, at the requisite level of generality, simply of what they *do*.[16] On these grounds, the incapacity of distinguishing a given apple from a given brick by some theory of Extension—say, Newton's theory—would presumably be reason enough for Spinoza to dismiss that theory altogether. Concomitantly, the absence of any real distinction between past and future, as well as among repeatable and in that sense identical experiments, would have made him deeply suspicious of these glorified methodological hallmarks of contemporary natural science.

Thus the laws Spinoza formulates in comments about the nature of bodies in part II take a form that accommodates such singularity. And to that extent, one may safely presume that the laws he was envisaging, the laws that the science of Extension—of "*motion and rest*" (E II p13sa1, emphasis added)—would eventually formulate, would not touch the singularity at issue. Nonetheless, a Spinozistic individual has its proper history, and, again in contrast to the way classical mechanics and physics in general treat the matter, an adequate science of Extension should account for this history. Thus, whereas classical mechanics and physics in general treat a singular apple or a singular brick as an undifferentiated token instantiating the relevant laws, a Spinozistic science of Extension should conceive it as coming into being when the unvarying relation identifying it is instituted, as evolving while surviving causal encounters with other individuals (and being transformed by them, in part retaining traces of the bodies it has encountered), and as perishing when its identifying relation is dissolved by some such causal encounter. Since singularity is connected with history in this sense, the laws of God's nature, unlike the natural laws to which we have been accustomed, should be *the laws of natural history,* whatever these might be. The science of motion and rest should tell us how individuals are composed and decomposed, how an individual evolves causally from state to state, and how nature as a whole evolves causally from state to state.

It is noteworthy that contemporary natural science has come up with chaos theory, which might help illustrate what an adequate science of Extension could have looked like for Spinoza. Put as simply as possible, chaos theory envisages the differential equation governing a physical system (a physical phenomenon) in a self-reflexive way by taking into account the equation itself, the range both of any parameters it includes and of the relevant set of initial and boundary conditions. Its purpose is to study how numerical variations in the values of parameters or of initial and boundary conditions influence the solution of the equation and thus the behavior of any physical system governed by it. In cases where small numerical variations in the parameters or the initial and boundary conditions yield disproportionately large variations in the solution of the equation, the behavior of the physical system is deemed "chaotic," for we typically do not expect apparently insignificant numerical differences to produce dramatic changes in overall behavior.

It is imperative to note that the differential equation itself has nothing inherently dramatic about it: once the values of the parameters and initial and boundary conditions are given, the solution of the equation is unique, which means that that the behavior of the physical system is fully deterministic (causal, in Spinoza's sense). Nonetheless, since these values are provided not by the theory but by experiments, observations, or estimations that cannot be sharply determined but must instead fall within a margin of error, the exact course of the physical system is not predictable, making chaotic behavior an expected outcome. The popular example is well known. The otherwise fully deterministic equations proposed to study weather

modes will be actualized sooner or later. Therefore, God's eternal mind contains the ideas of all naturally possible extended modes in that He knows that all of them will be eventually actualized in the determinate order in which His power will bring them about. And this order is the order of adequate ideas in God's mind, the order containing at their correct places the ideas of all the causes that will eventually bring about—will actualize—any given naturally possible extended mode.[19]

This allows me to honor the promise I made at the beginning of this chapter, namely, to clarify the sense in which we should understand Spinoza's assertion that "the order and connection of ideas is the same as the order and connection of things" (E II p7). To achieve this, I employ Spinoza's notion that an adequate definition consists in the idea that lays out the essence of what it defines: "the true definition of each thing involves and expresses nothing beyond the nature of the thing defined" (E I p8s2); equivalently, "the definition of any thing affirms . . . the thing's essence" (E III p4d), which entails that the intellect should be capable of inferring from that definition all properties that necessarily follow from that essence (E I p16d). And since "the knowledge of an effect depends on, and involves, the knowledge of the cause" (E I a4), the adequate definition of some particular thing includes the idea of whatever caused that thing to come to exist. Mainly for this reason, Spinoza's adequate definitions are often called "genetic definitions."[20]

So for Spinoza, an adequate definition of some particular thing is its genetic definition, or equivalently, the *identification* of the cause that has brought about that thing and the *expression* within the Attribute of Thought of the idea of that cause. Given this, and if we slightly extend the notion of cause so as to authorize the claim that the adequate definition of an idea is the cause of this idea in the Attribute of Thought (for its definition is precisely what engenders this idea in the deductive order and connection), then we may infer that the idea of the cause by which an extended mode comes into existence (i.e., the definition, obviously within the Attribute of Thought, of what causally engenders this extended mode) is the cause of the idea of that extended mode along the deductive order and connection. This, then, is why both ideas and things necessarily exhibit the same order and connection. And it is striking that Hertz, who on all accounts influenced Wittgenstein profoundly,[21] says almost verbatim what Spinoza says: "We form for ourselves [cf. "we make to ourselves" (TLP 2.1)] images or symbols of external objects; and the form which we give them is such that the necessary consequents of the images in thought are always the images of the necessary consequents in nature of the things pictured" (Hertz 1956, 1; see also Cohen 1956).

Hertz's appearance at this juncture as mediator between Spinoza and Wittgenstein prompts me to round off my discussion of Spinoza's physics before engaging Wittgenstein's way of understanding causality and its possible match with Spinoza's. To that end, I must further elaborate the relation between God's infinitude and the finite extended modes.

Again, Spinoza considers the causal chains interconnecting the finite extended modes to consist of an infinite number of links, with the infinite compass of these interconnections betokening God's immanence in each finite mode. That is, "God is the immanent . . . cause of all things" (E I p18), and hence of every single thing, because the infinite compass of the causal network relates each one of its nodes (each extended mode) to all the others and so to God's infinite causal power. God dwells in every extended mode, for this causal power, "his very essence" (E I p34), empowers that mode to connect causally with all the others.

At the same time, Spinoza was acutely aware of the distinction between the infinite character of God's causal power, on the one hand, and the finite character of the causes accessible to our finitude, on the other: "The idea of an individual thing existing in actuality has God for its cause, *not insofar as He is infinite,* but insofar as He is considered as affected by the idea of another thing existing in actuality" (E II p9, emphasis added). This implies that following a causal chain from finite mode to finite mode would never allow us to reach God or Substance as such; a causal gap separating finitude from infinitude is unavoidable.

The same causal gap can be witnessed from the opposite direction as well. The general characteristics of Substance and Attribute deductively entail neither the specification of the finite modes that exist nor the causal chains linking them. The infinitude of Substance and of Attribute cannot relate to the finitude of the modes otherwise than through the infinite compass of the causal network. Hence, if physics is the discipline exclusively concerned with the workings of finite extended modes, and metaphysics is the discipline mainly concerned with Substance and Attribute, the *causal* gap at issue manifests the *epistemic* separation of physics from metaphysics: physics, the natural history of bodies, cannot be deduced from metaphysics, and metaphysics cannot be deduced from physics.

But Spinoza does not leave things at that. Although he did not "intend writing a full treatise on body" (E II p13sa3l7s), that is, on physics, he did feel the need for reconciling, if not properly unifying, the metaphysical necessities associated with God's infinite causal power on the one hand with the physical necessities associated with ordinary finite causes on the other. Thus, as was noted in chapter 6, when pressed in this direction by his interlocutors, he brings in what has been called the "infinite immediate modes" and the "infinite mediate modes." To round off my examination of Spinoza's physics, I must try to clarify these "infinite modes."

Spinoza explains these infinite modes in replying to a query Schuler formulates in a letter already mentioned; the latter man writes, "I would like to have examples of those things immediately produced by God [the infinite immediate modes] and of those things produced by the mediation of some infinite modification [the infinite mediate modes]" (L 63). Spinoza answers, quite briefly, "Lastly the examples you ask for of the first kind are: in the case of [the Attribute of] Thought, absolutely infinite intellect; in the case of [the Attribute of] Extension, motion and rest. An example of the second kind [for the Attribute of Extension] is the face of the whole universe [*facies totius universi*], which, although varying in infinite ways, yet remains always the same. See scholium to lemma 7 preceding II p14" (L 64).

Ignoring the Attribute of Thought (i.e., what the "absolutely infinite intellect" is and why Spinoza does not give an example of an infinite mediate mode for the Attribute of Thought, which according to E II p7 should exist in correspondence with "the face of the whole universe"),[22] we may safely assume that Spinoza's brevity here, as well as his failure to enlarge on the infinite modes in the *Ethics* (simply alluding to their existence in E I pp21–23), constitutes yet another token of his reluctance to characterize physics in more detail than strictly requisite: the infinite modes should constitute the necessary interface between physics and metaphysics, and hence their content and status could not be formulated in the rigor demanded by the *Ethics* before or independent of the formulation of an adequate physics.

This is also to say that if such an adequate physics had been available to Spinoza, the axioms and lemmas forming his comments on the nature of bodies would not have been formulated as relatively independent material but would have rigorously followed from the content and status of the infinite modes themselves. To see why, note that the infinite immediate mode of the Attribute of Extension is identified with "motion and rest," and the first axiom of the "brief preface" states that "all bodies are either in motion or in rest" (E II p13sa1).

Since all other axioms and lemmas in this passage are intimately connected to that first axiom, we may speculate that in the full-blown physics Spinoza envisioned, the infinite immediate mode of motion and rest, as directly derived with the requisite logical rigor—that is, immediately—from Substance and as adequately spelled out in the process, would have constituted the "laws of God's nature" as expressed in the Attribute of Extension. The immediate infinite mode would thus have been equivalent to the determination of the natural necessities implied in motion and rest and in bodily movement generally.[23] It would have thus formed the first additional step in an epistemological account of the causal gap between the infinitude of Substance and the finitude of the standard extended modes. In terms of the previous example, the immediate infinite mode would have constituted the laws governing the collisions of the particles, the associated conservation laws, and the ways these particles come to form "complex individuals."

The same example allows us to understand Spinoza's mediate infinite mode. First of all, the "whole universe" referred to in Spinoza's example may be interpreted as the "universe" of the enclosure containing the particles moving in it. This whole universe has to be differentiated from the mere "what" of Substance as well as from the way Substance is expressed in the Attribute of Extension as Extension. Once the totality of the particles in the enclosure is taken as analogous to the totality of the extended bodies, this differentiation—a "modification"—makes the "what" constituting Substance appear to the perceiving intellect (E I d4) as the totality of the extended bodies that already conform to the infinite immediate mode and hence to the laws of God's nature. Thus the "face of the whole universe" is an infinite mode in that it concerns the totality of all particular extended modes; it is a mediate mode in that it mediates between the laws of God's nature and the particular extended bodies making up the whole universe.[24]

As a result, the way the whole universe appears, its "face," is perfectly analogous to the way the phase space of the proposed enclosure appears to us: at each particular instant of time, the totality of those particles is represented by (has the face of) a point in phase space that condenses the positions and velocities (the states of motion and rest) characterizing each and all particles at that instant. Concomitantly, phase space as such is nothing but the picture (the face) of all naturally possible positions and velocities of all particles in the enclosure—the face of "Nature as a whole" (E II p13sa317s). The position of this point varies continuously in phase space in perfect tune with the way all particles in the enclosure ("all the constituent bodies of Nature" [ibid.]) vary in position and velocity (in motion and rest). The phase space as a whole—the face of Nature as a whole—however, does *not* vary: given the ergodic hypothesis, the system of particles in the enclosure will necessarily go through every point of the phase space; that is, all naturally possible positions and velocities of all the particles will eventually be actualized. In sum, the face of the whole universe may be taken as equivalent to phase space as such, that is, to all the naturally possible states of motion and rest of all naturally possible extended modes whose formal essences are "contained in the Attributes of God" (E II p8) and that will therefore necessarily be actualized according to the proper causal order and connection.

Thanks mainly to Newton and Laplace, however, we are in a position to know something that Spinoza could not have known, namely, that in conjunction with the relevant mechanical laws (here just the laws of collision), the complete determination of just one point in phase space is sufficient for determining the state of the system at any instant of time. The whole natural history of that system is determined solely by the mechanical laws and the determination of just one of its possible states. Thus, although many decades had to pass before Laplace could arrogantly proclaim that, were he given the initial conditions of the universe, he could predict all its future,[25] my interpretation suggests that Spinoza had already formulated, with striking clarity, the fundamental insight according to which, epistemologically speaking, an adequate physics would require only one additional infinite mode to mediate between "the laws of God's nature" and the finite extended modes, with this mediate mode somehow involving the face that the actual world presents to finite human intellect.

Given that my interpretation is vague and sufficiently speculative to provide the requisite leeway, once we assume the ergodic hypothesis to hold even when the number of the dimensions of the enclosure and the number of particles in it become indefinitely large, we can conclude that, had Spinoza considered the possibility of atomic particles, the guiding ideas of his "brief preface" could have proudly stood up not just to the classical mechanics of Newton and Laplace but, even more impressively, to the main articulations of important physical theories much more contemporary to us.

Natural Necessity versus Logical Possibility

Having concentrated on Spinoza for so long, we should return to Wittgenstein so as to examine whether and to what extent his views on physics and causality match Spinoza's. Again, Wittgenstein showed little interest in either physical theory or the way particular physical theories might arrive at "finer," "simpler," or "more accurate" (TLP 6.342) descriptions of the world, focusing instead on the associated logical possibilities, that is, that "the world . . . *can* be described in the particular way in which as a matter of fact it is described" (TLP 6.342, emphasis added). Spinoza lacked the luxury of envisaging such possibilities; he could accept only what he considered to be the unique metaphysically necessary description of the workings of Nature.

This difference lies at the root of the striking divergence between the two writers' formulations: whereas Spinoza advocates that "nothing in nature is contingent, but all things are from the necessity of the divine nature determined to exist and to act in a definite way" (E I p29), and that "things could not have been produced by God in any other way or in any other order than is the case" (E I p33), Wittgenstein asserts that "outside logic everything is accidental" (TLP 6.3) and that "a necessity for one thing to happen because another has happened does not exist. There is only logical necessity" (TLP 6.37).

But such divergent formulations might not be the end of the matter. First, note that Wittgenstein does not hesitate to acknowledge that the scope of the physics of his day suffers from exactly the limitation that the previous discussion suggests would have alienated Spinoza: "We ought not to forget that any description of the world by means of mechanics [the established discipline of physics] will be of the completely general kind. For example, it will never mention *particular* material points: it will only talk about *any material point whatsoever*" (TLP 6.3432). As we saw, the structure of mechanics and of physical theories generally does not allow the treatment of singularity, that is, of particular bodies of indefinite variety, their causal interconnections, and their natural history. Contemporary physics does not add up to the natural history of the world and its inhabitants, for the "nets" with different shapes—"square," "triangular," "hexagonal," and so forth—that Wittgenstein invokes as metaphors for different possible physical theories that might "cover" the "surface" representing "the world" (TLP 6.341) have been established only at the cost of excluding the possibility of treating proper individuals, their coming to being, their development, and their passing away.

Within the limits set by this remark, Wittgenstein concentrates on causality as contemporary physics understands it by systematically conjoining it to the notion of a natural law. Thus he points out, first, that there can be no overarching law of causality governing things in nature, for "the law of causality is not a law but the form of a law" (TLP 6.32), and second and concomitantly, that "in physics there are [many] causal laws, laws of the causality form" (TLP 6.321). The phrase "law of the causality form" presumably refers to the mathematical structure—recall that "the form is the possibility of the structure" (TLP 2.033)—expressing a law; that is, it refers to the corresponding differential equation. As previously discussed, understanding causation in this way involves individuals (this particular apple, that particular brick) as items that can be distinguished only by the numerical values of a few well-defined characteristics considered to be fundamental and common to all bodies (e.g., mass)[26] and the conditions, initial and boundary, to which each item might conform. Such an understanding ignores all features or properties individuating natural bodies, and the natural bodies themselves become equivalent instances of the law's applicability. It is only under this essential restriction that natural bodies as different as apples are from bricks can obey "laws of the causality form" in the way physics (and Wittgenstein) lay them out.

Spinoza, however, refers cause only to proper individuals. For him, then, to invoke a causal connection is to say that some proper individual acts on some other proper individual and modifies it, where the cause, *in sensu stricto,* of the modification is merely this action itself. Concomitantly, the laws of (God's) nature are general descriptions of what bodies do to other bodies, descriptions that fully safeguard the singularity of the individual bodies caught up in the interaction.

To go on, then, note that Wittgenstein does not further consider how a causal law could concern proper individuals in a way that would have caught Spinoza's attention. He concentrates instead on causal laws generally, clarifying that since there are "many laws of the causality form," this phrase constitutes "a class name" (TLP 6.321). The reason Wittgenstein delimits the scope of his investigation in this manner is not hard to find: since "the exploration of logic means the exploration of *everything that is subject to law*" (TLP 6.3), the focus of his *logico*-philosophical treatise can be only the "position of logic relative to mechanics" (TLP 6.342), to physics and science generally and nothing else. With respect to the matter at hand, "outside logic everything is accidental" (TLP 6.3), and again, the accidental as such cannot present any issue for logic.

To proceed with the logical exploration of everything that is subject to law and to set forth the position of logic relative to scientific theories, Wittgenstein considers Newtonian mechanics as an example. He envisages that theory as "imposing a unified form on the description of the world" and goes on to illustrate this unified form in terms of an (evidently conceptual) net covering a "white surface with irregular black spots" that in turn is meant to picture the world (TLP 6.341). Once the shape of the mesh is chosen—for example, it might outline squares—if the squares are "sufficiently fine" (i.e., small), the "description of the world" has indeed been "brought to a unified form," for "every square is now either white or black" (ibid.). Considered as such a mesh, "mechanics determines a form of description" because it stipulates that "*all* propositions in the description of the world must be obtained in a given way" (ibid., emphasis added), that is, by following the layout of the net, while starting "from a number of given propositions—the mechanical axioms" (ibid.). In this sense, "mechanics is an attempt to construct according to a single plan all true propositions which we need for the description of the world" (TLP 6.343).

This illustration helps Wittgenstein determine the position of logic relative to physical theory, and by extension to scientific theory generally, because other, different such "nets" (i.e., formulations of physical theories) can be and have been constructed, all of them offering uniform descriptions of the world that are physically equivalent. Thus a "net with a triangular or hexagonal mesh" could have been "applied with equal success," which implies that the particular shape of the mesh, and hence the corresponding unified form, is of no concern to logic: it is "arbitrary" in this precise sense (TLP 6.343). Obviously Wittgenstein was thinking of the different, conceptually disparate but physically equivalent formulations of classical mechanics (the Lagrangian or the Hamiltonian formulations [H. Goldstein 1959], the one proposed by Hertz [1956], and so forth) available in his time. His references to the "law of least action" (TLP 6.3211) and the "law of conservation" (TLP 6.33) confirms that much.

Note, moreover, that Wittgenstein accepts the possibility that the surface he invokes to picture the world might be described "more accurately with a triangular and coarser" mesh instead of "a finer square mesh, or vice versa" (TLP 6.341). By taking such more drastic differences into account, Wittgenstein was no doubt also envisaging the radically novel uniform description of the world brought about by the STR. His assertion that there is no "passage of time" per se but only "comparison" between physical "processes," one of which might well be the "movement of the chronometer" (TLP 6.3611), is highly reminiscent of Einstein's discussion of time and simultaneity in his classic 1905 paper introducing the STR.

Be that as it may, since such conceptually disparate descriptions can be indefinitely many, while logic can be interested only in the *possibility* of such descriptions (in that "the world . . . *can* be described" [TLP 6.342, emphasis added] according to them), the particulars of any uniform description scientific research might yield can be of no concern to logic. Conversely, the bare possibility of describing the world in the terms of such a uniform description, of a "single plan aiming to include all *true* propositions" describing the world (TLP 6.343), has nothing to say about the world of "everything that is the case." As Wittgenstein has it, "That a picture like [the one portraying the world as a white surface with irregular black spots] be described by a network of a given form asserts *nothing* about the picture" (TLP 6.342), that is, about the world, for such a form of description is indifferent to the exact layout of the black spots on the surface (correlative to the configuration of actual facts out of all possible facts) and thus holds of any such configuration, that is, "of every picture of this kind" (ibid.).

Nonetheless, the exploration of "everything that is subject to law" (TLP 6.3) via logic does not end with this negative result. The uniform description in question offers something for understanding the world at the level of possibilities that logic demands, something characterizing the world by means of that description: "*this* characterizes the picture—the world pictured as the surface we are talking about— the fact, namely, that it can be *completely* described by a definite net of *definite* fineness" (TLP 6.342). This appears to mean that the possibility of describing the world by such a "net," the logical possibility of describing the world in its terms, the fact that the world "can be described in that particular way in which as a matter of fact it is described" (ibid.)—this possibility ensures that the net "speaks" about the world. To say this is to assert that the world is describable in such a manner. Wittgenstein states this explicitly: "through their whole logical apparatus [as expressed mathematically at the conceptual level] the physical laws still speak about the objects of the world" (TLP 6.3431), the "objects" making up the world's substance. It goes without saying that Spinoza would agree fully that the world is indeed describable at the level of Substance by one rigorous conceptual system.

In addition, to claim that the world can be "*completely* described by a definite net of definite fineness," or equivalently, that this net can provide in its own conceptual terms "*all* true propositions we need for the description of the world" (TLP 6.343), is to assert that the world is in principle knowable with no residue remaining, and knowable, moreover, according to a "single plan" (ibid.). It follows that such a "definite net of definite fineness" provides in principle a *sufficient* basis for the "one and only complete analysis" of propositions (TLP 3.25) setting forth "the total reality" (TLP 2.063), or to put it differently, the world of everything that is the case. All these observations imply that Spinoza would agree fully that the world is indeed knowable, with no residue left behind, according to a "single plan." Yet for him, this plan can be only one.

Since the possibility of describing the world by a "definite net of definite fineness" does not preclude the possibility of describing it by some other, conceptually disparate "definite net" of *different* "definite fineness," we may infer that, once we have laid down the relevant criteria, the world "can be described more simply by one system of mechanics [net] than by another" (TLP 6.342).[27] And this "too says something about the world" (ibid.), namely, that our knowledge of the world can be rendered conceptually more economical.[28] Note, however, that historical limitations would force Spinoza to diverge from Wittgenstein on this. As he was in no position to conceive a level underlying the conceptual, he could envision only one adequate description of the world, namely, its true description.

Anyway, this discussion demonstrates that Wittgenstein does not dismiss causation out of hand. His decision to call the conceptual nets at issue expressions of "laws of the causality form" even implies that he regarded causation as ineliminable from the proper philosophical understanding of physics and of science generally. Still, the logical possibility of indefinitely many such nets makes their particular ways of describing the world—their corresponding "laws of the causality form"—a function of the nets themselves. Wittgenstein fully grants the inference: "a necessity for one thing to happen because another has happened does not exist" (TLP 6.37), for the way in which one thing is causally connected to another thing depends on the way a given net/physical theory portrays the connection. For this reason "that the sun," a "particular" material point (TLP 6.3432), "will rise tomorrow is a hypothesis" (TLP 6.36311) associated with one or more such net, and hence we "do not *know* whether it will rise" (ibid.). It follows that there is no causal necessity connecting things in the world and hence that there can be "only logical necessity" (TLP 6.37). Talk of causality can never be talk about the world as such but only talk about the ways of describing it.[29] As Wittgenstein puts it, such talk "treats of the network and not of what the network describes" (TLP 6.35).

Concomitantly, anyone tempted to subsume all causal laws under a unique law of causality could not state anything more substantial than "there are natural laws": "If there were a law of causality it might run: There are natural laws" (TLP 6.36). Such a statement, however, could be issued only from a vantage point overarching the world and all scientific (conceptual) possibilities of knowing it and thus would be mere nonsense. Hence "that there are natural laws clearly cannot be said but shows itself" (ibid.). And it shows itself precisely through the existence of effectively functioning scientific theories and their different formulations, through *the fact* that the world "can be described in that particular way in which as a matter of fact it is described" (TLP 6.342) by means of these formulations. In short, Wittgenstein has no issue with causality and causal laws once causation is understood as function of the relevant conceptual frameworks and hence as residing at the conceptual level, thus remaining subject to the logical possibility that scientific research might create indefinitely many and conceptually disparate frameworks of this sort.

Wittgenstein specifies further that since any "law of the causality form" merely expresses one logical possibility among many, it is an "illusion" to view "the so-called laws of nature as the explanations of natural phenomena" (TLP 6.371). After all, an explanation can be put forth and do its work only at the conceptual level. But again, indefinitely many conceptually disparate formulations of physical theories associated with causal laws are logically possible. It follows that each such formulation can carry only the explanations proper to it, and hence explanation in general is always a function of the corresponding conceptual framework. Thus no law of nature could possibly explain anything exceeding the particular conceptual framework proposing it, and hence no law of nature can constitute the kind of explanation that could bring a logico-philosophical investigation to a justifiable close. Far from constituting such a close, considering natural laws to be the ultimate terminus of explanation occludes the whole realm of logical possibilities with which Wittgenstein is exclusively concerned, the realm of logic as such. And since, in coming "before the How" of the world, logic expresses without mediation the world's substantive "What" (TLP 5.552), to hide logic in this way is to hide the substantive "what" itself, the only proper terminus of logico-philosophical inquiry.

Yet, says Wittgenstein, "people today" (TLP 6.372) do not understand that they are victims of this illusion. On the contrary, the illusion in question "founds the whole modern view of the world" (TLP 6.371), making scientists and philosophers "stop short at natural laws, treating them as something unassailable, as the ancients stopped short at God and Fate" (TLP 6.372). Wittgenstein indicates that in this respect, "the ancients" and the moderns "were both right and both wrong" (ibid.). They were both right, presumably, in holding that no investigation could go on indefinitely and that all must stop somewhere. The ancients presumably erred, however, because they lacked the conception of a natural law that science later elaborated. Hence they were in no position to carve out the realm of logical possibilities and consider the level of logic in its own right. Of course, this is the historical limitation that blocked Spinoza and differentiated his approach from Wittgenstein's. Despite this historically determined limitation, however, Wittgenstein acknowledges that, in contrast to the modern "system of thought, which makes it appear as if *everything* were explained" (ibid.), the "ancients were clearer in that they recognized one clear terminus" in God or Fate.

Even if Wittgenstein might not have had Spinoza in mind when referring to "the ancients," his acknowledgment of the superior clarity of their conception can be seen as a respectful bow in Spinoza's direction. As we saw, Wittgenstein expressly admits that laws of nature as understood in his day could not deal with proper individuals, as the prospective laws of God's nature would do for Spinoza; beyond that, the clear terminus in God to which Wittgenstein pays tribute is the most salient feature of Spinoza's whole approach. Given that God is for Spinoza what substance is for Wittgenstein, the difference between the two authors with respect to causation involves nothing more than the distinction between the conceptual (or the grammatical) and the logical. Even with regard to the issues of necessity and contingency, where the disagreement between the two appears starkest, the distance separating them proves minimal.

To seal off this final agreement, we may once more appeal to the mediation of Hertz. Thus, where Spinoza maintains, "It is the nature of reason to regard things not as contingent but as necessary" (E II p44), Wittgenstein can reply, "In the terminology of Hertz we might say: only connections that are *subject to law* [*gesetzmäßige*] are *thinkable*" (TLP 6.361), for "what is excluded by the law of causality cannot be described" (TLP 6.362). I have shown that the thinkable, and hence the describable, is precisely what is in reason's nature to regard at the conceptual level. Thus, once the historical distance separating the two ways of tackling cause and natural law is taken into account, the statements summarizing the positions of the two authors prove nearly identical. It would be almost indecent—hubris toward history—to hope that the effective accord between two philosophers divided by epoch-changing events and almost three hundred years might have been closer.

Logic in God, Logic of God

It is time to sum up, to review the conversation between Spinoza and Wittgenstein by presenting a synopsis of achievements. Let me begin, then, by outlining how the overall structure of the work of the one may match the overall structure of the work of the other.

First, recall a couple of equivalences: Spinoza's notions of God's body and God's mind are much the same as Wittgenstein's notions of the world as body and the solipsistic mind of the metaphysical subject, respectively. Recall also that psychophysical parallelism plays the same role for Spinoza as the identification of solipsism with pure realism does for Wittgenstein, rendering the body and the mind of God, on the one hand, and the world considered as body and the solipsistic mind of the metaphysical subject, on the other hand, as the two facets of the all-encompassing "thing" that is God or Nature or Substance for Spinoza and just substance for Wittgenstein. One way of summarizing all these identities is to say that the God of Spinoza is the absolute *bodily* solipsist of Wittgenstein. At this most fundamental level, the overall structure of the *Ethics* does match the overall structure of the *Tractatus*.

In addition, the structure of the *Ethics* makes Spinoza go directly from the "fact" that God or Nature simply is—from the mere "what" of Substance (TLP 5.552)—to the infinitely many Attributes wherein God's power unfolds. In Wittgenstein's terms, this would be equivalent to saying that God's unfolding His power makes up the "how" (ibid.) of the world. Now note three things. First, all that I have been saying implies that the *Tractatus* comprehends its subject matter in the atemporal mode proper to logic, which is identical to comprehending it "under the form of eternity" (TLP 6.45; E V, passim). Comprehending things under this form presupposes that "the one and only complete analysis" (TLP 3.25) of each proposition has been completed and hence that all the associated grammatical hinges or "silent adjustments" (TLP 4.002) have been revealed for what they are and addressed accordingly.

Second, construing Spinoza's positions in the *Ethics* under the form of eternity compels us to consider God's power as having been unfolded and thus the "how" of the world as lying completely open to His gaze, or equivalently, to the gaze of Wittgenstein's metaphysical subject. Whether and to what extent the overall structure of the *Ethics* matches that of the *Tractatus* must be answered in terms of this atemporal picture. Third, note that given this, the structure of the *Tractatus* does not make Wittgenstein go directly from the "what" of substance with its indefinitely many objects to the "how" of the world; to arrive there, he passes through logic, which comes before the "how" (TLP 5.552). Some account must be given of this important discrepancy in structure.

One facet of the "how" is the realm of facts (for Wittgenstein) or the Attribute of Extension (for Spinoza); its other facet is the realm of thoughts considered as logical pictures of the facts (for Wittgenstein) or the Attribute of Thought as inhabited by ideas of their objects (for Spinoza). Earlier I proposed calling the first facet the *realm of the factual* and the second the *realm of the conceptual* for both authors. We saw that the inhabitants of either realm (facts and extended modes, thoughts and ideas) match their counterparts in the other and that both authors take one unique "order and connection" or one unique "form" and "structure" as organizing the inhabitants of both the conceptual and the factual realms.

I argued that the structural discrepancy just noted arises because, on the one hand, Wittgenstein carves out this common organizing principle, identifies it with logic, and reserves for it the place or role of go-between mediating the "what" and the "how" of the world. On the other hand, Spinoza leaves logic to work in parallel within both the conceptual and the factual, though he does not acknowledge this as such. We could say that logic, the realm of bare possibilities, comes before the "how" for Wittgenstein, for it conditions (Spinoza would say "modifies") both the conceptual and the factual, which is why he calls logic transcendental (TLP 6.13): it is the condition of possibility for thought (and of language) as constitutively thought *of* (and language *about*) the world, the condition of possibility for the conceptual in its constitutive correspondence to the factual. That Wittgenstein was in a position to mark out logic in this manner measures the historical distance separating him from Spinoza. The structural discrepancy at issue thus becomes simply a matter of historical distance.

Nonetheless, this distance produces a philosophical difference. Wittgenstein worked in a period when major paradigm changes were in the process of modifying the "silent adjustments" (TLP 4.002)—the underlying grammatical "assumptions"—that determine the use of otherwise rigorous scientific concepts. Accordingly, to go to the logico-philosophical bottom of things, he had to work "beneath" the conceptual (the grammatical) and tackle the level of logic itself. In contrast, Spinoza worked at a time before the scientific revolution had fully developed. There was nothing to indicate that the new science of motion would establish anything that might be subject to conceptual (grammatical) revision. There was nothing, that is, that might have made him envisage the possibility that the grammatical could be disengaged from the logical and that philosophical (elucidatory) work beneath the conceptual might be possible and eventually required. Spinoza's work was forced to remain at the level of the conceptual for precisely this reason.

When read under Wittgenstein's historically clearer lights, however, Spinoza's Attributes of Extension and of Thought carry a conceptual part that can be taken as tantamount to the grounding conceptual layout of the corresponding scientific perspectives (natural science, on the one hand, and psychology or psychoanalysis, on the other) and a logical part related to the fact that each of us is an indissoluble congeries of body and mind always already thrown into the world. Once the logical is thus disengaged from the conceptual, the indefinitely many Wittgensteinian objects can compensate for the infinitely many Spinozistic Attributes, for scientific perspectives (as I took Spinozistic Attributes either to evoke or to be) can be indefinitely many. Hence, once such disengagement of the logical from the conceptual has been effected within the Spinozistic Attributes, the structure of the *Ethics* comes to match the structure of the *Tractatus* impressively.

In addition, the fact that we are always already thrown into the world as only indivisible body-minds is expressed by the fact that we always already have thought, possess language, and come to perform linguistic acts. Again, the two works match almost perfectly in all those respects.

Further, recall that Spinoza provides an axiom, presented deliberately in the active voice, to the effect that "man thinks" (E II a2), and Wittgenstein uses an equally axiomatic form and the active voice in stating: "we make to ourselves pictures of facts" (TLP 2.1). The active voice implies that thinking (making to ourselves pictures of facts) forms a human activity, which shows that both authors posit human agency in the world. At the same time, both authors see purpose or intention as preceding and presiding over any activity, even if it is misapprehended or remains largely unacknowledged by the one undertaking the activity, and purpose or intention (Wittgenstein's "will") is what bears or supports the ethical. Both Spinoza and Wittgenstein conceive the ethical as an ethics of responsibility for precisely this reason.

It follows that once the ethical precedes and presides over any human activity, it forms the condition of possibility for human agency, which is why Wittgenstein calls it transcendental (TLP 6.421). Ethics is transcendental because purpose or intention, the hallmark of human agency, cannot be eliminated, contrary to the claims of the more naive forms of naturalism, which is to say that the mental (or the psychological) includes a logical core that cannot be eliminated by reduction.

As we saw, Spinoza endorses this position with a vengeance through what he says about conatus.

Excepting the historically determined philosophical difference just reviewed, then, the overall structure of Spinoza's work matches Wittgenstein's in all essentials. Nor is the match merely one of structures. As I have shown throughout, philosophical strategy and philosophical content are inevitably implicated in the match as well.

The developments witnessed since the times of Spinoza allowed Wittgenstein to rely on and develop the symbolism of his contemporary logic so as to demonstrate, in a way he considered "unassailable and definitive," that is, *logically* compelling, that the philosophical perspective of radical immanence is logically necessary and hence that the philosophical perspectives opposing it are logically impossible. If so, there can be no philosophical perspectives to begin with, and there can never be a question of espousing one. Accordingly, Wittgenstein's demonstration, if successful, uses logical rigor to destroy the very possibility of doing philosophy and thus to ensure philosophical silence. Again, this demonstration has to proceed along two strategic movements unfurling simultaneously. The first straightforwardly engages in philosophical activity, while the second undermines and in the end thoroughly erases the philosophical content that the first advances. The result is the performative obliteration of Wittgenstein's treatise as a whole and therefore of philosophy in its entirety; the means for carrying out the demonstration is the judicious employment of telling nonsense.

The historical conditions under which Spinoza worked blocked him from expressly conceiving the perspective of radical immanence as logically necessary. His aim was thus restricted to establishing this perspective in terms of a rigorous philosophical theory, the theory Spinoza considered to be the only true one. His strategy for achieving this goal could accordingly unfurl, at least on the face of it, along only one movement, proceeding within philosophical activity as practiced during his time. Yet the logical rigor with which Spinoza reasoned made him compose a text that comes almost to the point of undermining itself, *à la* Wittgenstein, silently deploying a corresponding second strategic movement that he recognized as far as he possibly could and all but expressly acknowledged.

Given that Spinoza and Wittgenstein share the same philosophical perspective (ignoring for the moment the logical impossibility of espousing any philosophical perspective), and that their reasoning exhibits the same logical rigor, I argued that the philosophical content each advances in the first movement of his strategy should fundamentally match the other's, which turns out to be the case. The philosophical content Spinoza advances matches, in both scope and depth, the philosophical content Wittgenstein advances along the first movement of his strategy.

Finally, with respect to the difference separating Spinoza from Wittgenstein, history seems to play a central role in that the philosophical protagonists in the two periods were apparently very different. Again, Spinoza largely ignores logic as such, leaving it to work hidden from view within the Attributes, the infinite modes, and the finite modes and their causal or deductive interconnections. We might thus say that for Spinoza, logic works indiscriminately in all the expressions of God, infinite as well as finite, and in all His workings. For Spinoza, then, logic is *in* God and remains concealed there in that precise sense.

Wittgenstein, however, places logic "before the How" of the world as regards both the conceptual and the factual realms, "absorbing" the logical part of the two Spinozistic Attributes. Seemingly in tune with what his contemporaries were advancing, Wittgenstein elevates logic to a position preceding the workings of the world. By that movement, Wittgenstein connects logic to the mere "what" of substance, with no mediation whatsoever, which is to say that this mere "what" provides the substantive anchorage—through the objects constituting it—of the bare possibilities of which the "facts" of logic (TLP 2.0121) consist. Since Wittgenstein's mere "what" of substance matches Spinoza's God, we may take the same "what" as amounting to Wittgenstein's God. Accordingly, since this "what" (both Wittgenstein's "what" and Spinoza's God) constitutes the sole anchor for logical possibilities (i.e., logic as such), logic becomes expression of God for Wittgenstein, the one single expression that mediates God's relation to the world. If logic is not itself God, for it requires the substantive "what" as an anchor, it is *of* God in being the only thing that comes *before* the world.

Saying that logic is of God in that it precedes the "how" of the world can be misleading, however, for it might then be taken as overarching the world, as most of Wittgenstein's contemporaries held to be the case, though perhaps unreflectively. Much of the present work has been devoted to showing why this is not the case, for on the contrary, *the whole point* of the perspective of radical immanence, which Wittgenstein pursues relentlessly, is that neither logic nor anything at all can come to occupy such an external position. Logic, be it *of* Wittgenstein's God, is thus *in* the world (in Nature or in Spinoza's God) simply because there is no outside. It follows that to say either, with Spinoza, that logic is *in* God or, with Wittgenstein, that logic is *of* God is fundamentally to say the same thing. The distinction between the two clauses is simply another way of marking the historical distance separating the two authors, this time by invoking the protagonists of their two works—God for the one and logic for the other—and pinning down the position of each protagonist with respect to the other.

Thus, despite the enormous social, political, scientific, and cultural divergences separating the eras when the two authors set out to do their work; despite the attendant profound dissimilarities in intellectual climate and general ideological outlook; despite the huge differences in philosophical priorities, agendas, and methods of approach; despite the almost unbridgeable disparities in philosophical vocabulary and forms of formulation—despite, that is, all that happened in the world, in civilization, in science, and in philosophy during the three centuries separating the two authors and their works—the structure, strategy, and content of the *Ethics* match the structure, strategy, and content of the *Tractatus* to a really extraordinary extent and degree, if not almost perfectly.

The discussion to this point remains far from having exhausted everything that could be said on the subject, so that the closing word at this juncture can be only that although I prompted this "conversation" between Spinoza and the young Wittgenstein, even I was surprised by the extent to which they reached agreement at this point in the dialogue, a reaction likely shared by most of those who have taken the trouble of following it.

Exodus: Toward History and Its Surprises

> Whoever reaches one's ideal transcends it *eo ipso*.
> —Friedrich Nietzsche, *Beyond Good and Evil*, §73

Evidence provided by his biographers shows that Spinoza had neither the opportunity nor the inclination to return to the *Ethics* after finishing it. As is well known, he had interrupted its composition to write his *Theologico-Political Treatise*, and after going back to complete the *Ethics*, he started composing the *Political Treatise*, a work his death interrupted abruptly. I have characterized both these works as more "applied," implying, among other things, that they do not put the overall philosophical picture provided by the *Ethics* into question. Consequently, the fundamentals of Spinoza's mature philosophy seem to be those put forth by the *Ethics*, with nothing more to add on the matter.[1]

In contrast, Wittgenstein notoriously had both the inclination and the opportunity to return to the *Tractatus*. This return seems to initiate a mode of philosophizing very different from that of his younger self, though the sense of this difference and the salient characteristics of the novel mode continue to tax Wittgenstein scholars. I promised a few words on these issues, and we are now at a point where I can honor that promise. Specifically, I will outline how a reading of Wittgenstein's later work might connect to the arguments I have made with respect to the *Tractatus*.

First, Wittgenstein's later work retains the fundamental insight of the *Tractatus*, namely, that there can be no position outside the world, thought, or language overarching them as wholes. His remark that his later thoughts "could be seen in the right light only . . . against the background of [his] old way of thinking" (PI Pr ¶5) can be read as implying, among many other things, that his return to philosophy does not impugn the perspective of radical immanence. If this is the case, then the conception of philosophy entailed by this perspective remains similarly unaffected: philosophy continues to be not a theory but an activity aiming at the elucidation of thought and thereby at silencing philosophical worries for good, at least as far as this can go, given the indefinitely varied contexts wherein philosophical worries tend to crop up. Thus a fundamental line of continuity between the *Tractatus* and Wittgenstein's later work is thoroughly preserved.

What changes, and this drastically, is the overall mode of doing philosophy: Wittgenstein forsakes the form of the logico-philosophical treatise he uses in his earlier work. Instead, the older Wittgenstein presents sets of loosely interconnected remarks, often in the form of internal dialogues, that set up and handle philosophical puzzles, many of them of Wittgenstein's own making, from various angles. Wittgenstein openly acknowledges "grave mistakes" (PI Pr ¶6) in the *Tractatus*, but the first question to address is why he proceeds precisely as he does.

The general contours of the answer may be drawn once we take Wittgenstein at his word: while proceeding to his novel way of engaging in philosophical activity, he admits that he now "let[s himself] off the very part of the investigation that once gave [him] most headache, the part about the *general form of proposition* and of language" (PI §65). He later explains that he lets himself off the hook this way because "what we call 'sentence' and 'language' *has not* the formal unity that [he had] imagined" (PI §108, emphasis added) and put forth in the *Tractatus*. The investigation should be involved not with the "phantasm" that this "general form" constitutes but only with the actual workings of language, with "the spatial and temporal phenomenon" (ibid.) that is language.

If language lacks the formal unity Wittgenstein proposes in the *Tractatus*, and as a result no general form of a proposition exists, then there is no working principle encompassing all aspects of language prone to philosophical scrutiny. Language should be considered instead as an assortment of unclassifiable, loosely connected "language games" (PI §7), inseparable from corresponding "forms of life" (PI §19). Such "forms of life" should be deemed the "accepted, the given" (PI §226)—the flesh, perhaps—of language usage generally. This implies, among other things, that language games should not be thought to involve just words; all kinds of bodily gestures form essential parts of language games, actively contributing to the determination of sense or meaning called for in particular circumstances.[2] Thus, Wittgenstein's later work not only preserves the previously highlighted connection between language and the live human body but makes it more pronounced.

If, moreover, propositions possess no general form, then the "complete analysis of the proposition" down to the level of "elementary proposition," which the *Tractatus* proposes, cannot make much sense. And if such an analysis does not make sense, then there is no principle allowing philosophy to penetrate the "disguise" constituting colloquial language so as to lay bare the "body of thought" (or lack thereof) covered by that disguise. If we abandon the term *disguise,* however, for it obviously relates to the mistaken formal unity and the correlative unadorned "body of thought," philosophical puzzles and the attendant worries nonetheless continue to arise from linguistic usage, from the "bewitchment of our intelligence by means of language" (PI §109). But if there is no complete analysis of the proposition and hence no guide for confronting such puzzles, these cannot be brought under subjects of standard philosophical concern. They cannot be solved or dissolved—even if only "in essentials"—in one fell swoop by a complete (logico-)philosophical treatise.

Since philosophical puzzles arise from linguistic usage, they can crop up anywhere at all, and given that language is a "spatial and temporal phenomenon," such puzzles might differ according to the particulars of such usage. Because there is no general way for confronting them, these puzzles can be tackled only locally, one at a time, and from various angles, because the role that linguistic usage plays in their formation entails that the puzzle is usually a function of that angle. In his later work Wittgenstein uses this approach, more or less, to confront particular philosophical puzzles, either as formulated by others (e.g., Augustine, discussed in the *Philosophical Investigations,* or Moore, discussed in *On Certainty*) or as concocted by himself for illustrative, instructive, or self-pedagogical (or, if you like, homeopathic) purposes. These considerations explain Wittgenstein's choice of form for his later philosophical activity.

After acknowledging that language lacks the "formal unity [he] had imagined," which implies at least a downgrading of all formal aspects to which the *Tractatus* had given pride of place, Wittgenstein proceeds to the appropriate diagnosis: he was led to overemphasize such aspects not as "a result of investigation, but [by] a requirement," namely, that logic possess "crystalline purity" (PI §107). This requirement proved illusory, no better than a *"preconceived idea"* (PI §108) that placed Wittgenstein's way of thinking "on slippery ice where there is no friction and so in a certain sense the conditions are ideal, but also, just because of that, we are unable to walk" (PI §107).

This diagnosis leads directly to a question crucial to everything concerning the *Tractatus*: "But what becomes of logic now? Its rigor seems to be giving way here.—But in that case, doesn't logic altogether disappear?—For how can it lose its rigor?" (PI §108) Wittgenstein does not reply to this directly. He proposes instead a different path for the philosophical investigation, the only path that can "remove" the preconceived idea lying, presumably, at the roots of the "grave mistakes" in his early work: "the axis of reference of our examination must be rotated about the point of our real need"; that is, the examination must "turn round" (ibid.) while keeping fixed at that point. Such a rotation aims to direct the investigation "back to the rough ground" (PI §107), where the friction necessary for proceeding is to be found.[3]

This is presumably the rough ground of unclassifiable and incessantly flowing language games together with their corresponding forms of life, the rough ground whose very bedrock can be eroded, thus allowing for radically novel linguistic usages, the rough ground where nothing can remain perfectly stationary, where the ground itself always remains open to change. This is the rough ground of grammar as I have been using the term. The rotation leading to such rough ground is thus the one that raises grammar to a position of prominence at the expense of logic, the one that leads Wittgenstein to state that his "investigation is therefore [not a logico-philosophical but] a grammatical one" (PI §90). Nonetheless, in chapter 3 we saw how such rotations do not censure logic by "bargaining any of its rigor out of it" (PI §108): even after the most radical grammatical break, logic becomes reinstated.

On this basis, we can sketch the general contours of the philosophical strategy Wittgenstein adopts in his later work and contrast it to the general contours of the strategy of the *Tractatus*. In the earlier work, Wittgenstein tried to satisfy our "real need"—that is, to take care of philosophical worries arising, then as now, from *everyday* linguistic usage—by leading the philosophical investigation toward the "slippery ice" of the formal (the logical and the mathematical) so as to clarify things at this level. His "preconceived idea" led him to think that this was the proper way to elucidate and thus solve or dissolve, at least "in essentials" but at any rate all at one go, all philosophical problems by subsuming them under subjects of standard philosophical concern. His later strategy, where we find him "rotating the axis of reference" of the philosophical investigation while keeping the point of "our real need" fixed, has him concentrating on the grammar of everyday linguistic usage and trying to dissolve, locally and one at a time, particular philosophical puzzles, though they are still considered to arise from linguistic confusion.[4] In developing his strategy, he leaves the formal more or less alone but nevertheless tries to examine (as some of his lectures and notes reveal) how parts or aspects of mathematics connect to grammar and to everyday linguistic usage and what is particular to them in this respect. In other words, the early work's axis of reference is directed to the formal based on the preconceived idea that the goal there can be reached at one go through the composition of a complete logico-philosophical treatise. The axis of reference of the later work is rotated in that it is directed *away from* the formal and toward everyday linguistic usage. "Our real need" remains the same—the rotation keeps that point fixed—but this need can be satisfied now only partially by loosely connected grammatical investigations. The formal itself, or at least some aspects of it, can be tackled along the same grammatical lines.

• • •

I have been arguing that an appeal to the grammatical aspects of radical scientific change can contribute to our understanding of the *Tractatus* even if that treatise does not tackle the issue frontally. I have also been arguing that the germ of what Wittgenstein would later call grammar already exists in the *Tractatus* and that this presence justifies, at least up to point, borrowing both the idea and the term to further this understanding. I can now complete this discussion by focusing on the perhaps most salient feature of the rotation in question, namely, the elevation of grammar at the expense of logic. At the same time, both the connections and the differences between the *Tractatus* and Wittgenstein's later work will appear in clearer light.

Recall that a *logical* standoff (e.g., a bona fide contradiction) often lies at the roots of a crisis situation leading to radical scientific change. A new paradigm then overcomes the standoff through a *grammatical* break. In widening the grammatical space available, this development makes sense of some conceptual distinction whose possibility had hitherto been *logically* foreclosed. Such widening makes the standoff disappear as such; logic is reinstated ex post facto. Before the event, no logical room was available for formulating the conceptual distinction at issue, while after the event, such room not only becomes available but also appears to have always already been available.

I have been supposing that the young Wittgenstein would rely on this last clause to maintain that the logical possibility appearing only ex post facto is a pure logical possibility as envisaged in the *Tractatus*. This is a pure logical possibility "underlying" the conceptual and thus independent of it, a pure logical possibility that really was there all along: if the propositions involved in the logical deadlock were analyzed down to the "concatenation of names" forming "elementary propositions," we could confirm that the logical possibility seemingly brought forth by the grammatical break neither constitutes a novelty—for the Wittgenstein of the *Tractatus*, at least, no novel logical possibilities can arise, for there can be no "surprises in logic"—nor lay dormant or remained idle before its "discovery." On the contrary, this is a pure logical possibility that, along with all the others, has always already been at work, determining from below all conceptual distinctions logically permitted as well as all logically permitted directions for scientific investigation.

On this basis, the young Wittgenstein would presumably continue by maintaining that it was, as it always is, contingently up to us to realize what was really at issue in the ostensible logical standoff so as to come to pinpoint the conceptual distinction required, the one that an allegedly novel logical possibility appeared as legitimizing only ex post facto. Before the event, before we pinpointed it, this logical possibility, although always at work, had simply been hidden from our view, for colloquial language had hidden it. In sum, according to this line of thought, logical standoffs preceding radical scientific change are not real but only apparent; they seem to be logical impasses only to the extent we are unable to reach logical space itself, where such appearances are removed. Concomitantly, the attendant philosophical worries would be relieved as well, for "the one and only complete analysis" is in principle capable of addressing all such worries.

In contrast, the rotation Wittgenstein performs with his later work seems to reverse the attribution of apparent and real in the case at hand. Thus the reality that the young Wittgenstein would have attributed to the pure logical possibility underlying the novel conceptual distinction would for the older Wittgenstein be only an illusion, a byproduct of his preconceived idea that language possesses formal unity and logic exhibits the crystalline purity of possibilities unfettered by linguistic usage. Conversely, the logical stalemate that young Wittgenstein would have considered apparent the older Wittgenstein would consider real, for there is nothing below the rough ground (actual linguistic usage) where the standoff cropped up.

This reversal of attribution follows directly from the rotation in question: since language does not possess formal unity, making a complete analysis of the proposition impossible, there is no way to reach logical space per se in its own "crystalline purity." Any results that Wittgenstein took himself to have reached using this approach were illusory, for, to continue the metaphor, he supposed himself able to walk on slippery ice where there is no friction, where, despite all efforts or appearances to the contrary, one remains immobile. The very idea of a space of pure logical possibilities underlying the workings of language (and of thought and of the world) is itself a side-effect of presuming that logic possesses purity, which was the preconception that guided Wittgenstein toward the formal, that led him to the initial requirement for the formal unity of language. Thinking in this way might have been attractive in laying out "ideal conditions" for philosophical investigation, but it is finally no better than a sourcebook of "phantasms" offering only illusory solutions to the puzzles confronting us and hence only illusory relief to our philosophical worries. In contrast, real solutions and real relief can be forthcoming only from the rough ground of that which is *actually there* in linguistic usage, only from grammar.

To the extent that they are phantasms, the space of pure logical possibilities and the complete analysis of the proposition supposed to lead there simply do not exist. There is nothing below our actual linguistic practices and no way to go beyond them so as to clarify what they produce. Hence no *effective* (i.e., not idle) distinction can be drawn between linguistic "disguise" and the naked "body of thought" covered under it; the logical standoffs we encounter in our linguistic practices cannot be stripped of their clothes, allowing us to uncover a pure logical form or to acknowledge that we have been deceived by the semblance of one.[5] Even if the logical deadlocks preceding radical scientific change appear ex post facto to have been hiding previously unrecognized conceptual distinctions within their linguistic folds, nothing permits us to maintain that such distinctions (as well as the pure logical possibility underlying them) were already there all along, for there *never was* a "there." There is no space of pure logical possibilities to underlie linguistic usage either before or after the grammatical break and hence to underlie the ex post facto as such; logic cannot subsume the ex post facto. In this precise sense, then, the ex post facto is irreducible to logic, or alogical, and the idea of a formal "there" (logical space per se) underlying linguistic usage across the grammatical break (and hence the ex post facto) is an illusion the ex post facto itself creates, doing so almost irresistibly.

We might try getting the measure of the older Wittgenstein's position on the connection between logic and grammar in cases of radical scientific change: that the reinstitution of logic after some logical standoff is overcome keeps logic's rigor intact, without "bargaining" anything "out of it," as Wittgenstein is careful to demand. Hence the idea that there are no surprises to be expected from logic continues to stand. But the notion that logic becomes *re*instated after a grammatical break withdraws the absolute predominance the *Tractatus* had attributed to it. Consequently, the fact that logic is reinstated *after* some standoff of logic as such has been effectively overcome makes the per se of logic an ex post facto recovery, that is, a recovery manifested as reinstatement.[6] We might call such a reinstatement an effect of the grammatical ex post facto. In other words, the per se of logic as given, the presently available space of logical possibilities, results from, is the outcome of, the grammatical break. And this is the case even though this outcome is simply a reinstatement, a result recovering the selfsame logic and thus spontaneously producing the illusion that some pure logical space per se persists throughout, always already lying there. This presumably is how logic (the per se of logic) loses its preeminence to the profit of grammar (the ex post facto of grammatical change) without it or its rigor being jettisoned in the process. To recognize the genesis of this mirage is to avoid being duped by it.

On this basis, no philosophical worry should arise from mirages created by the ex post facto. The logical standoffs at issue are as real as our linguistic practices are, retaining the force of such reality (e.g., effectively disallowing the inquiry to proceed) up to some grammatical break. When it takes place, such a break overcomes the impasse by introducing a grammatically novel distinction while retroactively shedding light on the way that distinction can appear ex post facto as having hitherto been logically foreclosed and conceptually untenable. Although the reinstatement of logic forbids us to say that logic changes in the process, it is misguided to ask whether the logical possibility that ex post facto appears to have been underlying the distinction in question was really there all along—whether, that is, the logical standoff is real or only apparent. Demanding a simple yes or no answer to this question obscures the play between the per se and the ex post facto, or the dialectic of reinstating or recovering the logical per se,[7] the dialectic making the per se of logic a function of the ex post facto of grammatical breaks.

It should be obvious that none of this can help us to tackle the logical standoffs that crop up in our scientific investigations or linguistic practices. Each such standoff is always singular, irremediably tied to the particular circumstances of its formulation, which are themselves always singular as well. Hence there can be no theory subsuming them all, no method for addressing them systematically. Although the invariable reinstatement of logic after the corresponding grammatical break attests that there are "no surprises in logic," the lack of these surprises, which is visible only after the grammatical break and hence only ex post facto, does not enable us to anticipate, much less forestall, the very real surprises that such a break invariably produces. And with regard to the ultimate source of such surprises, perhaps they are always lurking in the shadows simply because, to use a metaphysically loaded term, our practices and the way they evolve cannot ever exhaust the real.

• • •

It follows that the real as presently available, the real (and the imaginable)[8] presently accessible by concepts and conceptual relations that present no discernible logical deadlocks, is a function of the grammatical ex post facto. The argument is simple, at least initially. To say that concepts and conceptual relations present no logical standoffs is simply to say that they comply with logic as it is presently at work. But if the per se of logic is subject to the dialectic of recovery and reinstatement, then the concepts themselves, the distinctions they allow or forbid, the propositions they entail and presuppose, their *logical identities,* are a function of the same dialectic. Concepts have been established after past grammatical breaks and remain at the mercy of any grammatical breaks that might occur in the future. Thus the identity of any concept—of the conceptual per se—is a function of the grammatical ex post facto, and since the real cannot be accessed independently of concepts (or as McDowell [1996] would say, the conceptual goes all the way down), the real presently accessible is a function of the grammatical ex post facto as well.

To approach the same issue from a slightly different angle, recall that "the reality" in the *Tractatus* is the "existence and nonexistence of atomic facts" (TLP 2.06) encompassing the actual world together with all imaginable worlds. Within that book's framework, reality is the substantive "correspondent" or "partner" of the formal space of pure logical possibilities, the substantive anchorage this space requires if it is not to float freely as a pure form. But if the space of pure logical possibilities is in fact a phantasm, a byproduct of a misleading preconception, then the *Tractatus*'s "reality" should be a phantasm, too. If there is no "there" of logical space, then there is no reality to constitute a substantive "there," no reality per se. Reality per se is an illusion, too, a mirage created by the ex post facto. It follows that the only real that makes sense, the only real to which we can have any kind of access, even in imagination, is the real encountered in and by our practices,[9] the real that is a function of the grammatical ex post facto.

But this seems not to exhaust the issue. In chapter 3 I argued that a grammatical break within scientific practice both widens and modifies the grammatical space available, making our grip on the real better: some features or facets of reality previously considered to exist are totally discarded; others appear ex post facto as retaining a core of their identities, while this core reacquires its status ex post facto as well; and altogether novel features or facets sometimes come to the fore. What is discarded in this way is discarded for good, and what is novel stays in place at least until the next relevant grammatical break. Hence such change appears as progress: reality seems to become fuller, as it were, after the grammatical break. Unlike what happens with logic, then, the grammatical break does not simply reinstate reality; it enriches it.[10] If there is no reality per se, however, what is the source and substance of this enrichment?

First, such enrichment should not be understood as a process that renders the antecedent real rich*er* by adding something to it or modifying it. If that were the case, logic would underlie the addition or modification as such, making it comply with logic. But this cannot be the case: the prebreak real—or rather, in all cases, its conceptual rendering—is separated from the postbreak real (its conceptual rendering) by a logical standoff that cannot be negotiated so as to allow some addition or modification. There is no logical room on which the enrichment as such could be based; it becomes amenable to logic only ex post facto, by the reinstatement of logic. In other words, the enrichment by itself and on its own has no logical basis; it is alogical, for there can be nothing logical connecting logic's prebreak giveness to its reinstatement after the break.

Furthermore, since the prebreak real (its conceptual rendering) is separated from the postbreak real (its conceptual rendering) by a logical stalemate, there is no logical room for a relation between the two. Although the postbreak real is indeed enriched,[11] it does not by the same token become richer than the prebreak real. There can be no richer/poorer relation[12] or for that matter *any* relation between the prebreak and the postbreak reals and hence no logical room for any unbiased comparison between the two.[13] To put this differently, since logic becomes reinstated after the break, both the prebreak and the postbreak reals comply with the selfsame logic, logic tout court: each (the conceptual rendering of each) is logically consistent in its own right, and hence at the level of logic our cognitive relations to both are exactly the same,[14] even if they differ dramatically at the conceptual (grammatical) level. In all such cases, the radical surprise invariably associated with a grammatical break eventually wears off, the effects of the break become domesticated, and the associated novelties in both the conceptual (or grammatical) and the real are assimilated in and by our practices. We soon become as comfortable with the enriched real as we had been with its predecessor; philosophical worries either do not crop up or become assuaged through the progression of scientific inquiry.[15]

This still does not exhaust the issue, however, for the question persists: if reality per se is a mirage, what constitutes the source of the enrichment? To see the answer, first recall the obvious. Logical standoffs form the dead end of inquiry, the "bounds"—which are not bounds, for there is no outside to them—against which thinking (and imagining) bounces and comes to rest. That is, a logical deadlock can not be overcome from the prebreak vantage point, for anything "beyond" the deadlock is by definition nonsensical. A radical grammatical break, however, does manage to overcome some such standoffs by introducing a conceptual distinction hitherto logically foreclosed. In doing this, it widens the grammatical space available, offering a grammatical basis for a novel postbreak vantage point both descriptive and normative that permits, among other things, the retrospective reassessment of the standoff itself.

Given this, to answer the question we must clarify the peculiarities of this novel postbreak vantage point. First, recall that the grammatical break wipes out the logical impasse through the introduction of a hitherto logically inconceivable conceptual distinction. The prebreak concept on which this distinction directly bears does not change on its own, however, merely acquiring, as it were, a novel postbreak limb. The introduction of the distinction disrupts all the earlier concept's presuppositions and entailments, its logical and grammatical identity, thereby upsetting all the other prebreak concepts to which this one is grammatically connected. The overall outcome is the creation of the postbreak grammatical space (i.e., the postbreak conceptual) and the novel vantage point based on it. But since the grammatical basis of this vantage point arises solely from the radical restructuring of the prebreak grammatical space, it inevitably—by its very construction, so to speak—loses sight of the prebreak conceptual's grammatical autonomy. And if the prebreak conceptual can have no grammatical autonomy when viewed from the postbreak grammatical, there can be no relation of width between the two. If enrichment of the real cannot be a relation between the more and the less rich, neither can grammatical widening be a relation between the more and the less wide.

Concomitantly, if from the vantage point of the postbreak grammatical the grammatical autonomy of the prebreak grammatical has been lost from sight, then the very index of that autonomy, namely, the logical standoff bounding the grammatical space and thereby marking out this autonomy by tracing the corresponding boundaries, is lost from sight as well. The issue here flirts with paradox: we can determine whether an impossibility is a logical standoff only ex post facto, from the novel vantage point created by overcoming it, but since this new vantage point reinstates logic, the ostensible impossibility no longer appears to have been a logical standoff. To put it more succinctly, the postbreak grammatical has been instituted by overcoming a logical stalemate in the prebreak grammatical as just that, a logical stalemate, but because overcoming the stalemate in this way reinstates logic, whatever was overcome could not have been a logical stalemate after all. The vantage point created by overcoming the logical standoff eliminates from view the very reason of its own creation.

To explicate what is at issue, we might say that there seem to be two "classical" options for a way out. On the one hand, we could espouse some kind of logical relativism by attaching a qualifying index to logic (based, e.g., on sociohistorical relations) and hold that the standoff in question was indeed logical standoff but is not one anymore. This would clearly be anathema for Wittgenstein throughout his career. On the other hand, we could go along with the *Tractatus* in maintaining that the standoff had only appeared to be logical because colloquial language made it look that way. We have seen why Wittgenstein's later work bars this option. The previously outlined position is thus all that remains: if the same logic underlies both the pre- and the postbreak conceptual, with logic being reinstated after the grammatical break, then the "re-" of this reinstatement should be manifested at the grammatical level as well. And so it is. The "re-" in question is manifested at the grammatical level by the mere widening of the grammatical space (a widening, however, that admits no relation between the more and the less wide), by the mere fact that the grammatical space has been widened to accommodate a conceptual distinction hitherto foreclosed by logic, by the mere fact (or factum) evoked by the ex post facto itself.

It follows that what the novel vantage point necessarily loses from sight is simply the recuperation resident in the ex post facto reinstatement of logic, a restitution that escapes logic (and *a fortiori* grammar) because the ex post facto is not subsumable to logic (and *a fortiori* to grammar), because it is alogical (and therefore agrammatical).

We can now return to the real so as to gather the residual means for answering the question about the source of its enrichment. To that effect, note first that the novel conceptual distinction would be nothing but a play of words if it did not have some content, if it did not capture some feature or facet of the real. It follows that logical deadlocks constitute a limit of the prebreak conceptual realm in its relation to the real, a barrier forbidding the content of the novel conceptual distinction to be captured in the terms of the prebreak grammatical space. That is, the logical standoff forms a point where the real's passive[16] resistance to the prebreak conceptual manifests itself.[17] By the same token, to overcome the logical standoff by the novel conceptual distinction is to overcome the resistance of the real—to appease this resistance—by using that distinction to capture the feature or facet of the real that enriches it. The novel conceptual distinction is logically aboveboard and the capture logically unproblematic, for logic is concomitantly reinstated, which is to say that the enriched real complies with it fully.

Since, however, reality per se is a "phantasm," there never is, there never was, and there never will be anything more to the real than that which our concepts can *really* capture on the basis of the grammatical space subtending them, nothing more than the objects of our *effective* cognitive relations.[18] After the passive resistance it exhibited has been appeased and the turmoil of the grammatical break has subsided, the real calms down, so to speak, to comply with logic, which has been reinstated by the same token. It thus becomes again "in itself" as serene and gravely silent as it was before the commotion started, amounting once again merely to what our concepts say it is. On its side, logic comes concomitantly to seamlessly underlie the whole of the real, which is now just the postbreak real.

None of this, however, implies that the postbreak real, even if enriched, has been brought closer to illusory reality per se. If both the pre- and the postbreak real are that which our cognitive relations effectively concern, then nothing we say about the real can go beyond this mere "what." After its resistance has been appeased, the whole of the real is just whatever the postbreak conceptual says it is, and that is that. Certainly, the vantage point of the postbreak grammatical seems to license saying that the prebreak real had been misconstrued by the prebreak conceptual in some particular fashion. Nonetheless, such a retrospective appraisal (or *re*appraisal) presents additional peculiarities to be discussed a bit later.

We have reached a point where the question about the source of the real's enrichment can be answered quite simply: even if reality per se is a phantasm, even if the only real to which we can have access is the real encountered in and by our practices, the enrichment at issue *does* come from the real. The real is inexhaustible precisely because it is always capable of resisting the conceptual through a logical standoff, because our practices fall sometimes short of it, and because our concepts at times prove inadequate to it, though this is always ascertained ex post facto.[19] And no undue metaphysical weight is to be carried by the appeal to such inexhaustibility simply because no further philosophical discourse on the phantasm of reality per se is possible. Saying that the real is inexhaustible does not lead us out of the "rough ground" of our practices, for it merely names the fact that our practices sometimes encounter logical standoffs leading to radical grammatical breaks.

To condense this line of thought, we might say that while a grammatical break is "governed" by the dialectic of reinstating or recovering logic, it is governed by the dialectic of enrichment, too, with no relation between the more and the less rich or by the dialectic of resistance and appeasement. That the real is inexhaustible means only that it can always manifest its passive resistance to the conceptual, while the possibility of such appeasement means that the real is knowable by the conceptual through our cognitive practices. In addition, since the enrichment of the real invariably catches us by a fully shocking surprise, by a bolt coming out of the blue, the enrichment as such, by itself and on its own, amounts to the alogical whiff of such a pure radical surprise.[20]

• • •

The discussion thus far has concerned how a grammatical break in scientific development affects our cognitive relations to the real. It might be useful to consider as well how such a break affects the way we tend to or should perceive the history of those relations. Whereas the first area of focus is concerned with the way science works, the second is concerned with the way its historiography tends to or should work.

In chapter 3 I argued that we cannot minimize a conceptual distinction that overcomes a logical standoff, shoveling it back into the nowhere of its prebreak foreclosure and continuing to conceive the real as if that standoff had not been encountered and overcome. Precisely because the deadlock has been overcome and logic has been reinstated, our postbreak cognitive relations to the real are epistemically compelled to abide by the terms of the widened grammatical space, to comply with the corresponding norms, and to anchor themselves in the novel standpoint the break has produced. We do not get to choose whether to submit to the postbreak conceptual; it is objectively binding. It follows that the present account entails no relativism.

Saying that the postbreak conceptual is objectively binding is equivalent to saying that it enforces its own norm on all thoughts concerning the real, obliging them to abide by that norm. This implies that any way of looking back to those prebreak thoughts will similarly be subjected to this postbreak norm, which will enforce their *re*appraisal.[21]

So the postbreak conceptual enforces its own norm on the prebreak real and the prebreak conceptual, making the former conform to it and forcing a reappraisal of the latter, but in doing this, it produces wrongheaded effects in the way we tend to perceive the history of the postbreak conceptual, the tale telling how it came out of the prebreak conceptual and how the grammatical break brought it about. These effects follow nearly irresistibly from the peculiarities of the vantage point that the grammatical break institutes.

Thus, within the process by which the postbreak conceptual comes to subject the prebreak real to its own norm so as appease the associated resistance, the alogical character of the break tends to be lost from view, the associated lack of any logical relation between the (conceptual rendering of the) prebreak real and that of the (conceptual rendering of the) postbreak real tends to be forsaken, and the resulting impossibility of comparing the two tends to be altogether ignored. Concomitantly, the prebreak cognitive relations to the real tend to be perceived as ridden with an inborn index of cognitive inferiority: they were missing the conceptual distinction central to the break, they had not managed to discover the corresponding feature or facet of the real, and more generally, they had fallen short of exploiting our conceptual capacities to their fullest. In this way the grammatical autonomy (and sometimes even the logical consistency)[22] of the prebreak conceptual is obliterated, the grammatical asymmetry and the conceptual incommensurability between the prebreak and the postbreak conceptual (as discussed in chapter 3) are seamed over, and the enrichment produced by the paradigm change appears to be a relation between the more and the less rich to the profit of the postbreak conceptual in virtue of its position after the break.

In this way, the apparent recuperation of logic in its "reinstatement" is eradicated, and the historical character of the break as an event—when, where, and how the corresponding logical standoff was encountered and overcome—is reduced to a logical and grammatical nothing, to a subject of interest to "mere" historians of science whose job is solely to show how science managed to transform the middling prebreak conceptual into its glorious postbreak counterpart. The overall result is an occlusion of *real history,* the history taking place on the rough ground of our practices wherein logical consistency can never be at risk (*pace* propaganda, ideological rhetoric, and the like) even though grammatical autonomy may subtend an indefinite variety of past, present, and future language games; this history is lost, to the profit of the unremitting advancement by which our conceptual capacities increasingly come into their own. At least in matters related to scientific change, what has been dubbed the "Whig conception of history" thus comes to reign supreme, entrapping in its grip virtually all practicing scientists.

One might be tempted to argue that the Whig conception of history could be justified through an appeal to the *Tractatus*'s conception of reality and its formal partner, the space of pure logical possibilities, both taken to be lying in wait, ready to welcome the discoveries that, according to this conception of history, are bringing us closer and closer to reality per se. But no philosophical worries related to history can bother the *Tractatus,* for it relegates not just history but everything concerning the workings of the conceptual to the disguise it says constitutes colloquial language, leaving all that to the care of "the complete analysis of the proposition." Thus the young Wittgenstein would presumably have responded to such an appeal by insisting that neither reality nor its formal partner is prone to any kind of discovery, for neither is subject to history at all. They are put in place simply to assist the idea that language possesses formal unity, an idea that, once spelled out, suffices to let us rid ourselves, once and for all, all at one go, and at least in the essential elements, of all philosophical worries arising from the workings of history or anywhere else.

Wittgenstein does not directly confront history or the history of science in his later work either. But the "rotation" this later work performs does away with the point of view that bars the *Tractatus* from addressing history. Thus, if the older Wittgenstein had accounted for grammatical breaks in scientific practice more or less in the manner outlined here, he might have found room for the history of science. For one thing, it is undeniable that the history of science is punctuated by grammatical breaks.[23] As we saw, however, a grammatical break brings about a vantage point from which the grammatical autonomy of past cognitive relations tends to vanish. Therefore, the first duty of the historian of science is to resist this tendency by acknowledging the distinction between the properly cognitive dimension of enrichment and widening, on the one hand, and the properly historical dimension based on recognizing past grammatical autonomy, on the other. Acknowledging this distinction is tantamount to raising a barrier shielding past grammatical autonomy from the anachronisms that are ultimately due to the cognitive dimension in question.

Coming to shield this autonomy fully would involve thinking in terms of the corresponding prebreak grammatical. But such an exploit is, strictly speaking, impossible. *All* our beliefs about the real are based on the postbreak grammatical, which means that historians of science cannot force themselves to think in terms different from those in which they do think by forgoing the grammatical changes that have occurred in their objects of study.[24] It follows that no recovery of prebreak grammatical autonomy can be perfect, exempt from reconstruction. And this implies that the room wherein the historian of science is compelled to move is restricted by the play between recovery and reconstruction, a space that allows different interpretations of the past. It should nevertheless be obvious that, despite acknowledging cognitive limits to what the history of science can achieve, this account completely disqualifies anything like the Whig interpretation of history.

Still, the cognitive and the historical are not as foreign to each other as this conclusion might seem to imply. On the contrary, not until the cognitive feat giving rise to the postbreak conceptual occurs can we see that the prebreak conceptual can be appraised as cognitively inferior and assessed as incommensurable with the postbreak conceptual, enjoying grammatical autonomy in its own right. If there can be no logical relation between the pre- and the postbreak conceptual, then the latter has become severed from the former, hence the use of the term *break*. This severance is brought about by the act or deed introducing the novel conceptual distinction that, in overcoming the logical impasse, constitutes the postbreak conceptual (and grammatical), with this act or deed being the cognitive feat in question.

By the same token, however, this act or deed retroactively, (i.e., ex post facto) constitutes the prebreak grammatical as the space that previously ruled out the distinction introduced. Because this foreclosure constitutes the cognitive "bounds," or horizon, of the prebreak grammatical, constituting it ex post facto amounts to retroactively tracing those bounds and thus marking out its grammatical autonomy. The prebreak conceptual can become a well-delimited object of study (always carried out only via the postbreak conceptual under the limitations specified) only because of the cognitive feat of the grammatical break, even if we cannot come to think fully in terms of the prebreak conceptual (grammatical) itself. It follows that if the historiography of science is to be more than a "repository" of "anecdote and chronology"—to borrow Kuhn's opening phrase in *The Structure of Scientific Revolutions* (1962)—it has to bow to the determinative role of the ex post facto as regards not just its cognitive limits but also the very possibility of its constituting an identifiable discipline.

Thus historians of science face a plight. On the one hand, their job is fundamentally to pin down the grammatical autonomy of some of previous prebreak cognitive relations to the real while shielding this autonomy from the subsequent events that would eventually disrupt those cognitive relations. Since they are confined to their own conceptual schemas, however, they can carry out this task only on the basis of the present postbreak vantage point and with the means provided by the present postbreak conceptual. Historians of science must adopt the postbreak vantage point as a matter of course, thereby introducing, even if unknowingly, the inherently biased celebration of the postbreak conceptual at the expense of the prebreak conceptual being studied. But this is a strange kind of bias, for the postbreak grammatical, and hence the postbreak conceptual, is objectively binding, compelling *all* our cognitive relations to the real, present as well as past, to comply with its norm. It follows that from the presently incontrovertible postbreak vantage point, which is necessarily that of the historian of science as a historian, our past cognitive relations to the real must be taken as inferior (even if only implicitly) on *objective* grounds. Thus, even as they force themselves to pin down and shield the grammatical autonomy of some of our earlier beliefs concerning the world at large, historians of science are simultaneously compelled to cope with, to use an oxymoron prompted by the ex post facto's alogicality, an objectively binding bias that works *against* such autonomy. The practice of historians of science (or anthropologists trying to account for unfamiliar forms of life) is there to manifest the methodological particulars of such coping.

. . .

If the preceding characterization of paradigm change holds water, then the rotation Wittgenstein performs with his later work, raising grammar to a position of prominence with respect to logic, involves history directly, though the work barely mentions it. And it involves history decisively, for the rotation leads Wittgenstein to all but explicitly formulate the principle "governing" it.

Even if grammatical breaks in scientific development represent the perhaps starkest changes in our ways of conceiving things, analogous changes might come from different directions as well. In chapter 3 I mentioned major social and political upheavals in that respect. But I have ultimately attributed such changes to the "inexhaustibility of the real," which is manifested by the alogical character of the ex post facto. It follows that all major changes in our cognitive relations to the real— that is, the *history* of these relations—are governed by the ex post facto. But since the ex post facto is irreducible to logic, or alogical, it is of equal rank with it. This warrants the claim that the ex post facto is the fundamental principle governing the history of our relations to the real, the principle that underlies the grammar of that history. And since history tout court is inseparable from this narrower history, we might generalize by saying that the ex post facto constitutes the fundamental principle, with a rank equal to that of logic, that underlies the grammar of history throughout and consequently the discipline of historiography overall.

As this might sound too exalted, we should try to circumscribe the issue as accurately as possible. Recall first that in the *Tractatus,* reality is the endpoint reached in principle after all possible grammatical breaks have been accommodated (again, in principle). But the later Wittgenstein recognizes that this way of thinking allows us to envisage only phantasms, only illusions or mirages with no real consistency, rendering the reliance on the formal akin to walking on slippery ice. This way of thinking takes no notice that logical standoffs do crop up in scientific practice in totally unexpected places and are impossible to preempt by any analysis whatsoever; it disregards the logical status of these standoffs, which before the grammatical break occurs *arrest* logic, effectively forbidding us to envisage what might lie beyond them. This way of thinking, in other words, fails to address the real issue of a logical impasse effectively encountered in scientific practice, for, among other things, it ignores the real work necessary for overcoming it.[25] In downplaying the rough ground of our scientific and, more generally, linguistic practices, this way sacrifices the real of those practices to the profit of illusory reality per se, to the profit of the mirages of the per se.

The only real to which we can have access (which includes all that we can imagine) is the real we encounter on the rough ground of our practices, even as these practices change incessantly in indefinitely many ways. Efforts to go beyond this real to reach reality per se, even in principle, and thereby relieve philosophical worries is fed by an illusion of reality per se. Hence if the only real there is and can be conceived is the real encountered in and by our practices, if this real can be accessed only through our concepts, and if these concepts are indeed a function of the grammatical ex post facto, then we are entitled to maintain that the real per se, the per se of the real, is a function of the grammatical ex post facto as well.

On this basis the principle just mentioned can be formulated more completely. We saw why the per se of logic is a function of the grammatical ex post facto, we saw why the logical identity of concepts—the conceptual per se—is a function of the grammatical ex post facto, and we saw why the per se of the real is a function of the grammatical ex post facto. Hence we may generalize by saying that *there is no per se except of an ex post facto,* with the actual workings of the history of science, and of history generally, bearing direct witness to this.

This principle is not of my own making. Monk (1990, 305–6, 579) adduces compelling evidence that Wittgenstein considered a venerable expression of fundamentally the same principle as the maxim apt to summarize the whole of his later work. This is Goethe's famous "Am Anfang war die Tat" (In the beginning was the deed). The principle at issue here is clearly just another way of formulating Goethe's dictum: the deed that was in the beginning, the deed preceding even logic's having a say, is the deed of the irreducible alogical grammatical break, the fact (or factum) invoked by the ex post facto itself. This is the deed or fact that retroactively (i.e., ex post facto) creates the space, both logical and real, wherein that deed or fact is constituted as a deed or fact per se. In other words, this is the deed or fact that ex post facto creates the very space of its own per se and thereby—since this deed or fact is in the beginning—the space of the per se throughout. This is what I mean when I say that there is no per se except of an ex post facto.

Nor can I claim the idea that this dictum expresses the fundamental principle underlying the grammar of history, where grammar involves not just phrases but acts or actions of the human body. Influential authors whose toil, in both word and deed, has been deeply involved with the workings of history have not failed to elevate Goethe's aphorism, or something very like it, to that outstanding rank. So, for example, Engels (Marx and Engels 1974, 232) explicitly emphasizes Goethe's saying, an opinion Lenin (Lénine 1962, 111) later endorses wholeheartedly. In addition, Althusser, although he earlier disagreed with the empiricist twist in Engels's employment of the maxim (Althusser 1996a), introduces something quite similar to it in his later work as a guiding idea of what he calls "*matérialisme aléatoire*" (Althusser 1994). Here Goethe's initiating "deed" takes the form of the radically contingent encounter.[26] The "materialist's" reaction to such an encounter assumes the arresting figure of one who embarks on a train while having no idea of its destination. The action can be justified (or not) only ex post facto, by the train's arriving—only provisionally—at its next stop, which is yet another contingent encounter.

Within the same configuration of ideas, a place should be reserved for use to which Walter Benjamin puts the Paul Klee painting *Angelus Novus*. For Benjamin (1969, 257), this is a pictorial rendition of no less than the "angel of history," that is, of the fundamental principle governing it: the angel firmly turns its back to the future and the counterfeit mirage of unbounded "progress" and looks toward the past, for the positive forces of history are "nourished by the image of enslaved ancestors rather than that of liberated grandchildren" (260). The initiating deed is here the revolution that will retroactively "redeem" the past, thereby establishing its own per se.[27] To round off this highly selective survey, we might add that the fundamental role played by the *Nachträglichkeit* or *après coup* (another name for the ex post facto) in Freud's and Lacan's approaches to psychoanalysis warrants the assertion that the principle in question is constitutive as well of the grammar of human subjectivity as it unfolds along its own retroactive history.

There is no per se except of an ex post facto: this might well serves as the motto summarizing Wittgenstein's later philosophy and encapsulate the principle behind the grammar of history. But we should beware the lure this outstanding rank might exercise on us. We should resist loading the dictum with undue metaphysical weight and hence trying to unearth its deeper philosophical meaning or implications. It should be taken only as the aphorism it is, as the concentrate of a sobering idea whose main function is to underscore our finitude. From this point of view, the only point of view Wittgenstein would license, the principle simply advises us to relinquish the arrogant belief that, through logic and philosophy, we are capable of forestalling, albeit only in principle, the radical surprises that even the most compelling and most intellectually rigorous of our practices might have in store for us. It urges us to give up the idea that we can arrest the future, antecedently harnessing the radical contingencies our lives encounter as coming from the inexhaustible real of nature and of our practices. Given our finitude, given our absolute dependence on such inexhaustibility, anything we take today as final might change, while even what we are might be transformed by what we will become.

• • •

I now return to the *Tractatus* and the *Ethics* for a final look informed by the distance gained. This will close my discussion and, at long last, let the reader off to other pursuits.

Both the *Tractatus* and the *Ethics* are developed under a general outlook that can be characterized as static. This is the outlook of the *sub specie aeterni* or *aeternitatis,* the outlook zeroing in on the ultimate per se, an outlook necessarily oblivious to the radical surprises coming from history, for seen from the standpoint of eternity, history can have no say, and the associated radical surprises are out of the question altogether. The standpoint of eternity is the ultimate per se, and the ultimate per se can admit no ex post facto.

Approaching philosophical issues from this angle does not lack justification. The present work has been devoted to showing how both the *Ethics* and the *Tractatus* present themselves as succeeding in tasks set for them even by their authors' most stringent criteria; for my part, I have portrayed this successfully completed task as the effort to defuse the metaphysical load traditionally carried by the per se by shrinking it to the mere "fact" that we are a part of the world, thrown into it as bodies and minds inseparably wedded. The equivalence between Spinoza's God and what I call the inexhaustible real and the reinstatement of logic after even the most radical grammatical break clarify this defusing and shrinking, as well as how the ultimate—the thinnest possible—per se should be considered. Spinoza and Wittgenstein can use the same considerations to justify excluding the ex post facto from view and downplaying history.

Thus downplaying history does not result from an oversight. On the contrary, establishing philosophy's final victory over history (to use the terms employed in chapter 3) or that of the thinnest possible per se over the ex post facto (to use the terms employed in this chapter) was the intent. Sacrificing history in this sense[28] was the price Spinoza and the young Wittgenstein were compelled to pay for carrying out the task they had set themselves. Given that they both carried out this task to the end while submitting their work to the most stringent criteria makes the argument I claim to constitute both the *Ethics* and the *Tractatus* perhaps the ultimate word on the perspective of radical immanence under the static outlook at issue.[29]

The principle defining this perspective, that no position can overarch the world, thought, or language, is widely considered to be the only viable philosophical outlook, for—if enough latitude is given the term—it is simply the naturalistic outlook many philosophers openly espouse. The argument making up the *Ethics* and the *Tractatus* therefore continues to be relevant and instructive. Outside philosophy, however, one of the major current issues seems to be fundamentalism of one or another variety, ranging from triumphal religious fundamentalisms to the more prosaic but hardly less insidious fundamentalism of the "free market." And since fundamentalism by definition entails raising some item or other to a position overarching everything we do or should be doing and everything we think or should be thinking—if not just everything there is or can be other than the item raised—the argument making up the *Ethics* and the *Tractatus,* even if it excludes the ex post facto and wipes out history, emerges as decisive for meeting fundamentalism head-on.

But I just said that the argument making up the *Ethics* and the *Tractatus* does not constitute the last word on the perspective of radical immanence. Wittgenstein's later work continues to hold fast to it, and in performing the "rotation" he announces in the *Philosophical Investigations,* he essentially abandons the exorbitant ambition of getting rid, once and for all and at one go, of the metaphysical load traditionally carried by the per se. But what he gains instead is no less than the *consummation* of that very perspective. Wittgenstein's later philosophy is such a consummation because, in confronting philosophical worries only locally and, in a sense, always provisionally, it acknowledges the impossibility of doing away with philosophy once and for all while remaining in tune with our finitude, with our subordinate position with respect to the inexhaustible real and the radical surprises coming from its direction, in tune with the eminently human modality of the ex post facto.

To round off this examination we might ask one last question: why did Wittgenstein have to pass through the ultimate per se before coming to realize the need to perform the rotation that raises the ex post facto to prominence as condensed by Goethe's dictum? Why did Wittgenstein go or have to go through his "early" self before coming to attain what has been hailed as his philosophical maturity?[30]

To see the answer, note first that no radical grammatical change appeared in the roughly fifteen years between the time when Wittgenstein composed the *Tractatus* and when he embarked on his later work. The fundamental (post-Kantian) philosophical techniques for tackling the big issues that would eventually emerge around theory change were more or less set into place at the time the *Tractatus* was written. Wittgenstein was thus offered both the historical opportunity and the philosophical tools for establishing the perspective of radical immanence along either of the two horns identified: he could pursue either a strategy focusing on the ultimate per se or a strategy aiming to tackle the issues surrounding radical theory change. My last question can therefore take a few forms. What induced Wittgenstein to focus on the ultimate per se in the first place? What made him look backward, toward an identification with Spinoza, rather than forward, toward Kuhn and the issues he would bring forth? Why did he zero in on the ultimate per se below the radical conceptual changes then sending shockwaves through his times rather than focus on those historical changes and surprises themselves, which might have led him to uncover the prominence of the ex post facto? Why was he attracted toward the "slippery ice" of the formal and away from the "rough ground" of human practices? Why did he espouse the static, ahistorical outlook we have been considering?

None of these questions, which are fundamentally the same question, admit a "deep" philosophical answer. Radical scientific breakthroughs, on the one hand,

and major social and political upheavals, on the other, constitute the main external motors of philosophical change. When Wittgenstein was first attracted to philosophy, both motors were at work but pulling in different directions. From the one side, the revolutionary changes in mathematics and physics appeared to undermine the Kantian framework and thus call for a novel philosophical foundation. We might call this the pull of the per se. Philosophical logic, linguistic analysis, Husserl's phenomenology, logical positivism, and logical empiricism, as well as Popper's critical rationalism, reflect efforts to respond to this pull.

From the other side, the social and political upheaval leading to the outbreak of World War I and the subsequent shattering developments, not the least being the Russian Revolution of 1917, involved to a crucial extent the ideas of Marx and the conditions of their effective application. Widely considered then as part of Hegel's lineage,[31] the work of Marx, Engels, and their followers reflected what we might call the pull of historicity. Concurrently, although Freud was ostensibly flirting with forms of philosophical legitimization that would admit the child of his labors into the exclusive club of the natural sciences, the practice of psychoanalysis was elevating another form of historicity to prominence.

These opposing forces are still pulling us today. Thus, one might reasonably maintain that the analytic tradition finds its roots at the pull of the per se and as a response to the "crisis of European sciences,"[32] while the Continental tradition finds its roots at the initial pull of historicity and as a response to the generalized social and political crisis erupting around World War I. Down the line, the twist Heidegger gave to Husserl's phenomenology tended to push it away from concerns of logic, science, and epistemology—the ultimate per se—and shovel it under what would be constituted retroactively[33] as the Continental tradition. Concurrently, Heidegger's involvement with the Nazis cast a damning political shadow on phenomenology altogether. The condescending view that the Continental tradition is political—or even literary or journalistic—rather than properly philosophical thrived on precisely such grounds, which have been and to an extent continue to be regarded as leaving room for the inadmissible intrusion of political or other immediate concerns into the dispassionate exercise of philosophy, whose only aspiration can be that of attaining everlasting peace, the eternal peace enjoyed by the ultimate per se. And here history must be altogether absent.

Appraising things from the distance gained, with ex post facto hindsight, we might claim that in the earlier twentieth century, these pulls were exerting unequal forces. From the one side, the philosophical stakes in the groundbreaking scientific developments were sufficiently important to attract philosophical attention nearly irresistibly; even Lenin (Lénine 1962) felt obliged to devote a book to such matters. From the other side, however, the stakes seemed far less central, with professional philosophers feeling the pull of historicity less strongly. After all, to assess either Marx's version of the historical or Freud's version of human subjectivity philosophically, one has to understand the corresponding endeavor in the first place.[34] But such an understanding could not then (and especially then) be fully acquired independent of a wholehearted involvement in the particular practices these endeavors required, in the streets or on the couch, with all their associated political divisions, ideological conversions, and heartrending splits played out either in the open air of political fight or in the privacy of one's relation to oneself.

At the turn of the twentieth century these endeavors were still brand new. The experience gained through participating in the corresponding practices had not had time to gestate, and no one yet had the distance necessary for philosophical assessment of them. As Althusser (1966) would put it much later, the novel forms of considering historicity at work were still in their "practical state." Thus, although the philosophical acumen of some of the protagonists involved made them gain glimpses of the ex post facto as the principle "behind" the "grammar" of history, at the time Wittgenstein was composing the *Tractatus,* these glimpses had not yet reached center stage or achieved philosophical maturity. Whether they have now achieved it remains an open question.

These circumstances presumably explain why the young Wittgenstein followed the horn of the per se rather than that of the ex post facto. Following the first rather than the second horn was not a matter of preference or choice, of discerning or failing to notice, of overestimating or underestimating. Rather, the prevailing historical conditions, the intellectual climate, and the theoretical environment with which the young Wittgenstein was involved conspired with his personal idiosyncrasies and the accidents of his life to make him take the road he took. And if this is indeed the answer to my last question, its prosaic and thus sobering character is noteworthy, for it implies that philosophy is not as independent of theoretical context and historical circumstance, as free of bias or of prejudice, as many philosophers would like to believe it to be. The availability of the appropriate philosophical tools does not entail that even the most penetrating philosophical mind will locate them, take advantage of them, and blaze a novel philosophical pathway. The ascendance of history over philosophy is thus manifested in yet another, perhaps simpler, fashion that once again stresses human finitude and thus once again pleads for the consequent humility.

The intrepid reader who has taken the trouble to follow me will no doubt have noticed that I never offered a solution—or dissolution—to any of the questions or puzzles I examined. They remain as open as they were at the outset. And this is as it should be. Any personal shortcomings aside, the eminently human modality of the ex post facto, all the future surprises with which history might stagger us, rule out the reassurance of a final closure. Thus this book does not lead toward novel philosophical lands, toward other issues that remain open: everything remains open and must always remain open, despite the most valiant and most necessary efforts to the contrary. Thus the final word should be Althusser's: the future indeed lasts forever.

Notes

Preface

1. Helpful overall have also been several collections: Garrett 1996, Greene 1973, Kashap 1972, Koistinnen and Biro 2002, Shahan and Biro 1978, Yovel 1999, and Yovel and Segal 2004.

2. Mention should also be made of the following monographs on the *Tractatus*: Anscombe 1971, Black 1964, Brockhaus 1991, Maslow 1997, Mounce 1985, Morris 2008, Nordmann 2005, Stenius 1996, and White 2006.

3. And more so with Hutto (2003), who tries to steer a middle course between theory and therapy.

4. In retrospect, my relations to my students should have been modeled on the relations Joseph Jacotot, the "ignorant schoolmaster," entertained with his. See Rancière 1987.

Coordinates of a Conversation

1. "Not true" because it is not simply "false." To enter into the logical details of this matter, one has to treat meaningfulness and meaninglessness, and hence propositional content, in conjunction with truth and falsity. Thus, relating the two couples with the standard "axiom" that every meaningful proposition is either true or false, it follows that a proposition that is not true is not by the same token false, for it can be meaningless as well. The proposition "a position outside the world (and thought and language) cannot be" presupposes the possibility of there being such a position; hence it presupposes what, by the proposition's own lights, is false. This is what I call its self-destructive character. But conversely, if the *presupposition* in question is false, the truth value of the initial proposition can be neither true nor false, at least according to Van Fraassen's (1968) treatment of presupposition: "A presupposes B" is identified with "if B is false, then neither A nor not-A is true." Furthermore, a proposition's not possessing a truth value through such dependence to a presupposition does not render the proposition meaningless. For example, given that the qualification "Frog" (as an epithet for a French person) may be highly injurious, the proposition "X is a Frog" is neither true nor false independent of the country of origin of X, for it presupposes in either case acquiescence to the injury. Among other things, this is important in court proceedings. For example, one cannot answer simply yes or no, as a prosecutor demands, to the question "Did you see a Frog close to the victim?" Nobody would consider "X is a Frog" to be meaningless, while a Russellian analysis of that proposition could hardly avoid being ideologically loaded one way or another, and to lift this load, one would have to enter into heavy metaphysics. For my part, I tend to believe that whether the proposition

"a position outside the world (and thought and language) cannot be" is or is not meaningless cannot be decided by such formal ways of reasoning, for the answer depends crucially on the context in which the proposition is used and the intention governing its enunciation. I owe the clarification to Aristidis Arageorgis.

2. I seem to be assuming here the "principle of bivalence," which is the "principle" that apparently fails in the case of propositions having false presuppositions. As I will show, however, logic for Wittgenstein can involve no principles by which propositions are obliged to abide or, more generally, no principles regimenting propositions from the outside in whatever manner. The immanence of logic in language and in thought implies that much. To a reader who is not satisfied by this short answer (admittedly rather cryptic at this stage) and persists in talking about the "failure" of the principle of bivalence in cases of propositions having false presuppositions (or in other cases—for example, in what has been dubbed quantum logic), I can add the palliative comment that the analysis of propositions the way Wittgenstein depicts it in the *Tractatus* wipes out all such problematic cases and comes to "save" the principle with a vengeance.

3. For a general study on confusion, but independent of Spinoza, see Camp 2002.

4. This brings to mind conundrums encountered in the context of ancient skepticism. In discussing Wittgenstein's methods in the *Tractatus*, Stern (2003, 139) refers explicitly to the relevant passages in Sextus Empiricus. For further discussion on the way Wittgenstein's strategy in the *Tractatus* might relate to skeptical ways of proceeding, see Reid 1998 and McManus 2004. For the overall relation of Wittgenstein to skepticism, the classic is Cavell 1999. Conant 2004 is also quite clarifying. For a thorough discussion of ancient skepticism and how it might be involved in the *Tractatus*, see Bourlakis 2006.

5. Grammar does not appear as such in the *Tractatus*. However, something much like it—the sum of the "silent adjustments" necessary for understanding colloquial language (TLP 4.002)—does form an important element therein, while this element, as clarified retrospectively through its association with grammar, suffices to let my discussion develop without jarring overmuch with my principle of reading, namely, to remain as close as possible to the two texts at issue. For the history of Wittgenstein's uses of the term *grammar*, see the corresponding entry in Glock 1996.

6. The cautionary quotation marks are intended as a warning that for Wittgenstein, *all* such possibilities are there permanently. They "remain" because nothing can take any away or bring any about.

7. Wittgenstein does not seem to have been impressed by relativity theory and the beginnings of quantum mechanics, while concepts and their grammar, let alone conceptual revolutions, do not enjoy a pride of place in the *Tractatus*. The reason for such disregard is presumably what I just said: the *Tractatus*, as a *logico*-philosophical treatise, focuses not on concepts and their grammar but on the logical possibilities underpinning all conceptual and grammatical issues.

8. For a discussion of the grammar of radical conceptual change in science with no reference to either the *Tractatus* or the *Ethics*, see Baltas 2004, 2007a, and 2009.

9. This formulation—expressing what philosophers of science call "pessimistic metainduction"—is anachronistic. In Wittgenstein's time, philosophers tended to account for theory change in terms of reduction and cumulative progress without paying close attention to the particulars. The work of Gaston Bachelard is a notable exception, but it had little influence outside France. Radical theory change came to the forefront in the late 1950s and early 1960s with the work of Toulmin, of Hanson—whose *Patterns of Discovery* (1958) relies decisively on Wittgenstein's later work—and of Kuhn (e.g., 1962, 2000), among others.

10. We might visualize the situation by considering the bare possibilities in question as lying at the "bottom level," below concepts and their grammar. This is misleading, however, for no strata can be involved here: logic for Wittgenstein must lie *on the surface*, on the same plane with language and thought. That it is immanent in them implies that much.

11. For overview of those matters from a very congenial standpoint, see Floyd 2005. See also Floyd 2000 and G. Pissis 2002.

12. Compare the late Wittgenstein's explicitly holding that "the human body is the best picture of the human soul" (PI, pt. 2, §4). Cavell (1999, passim) discusses this passage in a very illuminating way without, however, bringing Spinoza into the picture. See also Mulhall 2001.

Chapter One. Mutual Introductions

1. For Nietzsche's "discovery" of Spinoza's philosophy, see Yovel 1989 (104–35). See also Schacht 1999.

2. Deposing God as I intend it need not amount to the annihilation of religious *feelings*. Biographical accounts of Spinoza and of Wittgenstein attest to their deep respect for such feelings. We know that Novalis admired Spinoza's "intoxication" with the idea of God and that Wittgenstein entertained complex attitudes toward the spiritual aspects of religion. For a discussion of Wittgenstein in such respects, see Shields 1993, Phillips 1994, and Tessin and von der Ruhr 1995.

3. Nietzsche may have expressly acknowledged his affinities with Spinoza (Yovel 1989, 105), but his affinities with Wittgenstein are currently under discussion. See Cavell 1999, Conant 2000b, and Mulhall 2005. Nietzsche and Wittgenstein might be connected through the mediation of Schopenhauer; see Weiner 1992.

4. For a detailed account showing how the rise of natural philosophy relates to the social, political, and ideological upheavals of the period, see Israel 2001. Butterfield (1982) provides a more internalist account .

5. See Kline 1980. In contrast to the situation in Spinoza's time, when what we would today call social and political philosophy was interwoven with what we would call epistemology, philosophy of mind, and philosophy of science, in Wittgenstein's time the two branches tended to follow separate paths. Thus, apart from work related in one way or another to Marxism, philosophers concentrated on coming to terms with the radical novelties mathematics and physics were then pouring out. I will discuss the issue in the final chapter.

6. From the vantage point of the *Tractatus,* qualifying "the" logic in question as "new" should be taken with a large pinch of salt. In TLP 6.1251 Wittgenstein emphasizes that "there can *never* be surprises in logic," which implies that no "new" logic can ever spring forth. Not to overburden the reader, I keep this characterization in the present chapter.

7. As Coffa puts it in the sentence that opens chapter 1 of his work, "For better and worse, almost every philosophical development of significance since 1800 has been a response to Kant" (1993, 7).

8. Phenomenology should be included here as well, for, as shown by Dummett (1993), it bore intricate relations with the analytic tradition at the beginning of both. See Gier 1981.

9. Monk (1990) and Janik and Toulmin (1973) offer vivid accounts of how, in Wittgenstein's Vienna, the ethical was taken as indissolubly linked to austere linguistic expression. See also Lecourt 1981.

10. This is detailed in Deleuze 1981. See also Schipper 1993.

11. For Wittgenstein's biography, see Monk 1990, McGuinness 2005, and Waugh 2008; for Spinoza's, see Nadler 1999, R. Goldstein 2006, and H. Cohen 1972. Klagge 2001 includes an exchange between Monk (2201) and Conant (2001) on the constraints of a philosophical biography with particular reference to Wittgenstein.

12. If Spinoza risked secular imprisonment and eternal condemnation, Wittgenstein risked the ridicule of respected teachers and colleagues. Thus in his foreword to the *Tractatus,* Russell does not hesitate in resorting to typical Oxbridge irony with respect to Wittgenstein's distinction between saying and showing: "Mr. Wittgenstein manages to say a good deal about what cannot be said." Or even more pointedly with respect to the Wittgensteinian conception of method: "It is true that the fate of Socrates might befall a man who attempted this method of teaching, but we are not to be deterred by that fear, if it is the only right method" (ibid.). Also

with respect to showing, Ramsey (1931) writes in a similar spirit: "But what we can't say, we can't say, and we can't whistle it either."

13. Schouls (1980) provides an interesting discussion of this.

14. For a comparison of the two works taking account of the overall context determining this genre of writing, see Vokos 2002.

15. This is a kind of "rational reconstruction," in Lakatos's (1980) sense. I don't aspire to historical accuracy either as to the evolution of the views of Descartes and Spinoza or as to the way Spinoza had been conceiving his relation with Descartes at this or that period of his life. I am interested only in the core positions of their strategies as this is more or less typically understood today.

16. For an insightful presentation of the *Treatise,* see Grigoropoulou in Spinoza 2000b.

17. Spinoza's nominalism, i.e., his reluctance to envisage anything other than proper individuals, makes the very notion of "placeholder" difficult for him to uphold.

18. I will barely touch on Spinoza's political philosophy. For interesting recent appraisals, see Negri 1982 and 2004, Balibar 1985, Montag and Stolze 1997, Montag 1999, and Gavriilides 2000.

19. For the evolution of Descartes's thought, see Machamer and McGuire 2009.

20. Among the exceptions we can count, ironically, the late Descartes. See Machamer and McGuire 2009.

21. For a thorough discussion of method in Spinoza, see A. Garrett 2003.

22. We need not enter here into whether the conceptions of method at play in the seventeenth century differed from those at play at the beginning of the twentieth century.

23. Although I tried to avoid the confusion, it does no harm to stress that method differs from order: method for Spinoza involves what I have been saying it involves, while the *Ethics* is deployed according to the geometrical *order,* not according to some geometrical "method." See di Poppa 2006.

24. For an analysis of the relevant sense of immanence, see Deleuze 1968 and 2005 and Yovel 1989.

25. As Sellars (1991, 1) would have it, Nature in this sense includes "not only 'cabbages and kings' but numbers and duties, possibilities and fingers snaps, aesthetic experience and death."

26. Curley (1988, ch. 1) details how Spinoza separated himself from Descartes on such grounds.

27. The proposition E I p20 equates God's essence with His power to act, while E III p7 equates the essence of each individual thing with its endeavor—its conatus—to persist in its own being and hence, once again, with its power to act on its own.

28. This is Bergmann's felicitous characterization, later adopted by Rorty (1988).

29. Reading the *Tractatus* as challenging the view just sketched might appear as odd, for until recently (see Crary 2000a), the received view was the exact opposite: Wittgenstein was typically considered to be a protagonist of the movement elevating logic to that exalted status. At this stage, I simply note that the *Tractatus* does progress along the road I traced, but does so by following the *reverse* order: it starts from the world, passes from thought, and goes to language to end with silence. Perhaps this is neither a coincidence nor a mere literary device. The matter will be taken up in chapter 5.

30. Conant (1991a) offers a thorough discussion of the impossibility of even imagining logically alien thought.

31. If "all propositions in logic are tautologies" (TLP 6.1), if they are all "of equal rank," meaning that "there are not some which are essentially primitive and the rest deduced from these" (6.127), and if, to boot, "logic must take care of itself" (5.473), these are certainly weird "laws."

32. In such respects it is worthwhile to read Boghosian and Peacocke 2000 together with and in contrast to Mason 2000.

33. The way in which the *Tractatus* elucidates the workings of logic while respecting fully its immanence in language and in thought will be examined in the following chapters. We

will see that room is left for a correct logical symbolism *displaying* or *rendering manifest* logic's immanence in language and thereby in thought. One task of the *Tractatus* was to formulate an adequate symbolism in this sense.

34. "Blurred," meaning not clear and not distinct; or, as Spinoza would have it, "fragmentary (mutilated) and confused."

35. In this respect see Narboux 2001.

Chapter Two. Purposes and Ends

1. I use Heideggerian turns of phrase to suggest that Heidegger is not foreign to the perspective of radical immanence. Mulhall (1990, 2001) discusses Heidegger in connection to Wittgenstein.

2. Nehamas (1998) discusses this tradition as starting from Socrates without, however, referring to Spinoza, Wittgenstein, or Heidegger.

3. Gavriilidis (2000) presents a detailed analysis of the political implications of this thesis.

4. Even apparently aimless activities such as wandering around or dancing by oneself are driven by some purpose at a deeper psychological level that might not be directly accessible to the one engaging in the activity. For example, one may wander around for the "purpose" of letting some experience sink in or dance by oneself for the "purpose" of expressing a joyful mood. I owe this clarification to Evgenia Mylonaki.

5. See, for example, Spinoza's scathing criticism of Descartes in E V Pr.

6. Intentionality, as understood by Brentano and Husserl, may enter into the picture at this point.

7. Given that Lacan has acknowledged Spinoza's influence on his work, Lacanian psychoanalysis perhaps finds one of its roots here.

8. To the extent that the body is taken into account, these are exactly Spinoza's emotions, or "affects."

9. The purpose preceding our engaging in any activity might be characterized along three axes. We might engage it along the axis of will (willingly or unwillingly, in good or bad will), along that of thought (for a reason), or along that of emotion (for fear, ambition, anger, and so on) or according to any combination thereof. In his particular way, Wittgenstein tackles all three dimensions. Now will, thought (intellect), and emotion are the three compartments of the classical partition of the soul and thus of human agency. That Wittgenstein tackles them all attests to the *completeness* of the *Tractatus* as a (logico-)*philosophical* treatise.

10. Will is not an emotion for Wittgenstein, though it is for Spinoza. Wittgenstein, however, relegates "will as a phenomenon" (TLP 6.423), in company with the emotions, to the province of psychology, which did not form an autonomous subject in Spinoza's time. Hence, to the extent that Wittgenstein's "will" comes down to what I have been calling "purpose," the difference may be seen as secondary, at least with respect to what is at issue here.

11. This might point toward Kant, whose *Fundamental Principles of the Metaphysic of Morals* starts by appealing to "good willing" as a basis for considering the ethical.

12. Wittgenstein does not talk in TLP 6.43 of what the "happy" or the "unhappy" person feels but about the "*world* of the happy" and that of the "unhappy" (my emphasis).

13. Spinoza's ontology is exhausted by a unique Substance with infinitely many Attributes and by modes, i.e., particular things as they connect in facts.

14. In chapter 4 I will show how purpose can be thought *of* for Wittgenstein as well, even if it is not a thought properly speaking.

15. This is a logical—or rather grammatical—point inspired from Wittgenstein's "private language argument." I am indebted to Spyros Petrounakos for relevant discussions. Psychoanalysis ratifies the same point by bringing in unconscious motivations and the attendant impossibility of consciously assessing them. But psychoanalytic treatment might throw light on motivations—purposes— hidden from conscious view as well.

16. The particulars of such a self-referential property of responsibility have been worked out by Lacan with respect to the self-authorization of the psychoanalyst (Lacan 2006d).

17. Compare Cavell (2010, 321): "[Language] reveals human speech to be radically, in each uttered word, ethical. Speaking, or failing to speak, to another is as subject to responsibility, say to further response, as touching, or failing to touch, another." Thus "there can be no ethical propositions" (TLP 6.42) because *every* proposition bears ethical intent.

18. Wittgenstein calls ethics "transcendental" (TLP 6.421), a qualification he applies also to logic (TLP 6.13) but nothing else. The discussion in chapter 6 shows this to be consistent with the perspective of radical immanence.

19. The ethics of responsibility, as I intend the notion here, goes beyond moral judgment (Crary 2007a), involving all our relations to ourselves, to others, and to the world at large, while it is linked with political responsibility. In conjunction with Crary 2007a, see Revault d'Allones 2002; Butler 2005; Cavell 1999, 2010; and Cavell et al. 2008. For a synthetic presentation, see Spiliopoulos 2011.

20. And he admits this openly when not bound by the discursive strictures of the *Tractatus*. For example, in the famous letter to Ludwig von Ficker, his prospective publisher, he states that "the point" of the *Tractatus* is "is ethical" (qtd. in Monk 2005, 22).

21. The "color" in question is made visible by the ethical vocabulary McDowell uses when he talks about mind and world in his homonymous book (McDowell 1996), a work that has had a major, albeit indirect, influence on the matters discussed here. This is a vocabulary fittingly not elaborated, commented on, or even mentioned but left alone to *display* that the responsibility in question is being assumed *in deed*.

22. I am employing the term *frame* with most of the connotations implied by Diamond (2000b) but without endorsing her distinction between the frame and the body of the *Tractatus*.

23. Diamond's (1995a) "realistic spirit" and McDowell's (1996) "direct realism" presumably denote precisely this kind of realism.

24. Being clear and distinct is the mark of adequate ideas for Spinoza and most of his contemporaries, but Wittgenstein talks about exactly the same thing: "Philosophy should make clear and delimit sharply the thoughts which otherwise are opaque and blurred" (TLP 4.112).

25. I will discuss Spinoza's relation to the physics of his day in chapters 7 and 8.

26. Many scholars have addressed Spinoza's doctrine of the eternity of the mind without arriving at substantial consensus. See Bennett 1984, Curley 1988, Moreau 1994, Mason 1999, and Nadler 2006. What follows forms a bare outline that I do not pretend adjudicates the corresponding disagreements.

27. The notion that the idea of one's body is an eternal truth in the mind of God might be considered as having a "durational" analogue: the sometimes indefinitely persisting memory of someone's having passed through this world having achieved whatever he or she might have achieved.

28. The phrase in question goes like this: "For the eternal part of the mind is the intellect, through which alone we are said to be active." Spinoza himself refers to E III p3, where he holds that "the active states (*actiones*) of the mind arise only from adequate ideas," thereby securing the inference. The qualification "we are said to" indirectly addresses his contemporaries who did not conceive mind as the idea of the body (E II p13) and hence accept that mind and body always run in parallel as expressions of one's essence in the Attribute of Thought and of Extension.

29. This is why "expert knowledge" is distinguished from knowledge organized by concepts just as "knowing how" is distinguished from "knowing that."

30. The qualification is required for, say, expert drivers may well drive trucks expertly while their minds and parts of their bodies are devoted to other concerns. In such cases no feeling of deep contentment need arise. I owe the clarification to Kostas Pagondiotis.

31. Films of varying quality have portrayed how the toil necessary for mastering Eastern

martial arts may enhance the capabilities of a body-mind to amazing degrees. Such films exert this attraction presumably because they present the capabilities of a body-mind as being almost literally boundless, as well as showing that spectacular expert action involves body and mind in inseparable consort.

32. Thus Wittgenstein became a schoolteacher after he walked away from philosophy, while Spinoza, after completing the *Ethics,* devoted his time to more "applied" work.

33. This is not exactly true for either Spinoza or Wittgenstein. In the case of Spinoza, the more "applied" work subsequent to the *Ethics* was clearly philosophical. Nonetheless, its more "applied" nature may be taken as signifying that the "unassailable and definitive" results of the *Ethics* had silenced philosophy "in its essentials." To that extent, Spinoza's position on philosophical silence may be considered as being close to Wittgenstein's. Wittgenstein, however, did later return to philosophy even though he took himself to have silenced it for good. The sense of this return has taxed Wittgenstein scholars, but I will address it only in the final chapter.

34. This is not a mistake we make, and language does not push us in this direction because it was badly "designed." Psychoanalysis teaches that such marking out of the "I," with all the attendant illusions, is the price we have to pay in undergoing the tortuous process whereby we enter the properly human order, an order that is at once social, linguistic (or symbolic), and sexual, i.e., governed by the taboo of incest. See Lacan 2006e and Althusser 1976.

35. Compare Cavell (1989, 494): "For suppose my identity with my body is something that exists only in my affirmation of my body. . . . Then the question is: what would the body *become* under affirmation? What would it become of *me*? Perhaps I would know myself as, take myself, . . . as a universe." An earlier passage (Cavell 1989, 383–89) seems to support what I am trying to say here.

36. Gibson's (1979) "ecological" approach might be tied in here. See Damianos 2007.

37. I owe this clarification to Spyros Petrounakos.

Chapter Three. Grammar

1. This situation seems to change rapidly, at least as regards Spinoza's political philosophy. See the references in note 18 of chapter 1.

2. See Rorty, Schneewind, and Skinner 1984 and Garcia 1992. The work of Michael Frede is almost a solitary star in this respect.

3. The beginnings of systematic philosophy with Plato are deeply indebted to the emergence of mathematics. See Cornford 1970.

4. Mason (2003) offers an interesting discussion of understanding.

5. This motor is not properly external: conceptual advances taking place outside philosophy inevitably are already indebted to philosophy. See Althusser 1974b.

6. For Althusser (1972), "historical materialism" and psychoanalysis should be included here as well. As he phrases the same idea in Althusser 1976, Marx and Freud (and Nietzsche) had come to the world "fatherless."

7. For ways of connecting grammar with paradigm change, see Baltas 2000, 2004, and 2007a.

8. In the following I will try to moderate the obvious anachronism of the formulation.

9. See, for example, Mathews 2000 and Vosniadou 2002.

10. I am relying mostly on *On Certainty*, Wittgenstein's last work. See Morawetz 1978 and Rhees 2005.

11. The following section borrows heavily from Baltas 2004, 2007a, and 2009.

12. I use the term *normal* as Kuhn (1962) does when speaking of "normal science."

13. In the case at hand, these are the background "assumptions" fastening the classical definition of a wave. Resistance to their being questioned is encountered because those still entrenched in the grammatical space that grounds the understanding of classical mechanics cannot see that something beyond classical physics could come to drastically change some of

their most fundamental conceptions; therefore, they perceive those supporting the STR as trying, *per impossibile,* to tear apart the *analytic* relation defining the notion of a wave through that of a material medium.

14. Because, for example, understanding Newton's concept of force requires coming to terms with the idea that a body can act instantaneously on another body even though nothing lies between them. No everyday experience can vindicate this idea. For the relevant pedagogical research, see Vosniadou 2002 and Mathews 2000, among many others.

15. The material in this and the next section is indebted to Baltas 2004 and 2007a.

16. Ogden translates *"stillschweigenden Abmachungen"* as "silent adjustments," while Pears and McGuinness have it as "tacit conventions." Both renderings seem compatible with Wittgenstein's German, but the second translation can be profoundly misleading: it confers a conventionalist flavor on Wittgenstein's conception of language that the *Tractatus* does not support. Moreover, "adjustments" marries well with the "hinges" of *On Certainty.* In addition, "silent" should be preferred to "tacit," for it resonates with Wittgenstein's injunction in TLP 7 that concerning matters of which one cannot speak, *"muss man schweigen."*

17. In fact, Wittgenstein refers here only to forms in logic that supposedly could be invented, but his point can apply unproblematically to scientific discovery or invention generally.

18. J. Weiner 1990 examines Fregean "elucidations" as the means for allowing someone not initially familiar with the *Begriffschrift* to enter it and, so to speak, understand it from the inside. See also Conant 2000a.

19. I borrow this idea from Diamond 1995b and from a discussion with Andreas Karitzis. Diamond makes clear that no precise rules can be forthcoming in cases such as this, so that imagination should come into play in ways similar to those I discussed with respect to paradigm change. Compare Frege: "Science needs technical terms which have precise and fixed meanings, and in order to come to an understanding about these meanings and exclude possible misunderstandings, we give examples illustrating their use. Of course in so doing we have again to use ordinary words, and these might display defects similar to those which the examples are intended to remove. . . . In practice, however, we do manage to come to an understanding about the meaning of words. Of course we have to count on a *meeting of minds,* on others *guessing* what we have in mind" (cited in J. Weiner 1990, 230; emphasis added).

20. The elements making up such a ladder are close to the Fregean "hints" (*Winke*) that Weiner (1990) discusses in the last chapter of her book. See also Conant 2000a.

21. Phrases such as "I have the intuition that we should try proving this" or "It is my intuition that the experiment will turn out this way" are common stock within the practice of physics. How such intuitions may or may not resemble Kantian intuitions cannot be examined here. Much of the work of Gaston Bachelard (e.g., 1975) aims at building such intuitions with respect to twentieth-century natural science and mathematics. See Baltas 1986b.

22. Wittgenstein here refers such intuitions to mathematics, but the scope of his remark can cover situations such as the one presently being examined. At any rate, physics has been inextricably involved with mathematics since Galileo.

23. Lest I be misunderstood, I should specify that coming to feel such an effect *fully* is only a moment (the eureka moment) of a process that might involve long struggles to come to terms with a host of associated issues. Merely taking in Weinberg's definition is not enough, which is why I characterized it as only the first rung of the ladder at issue.

24. I am referring to what Bachelard (1975) calls *"obstacles épistémologiques,"* which can be explicated in terms of the silent adjustments discussed here.

25. This recalls but need not relate to Chomsky's conception of deep grammar. In any case, the bodily aspect of grammar in this sense cries out for more work.

26. Harris (1988) provide material interesting in this respect.

27. Althusser (1996a) has made much of such overdetermination; see also Lacan 2006b.

28. The idea of *his own* body: for Spinoza this would be nothing less than Wittgenstein's *mind* . . .

Chapter Four. Strategies

1. For future reference, note that Spinoza was certainly not more lenient toward nonsense. Many of the scholia of the *Ethics* use harsh words against the reigning confusions of his day, and in an almost side remark that has received less attention than it deserves (with the exception of the exchange between Savan and Parkinson I discuss in chapter 6), Spinoza speaks about the "so-called transcendental terms such as 'entity,' 'thing,' [and] 'something' [that]... signify ideas confused in the highest degree" (E II p40s). Yet he does not hesitate to employ such terms freely, without any means of protection, at the very foundation of his system. Almost exactly like Wittgenstein, he appears thus as totally unperturbed by this self-inflicted charge.

2. I am indebted to the way Diamond (2000b) highlights the importance of this point.

3. In his later work Wittgenstein makes this quite explicit: "Don't[,] *for heaven's sake,* be afraid of talking nonsense! But you must pay attention to your nonsense" (1984, 56e).

4. To repeat, "this limit can, therefore, only be drawn in language" (TLP Pr. ¶4).

5. This battery of metaphors is not intended as literary ornament. I am alluding to the bewilderment, sometimes accompanied by a deep sense of threat, exhibited by some commentators with respect to the *Tractatus*. After all, a philosophical approach presenting itself as fully self-sufficient and fully self-authorized, and thus as royally disdaining all "objective" assessment by its author's peers, might well be perceived as threatening others' self-images at the level of professional self-preservation. Similar considerations may account for the fairly common belief that Derrida aimed to demolish everything for his own sheer satisfaction, for "deconstruction" is not foreign to what I am trying to say here. See Staten 1984, Garver and Lee 1994, and Stone 2000. For Derrida's initial "statement" of deconstruction see Derrida 1976.

6. As Hacker (1986), among others, would have it.

7. At least in this respect, Conant and Diamond are perfectly right. See Crary 2000a.

On a lighter tone, an admittedly weak argument explaining why the distinction between "substantial" and "mere" nonsense cannot hold water is provided by the language of the Smurfs. In this language all words bearing content have been replaced by the string "smurf." Given the drawings, the expressions of the characters' bodies and faces, the time evolution modeled on the sequence of the pictures on the page, and the preservation of the skeleton of logical and grammatical structure as well as of the standard linguistic expressions of propositional attitudes, an exchange such as "I smurfed the smurf," "Good, for I feared that you had smurfed the smurf" makes perfect sense, or rather, a fuzzy class of compatible senses not very distant from one another. (The Smurfs might encounter difficulties in distinguishing "rabbit" from "undetached rabbit parts.") Given now the (probable) fact that those distinguishing "substantial" nonsense from "mere" nonsense would hold that the string, say, "She smurfed the smurf" is total gibberish, i.e., mere nonsense, the existence of the (admittedly rather poor) language of the Smurfs shows that total gibberish can be no nonsense at all. Hence, I cannot see what general criterion, covering all contexts, could possibly distinguish "mere" nonsense from "substantial" nonsense. And in any case, such a general criterion would itself be nonsensical, for it could be forthcoming only as formulated from a vantage point overarching all language.

In the same vein, consider the following commercial played some time ago by the Greek radio: "soccer is played with the hands; it doesn't need a pitch or a ball; it can be played with one player and you can win with a draw. The players know how to play." The commercial refers to soccer betting.

8. Conant and Diamond (2004) provide an excellent discussion of this point.

9. The last paragraphs might help to clarify how Wittgenstein proceeds in his later work as well.

10. All available evidence (Monk 2000) indicates that Wittgenstein was acutely aware of the inevitability of such resistance. Later he formulates such awareness succinctly and seems even to accept that his texts should offer themselves to their readers in precisely this fashion: "I ought to be no more than a mirror in which my reader can see his own thinking with all its deformities," while hoping against hope that, "helped in this way [the reader] can put [his thinking] right" (Wittgenstein 1984, 18e). Independent of Wittgenstein's later views, however, the "good will" necessary for getting rid of such resistance does seem to remain a fundamental prerequisite for understanding the *Tractatus*.

11. The deep unity of these two movements along the flow of the *Tractatus* is well laid out by Ostrow (2002), though he does not distinguish them as such. See also Ricketts 1996.

12. We might say that the work done in the *Tractatus* aimed at rubbing out the Cheshire Cat from inside it, as it were, not leaving even its smile. Wittgenstein undertook his work to arrive at the nothing of philosophical silence, a nothing not very different from the one to which Althusser led his philosophy. See Baltas 1995.

Chapter Five. Organizing Content

1. At the beginning of part III Spinoza uses the term *postulates,* but in formulating these postulates and while explicitly referring them to preceding propositions, Spinoza makes clear that the terms *axiom* and *postulate* can be used more or less interchangeably, even if postulates seem to be advanced in relation to a narrower field of study.

2. Given this interruption, as well as some alleged discrepancies in the flow of the text (e.g., the lack of prefaces in the first two parts), Negri (1982) maintained that these last parts endow the *Ethics* with a "second foundation," one less "metaphysical" and more directly political than that provided by the first two parts. My approach allows me to ignore such matters.

3. Russell, for one, found the demonstrations "tedious."

4. "I have written a book . . . *nobody* will understand," Wittgenstein wrote to Russell in 1919 (in Conant 2002).

5. See Janik and Toulmin 1973, Lecourt 1981, and Nordmann 2005.

6. As noted, Wittgenstein's demonstrations involve elucidation, showing, and nonsense.

7. Readers can multiply the number of layers at will and, if they have the courage, rewrite the *Tractatus* accordingly.

8. There is evidence that Spinoza initially meant to divide the *Ethics* in three parts. This brings out the similarity in organization between his and Wittgenstein's work even more clearly.

9. Stenius (1996) has remarked something like that, albeit rather hesitantly.

10. On many occasions Wittgenstein remarks that what readers can do on their own should be left to them.

11. By the end of the second reading, all propositions of at least the first category can be considered as having lost their metaphysical weight in the same manner. This is to repeat that everything regarding the sense of propositions depends on context and purpose. The propositions of the *Tractatus* can function as telling or as pointless nonsense in the context of the philosophical activity but also as innocuous everyday propositions with no metaphysical weight attached to them once the text's philosophical toil has been brought to completion. Compare: "What *we* do is to bring words back from their metaphysical to their everyday use" (PI §116).

12. The first chapter of *Discourse on Method* begins thus: "Good sense is, of all things among men, the most equally distributed." Potentially anyone could occupy the place of the philosopher.

13. For an analysis of Spinoza's relation to Descartes on the matters at issue here, see Curley 1988.

14. The requirement for such a "meeting of minds" is rarely discussed today. Many philosophers, particularly of the analytic persuasion, tend to see philosophical texts as inert and impersonal systems of arguments that pass the time waiting for further arguments to justify,

qualify, or refute them. Thus invoking, for example, the spirit of a text or asking how an author's intention might have manifested itself—or hidden itself—in his or her text is usually considered to be a matter lying beyond rational adjudication or even simple honest discussion. In such situations, talking about a communion of minds might well sound mystical.

15. Frege correspondingly asks for "a little good will and cooperative understanding, even guessing" (in J. Weiner 1990, 230). One might see this as an invitation to apply a form of the "principle of charity."

16. It expires as a *philosophical* proposition. It may well be used for this or that purpose in contexts not related to philosophy.

17. Recall Wittgenstein's injunction: "Don't *for heaven's sake*, be afraid of talking nonsense! But you must pay attention to your nonsense" (1984, 56e).

18. In fact, a clause such as "the electromagnetic field is mefkumsunus," that is, a clause involving a string making no sense whatsoever, might serve for naming an entity that only a leap into the ungrammatical can let us conceive. Compare the following passage, where Nasio (1998, 78) discusses what Lacan calls "object *a*": "What is object *a*? Object *a* is only a letter, nothing more than the letter *a*, a letter having the central function of naming an unresolved problem, or rather, of signifying an absence. What absence? The absence of a response to a question which is constantly repeated. Since we have not found the solution we hoped for and required, we mark it with a written notation—a simple letter—the opaque hole of our ignorance; we put a letter in the place of a response that is not given. Object *a* designates thus an impossibility, a point of resistance to theoretical development. With this notation, we can—in spite of our obstacles—continue our research without the chain of knowledge being broken. You see object *a* is finally a ruse of analytic thinking designed to circumvent the rock of the impossible: we circumvent the real by representing it with a letter." For more details on the connection of these ideas with grammatical change in science see Baltas 2004.

19. I have used TLP 1.1 only as an example; it could be replaced by any proposition belonging to the first category.

20. Though Wittgenstein asserts propositions that he himself categorizes as nonsensical, this does not indicate duplicity. In the *Tractatus,* propositions such as "the world is everything that is the case" can be used in mundane, nonphilosophical contexts to convey ideas one sincerely believes. Thus we might say that the propositions Wittgenstein advances along the first movement of his strategy express in nonphilosophical contexts ideas Wittgenstein does entertain. These propositions thus constitute standard philosophical activity in that subjective sense as well.

21. For it would be analogous to the way Pierre Menard rewrites *Don Quixote* in the famous story by Borges (1998, 88).

22. Although differing with the present reading, as well as between themselves, the approaches of Ostrow (2002) and of Friedlander (2001) converge in aiming to provide the general contours of such a reenactment.

23. Ricketts (1996), Floyd (2000), and Conant (1991b), among others, have worked meticulously on some such rungs reenacting the corresponding parts of Wittgenstein's demonstration.

24. The gist in question is just the idea behind the perspective of radical immanence. This idea should be considered now as *devoid of all metaphysical weight,* as merely the idea summarizing the secular or naturalistic attitudes at work in the everyday dealings of those who do not consider themselves, explicitly or implicitly, as subject to corresponding external "laws" or "forces," whether ontological, epistemic, or moral. I tried to explain how the *Tractatus's* propositions can become innocuous in nonphilosophical contexts, and this argument can be extended to cover the definition of the perspective of radical immanence.

25. I have maintained that indefinitely many ways of working out the perspective of radical immanence exist, with the live options depending on the historical and philosophical context in play. Hence not only Spinoza but also other philosophers could be brought to the game and

function as Wittgenstein's interlocutors in analogous conversations. I started by referring to Nietzsche, and I have mentioned still other philosophers who endorse the perspective of radical immanence in one form or another. Perhaps any two philosophers espousing this perspective could be brought to converse with each other in this way. In fact, conversations of this kind already exist in the literature. To refer only to Wittgenstein, Conant (1995, 1997) has made him converse with Kierkegaard; Mulhall (1990, 2001), with Heidegger; Staten (1984), Garver and Lee (1994), and Stone (2000), with Derrida; others, with Kant or Schopenhauer; and so on. Wittgenstein stands out among other philosophers of the same persuasion largely because of his commitment to rigor, and he appears closest to Spinoza because the two share what amounts to the same commitment. Such proposals of conversation, however, focus mainly on Wittgenstein's later work and not on the *Tractatus*. In the final chapter, "Exodus," I will discuss how the perspective of radical immanence is retained in his later work.

Chapter Six. Metaphysics

1. For a thorough analysis of this change as referred expressly to Spinoza, see Carriero 2005.

2. Macherey (1998) distinguishes two parts in the deployment of part I. The first (propositions 1 to 15) concerns what God is, while the second (propositions 16 to 36) concerns what God does given what He is.

3. In a slightly different context, Mason (1999, 68–69) draws the analogy with TLP 3.03: "We cannot think anything illogical, for otherwise we should have to think illogically."

4. All anthropomorphic characteristics of God have already been explicitly denied in the course of the text. For example, in E I p17s Spinoza holds that "neither intellect nor will pertain to the nature of God."

5. See letter 64 (Spinoza 1982). Varying interpretations have been proposed with respect to the distinction (and connection) between "the laws of God's nature" and those in the process of being established by science. We have profited from those of Curley (1969; see also Curley's comments in Spinoza 1985a), Matheron (1988), and Mason (1999).

6. On the difference between immanent and transitive causation, see Fourtounis 2002 and 2005.

7. Negri (1982) calls this approach in the historical context it was composed "savage anomaly." All sources attest that Spinoza was well aware that he constituted such an exception. Thus, in castigating his contemporaries while referring unmistakably to what himself was doing, he writes: "For they found it easier to regard [that] blessings and disasters befall the godly and ungodly alike without discrimination as one among other mysteries they could not understand and thus maintain their innate condition of ignorance rather than to demolish in its entirety the theory they had constructed and devise a new one. Hence they made it axiomatic that the judgment of the gods is beyond man's understanding" (E I Ap).

8. Mason (1999, 35) puts it thus: "[Spinoza] was not an ontologist in the sense of offering any account on what kinds or numbers of things exist. In fact, the tendency of his thinking is to take ontology—questions about what exists—out of philosophy, as we might understand it, altogether, leaving them for physics." And he goes on to specify that "it would be more accurate to say that his approach deprives philosophical ontology of any point. . . . The view that nature exists—that there is nature—leaves open any question of *what* exists in nature—of what exists: open for scientific inquiry, perhaps" (41). Without the "perhaps," this is precisely Wittgenstein's position as I have been rehearsing it.

9. Here, too, Mason (1999, 38) bears out my position from a resolutely Wittgensteinian perspective (although Wittgenstein is not mentioned): "There will be no external explanation for the existence of nature if we understand it as being without limits."

10. Recall Mason: "[Spinoza's] approach deprives philosophical ontology of any point."

11. These might have been conditions leading, for example, to Husserl's Cartesianism.

12. According to my account, the concept of a wave is analytically related to the concept of

a material medium within classical mechanics, and yet this analytic relation can be questioned by an act of imagination forwarding a conceptual distinction inconceivable in the grammatical space subtending classical mechanics. Furthermore, the possibility of challenging an analytic relation in this way was foreclosed to Spinoza, for historical conditions compelled him to run together the logically and the grammatically impossible. In considering the clause "one and only one" to be "very improper" when referred to God, then, he can be taken as invoking either a conceptual (grammatical) or a logical impossibility.

13. I thank Tasos Betzelos for bringing to my attention the letter to Jelles. In an interesting discussion of this passage, with respect to the larger issue of Spinoza's conception of God's unity, Mason (1999, 39–41) ushers into the forum Curley, Jonathan Barnes, Frege, Guéroult, Hampshire, Rorty, Lévinas, Aquinas, Hermann Cohen, and Pollock. To them we can add Macherey (1993).

14. A great part of Savan's (1953) text is devoted to unearthing and laying bare such logical inconsistencies.

15. Savan (1953, 68) refers to the *Cogitata Metaphysica*; to letters 12, 19, 50, and 83 (in Spinoza 1982); and to E I Ap and E IV Pr.

16. I have replaced Savan's rendering of the passage with Shirley's translation (Spinoza 1995). Spinoza's evocation of something like two movements in the flow of a text brings directly to mind the two movements of Wittgenstein's strategy.

17. Crary 2000a is informative in this regard.

18. I have already distinguished the figure of the ladder from that of the scaffold because the structure of the *Tractatus* appears to differ substantially from that of the *Ethics* particularly in respect to nonsense. But the difference between a ladder (with nonsensical rungs) one can climb off and a scaffold (with dispensable planks) one can stand atop to see Nature rightly seems to amount to very little. Indeed, even the slight difference remaining can be eliminated, for (inversely, so to speak) Wittgenstein's ladder can be profitably considered in some contexts as a scaffold with dispensable planks.

19. The idea that the figure of the ladder should subsume the figure of the treatise was perhaps hovering in Spinoza's mind: "We see also that reasoning is not the principal thing in us, but only a stairway"—a ladder—"by which we can climb up to the desired place" (ST II 26:6).

20. The only two exceptions (E I def 2 and 7), starting with "it is said to be," are of no consequence in what follows, for they too implicate persons, if only the form of "some people."

21. Demonstrating that equivocation is *necessarily* unavoidable is one of Derrida's main contributions. See Derrida 1981 for the case of Plato's *Phaedrus* and Derrida 1976 for the case of Rousseau's *Confessions*.

22. In the next chapter I discuss how Spinoza relates clarity and distinctness of ideas to truth. In any case, Spinoza shares with (or borrows from) Descartes the position that the mark of *certainty* is clarity and distinctness of ideas in one's mind.

23. As I implied in the preceding chapter, Spinoza provided scholia, prefaces, appendixes, and explications in order to meet halfway the readers who would engage the *Ethics* in *good faith*.

24. If that were not required, if philosophers could "express correctly what is in their minds" with the requisite rigor while their interlocutors were capable of not "misunderstanding" them, then philosophical controversies could not arise; everyone would agree with the true theses advanced. Although the premises of the corresponding argument are apparently very different, this is, notoriously, Wittgenstein's conclusion, too: "If one tried to advance *theses* in philosophy, it would never be possible to debate them, because everyone would agree to them" (PI §128).

25. That is, his "study of sign language" (TLP 4.1121), aiming, among other things, at coming up with a symbolism rendering perspicuous the logical impossibility of an external vantage point.

26. The historical distance separating Spinoza from Wittgenstein can perhaps be also marked by the reluctance or hesitation with which Spinoza introduces his discussion of the

"most confusing transcendental terms." Thus, although he previously "decided" not to "embark" on questions of language, not only because he had "set them aside for another treatise" (Shirley [1982, 89] identifies this with TEI), but also because he wanted to "avoid wearing the reader with too lengthy a discussion" on such matters, he nevertheless proceeds with a "brief" treatment of "the so called 'transcendental terms' so as to omit nothing that it is essential to know" (E II p40s1). Although this is wildly speculative, I am tempted to attribute this reluctance or hesitation at least partly to Spinoza's having realized that the self-destructive role that the confusing transcendental terms were playing in the *Ethics* might have required a closer scrutiny. The constraints imposed by historical context made it extremely difficult to draw the consequences of such self-destruction the way Wittgenstein did much later. In any case, the fact that he does inflict confusion on his own treatise by employing transcendental and universal terms without restraint, as well as the precaution of starting his definitions in the first person, attest that the implacable rigor with which he reasoned was in a position to shake even the pillars of the historical context in which he worked.

27. Since "the idea constituting the human mind" can have as object just the human body and *nothing else* (E II p13), all ideas of the human mind are ideas of the corresponding bodily modifications. And since "knowledge [hence the idea] of an effect depends on and involves knowledge [hence the idea] of the corresponding cause" (E I a4), the idea capturing the bodily modification in question must involve only the ideas of the bodies causally responsible for it, bodies that make up an infinite causal chain. The idea of the bodily modification in question involves the idea (or ideas) capturing, of course inadequately, the *entire infinite causal chain* responsible for this modification.

28. "When a liquid part of the human body is determined by an external body to impinge frequently on another part which is soft, it changes the surface of that part and impresses on it certain traces of the external body acting upon it" (E II post 5). This attempt at grounding is impressive given that Spinoza did not possess scientific information on the functioning of the brain and the central nervous system.

29. Many scholars have noticed affinities between Spinoza's and Freud's thought. See, for example, Yovel 1989 (136–66). But see also De Deugt 2004.

30. This is also Descartes's position, though for different reasons.

31. Spinoza expounds here in its proper order what I took as the basis of his method in chapter 1: "Who can know that he is certain of something unless he is first certain of it? What standard of truth can there be that is clearer and more certain than a true idea? Indeed, just as light makes manifest both itself and darkness, so truth is the standard both of itself and falsity. Truth is its own standard" (E II p43s). Much cogitation about theories of truth and the like would have crumbled like a castle of cards if philosophers had heeded these phrases, which also cut short any temptation to invoke metalanguage with regard to truth in almost exactly the manner Wittgenstein cuts short any such temptation with regard to logic.

32. This point, too, shows that Spinoza's metaphysics—here, his epistemology—is minimal.

33. Hence our memory is correspondingly limited too; unfortunately (or rather fortunately), it cannot achieve the feats of Borges's (1998, 131) Funes the Memorious.

34. Saussure (1968) similarly uses the notion of a coin to figure his linguistic "sign."

35. In another philosophical vocabulary this is Heidegger's Being, Being as shared by all beings (Heidegger 1962).

36. It cannot be a coincidence that Wittgenstein hypostatizes the "what" in the proposition at hand, marking the move by a capital. *Hypostasis* is the Greek for "substance," and obviously enough, if Wittgenstein can allow himself one such move, he cannot do better than reserve it for hypostasis itself—as well as for the "how" immediately and inseparably attached to hypostasis.

37. The accounts of Friedlander (2001, 34–46) and of Ostrow (2002, 23–33) offer a finer-grained discussion of these matters more or less along the lines I am pursuing here.

38. Friedlander (2001, 44) puts it thus: "An object is form insofar as it is inseparable from a space of possibilities; it provides content insofar as it occurs as a node in a specific configuration of that space."

39. Elsewhere (Baltas 1997) I have relied on the "what" in question to dissolve the opposition between scientific realism and antirealism. It refers, however, neither to Spinoza's nor to Wittgenstein's conception of Substance but to Lacan's "*le Réel*," which arguably lies close to such a minimalist conception.

40. Because Spinoza fails to offer any characteristic feature of his simplest bodies, many interpretations have been proposed about his intended meaning. Thus Macherey (1997, 132) tends to consider them as something like point masses, more or less of the Newtonian kind. Matheron (1988, 40), while not speaking explicitly of point masses, proposes a particular form of the "unvarying relation" I have been discussing that averages over the parts composing a complex individual by making the composing bodies relate to the composed body as point masses relate to their center of mass. Deleuze (2001) considers the simplest bodies to be some kind of infinitesimals.

41. No wonder! These are, of course, the defining problems of the differential and integral calculus, which Newton and Leibniz would later set up as working mathematical tools and would become mathematically rigorous and hence philosophically reliable only with the work of Cauchy, Bolzano, Weierstrass, Dedekind, and others as late as the middle of the nineteenth century. Interesting in this respect is Anapolitanos (1999), but in reference to Leibniz, not Spinoza.

42. Shirley (1982, 42) identifies in this respect Spinoza's *Descartes's Principles of Philosophy*, II, 2–3.

43. Macherey (1997, 130–33) offers an interesting discussion of this issue.

44. This is reminiscent of Lacan's "mirror stage" (Lacan 2006c). Given that the "I" in question must be an "imaginary" idea for Spinoza, it would be perhaps historically more accurate to say that Lacan borrows his mirror stage from Spinoza. See Kordela 2007.

45. That the "I" is "imaginary" in this way turns it into an unconscious construct in the sense Freud and Lacan use the term. Perhaps Lacan's "imaginary register" can be grounded here.

46. Wittgenstein's relations to Kant, noticed by many, find their perhaps clearest expression here.

Chapter Seven. Matching Content

1. Recall Della Rocca's (1996) discussion of language: despite the epistemic isolation at issue, "neutral"—and hence confusing—universal and transcendental terms are unavoidable once we employ language to examine an Attribute. In this sense, no language can ever be purely an Attribute's own.

2. Spinoza is certainly not an empiricist. In contemporary parlance he would hold, together with Hanson (1979), that any act of perception is "theory laden." See also Pagondiotis 2005.

3. An ample discussion has been going on regarding how Spinoza's Attributes should be understood in connection to whether he should be considered a monist or dualist. See Della Rocca 1996, Guéroult 1968, Macherey 1998, and Mason 2001. Lennox (1976), too, offers interesting material. On the present account, Attributes are neither objective nor subjective in any clear-cut sense; concomitantly, Spinoza is neither a typical monist nor a typical dualist, echoing Wittgenstein's comment "there is no philosophical monism or dualism etc." (TLP 4.128).

4. I owe this remark—and a discussion of the vista it might open—to Andreas Karitzis.

5. Curley (1969, 144–54) discusses the issue in clear terms. Although his approach is congenial to mine, I disagree with his finally opting for—reluctantly, to be sure—the first of the two answers I will be discussing. As I will try to show, we might perhaps do better.

6. Or rather conceived, for perception is passive and there is no passivity in God.

7. I am gesturing toward an ideally systematic reconstruction of Freud's theory as elaborated by Lacan and as enriched by all other contributions that can prove consistent with it.

8. Fink (1995, 139–46) offers a clear presentation of Lacan's conception of psychical causality.

9. Elsewhere (Baltas 2005) I discuss psychoanalysis along those lines, but with no reference to Spinoza's Attributes. See also Gourgouris 2010.

10. Given Althusser's (1974a) confessed Spinozism, perhaps the very idea of "scientific continents"—including, apart from HM, the "continent" of mathematics, that of natural science and, more hesitantly, that of psychoanalysis—found its inspiration in the Spinozistic Attributes. I exclude mathematics from the discussion here, for it would take us too far afield to try spelling out how mathematics could be considered a "scientific continent" or a Spinozistic Attribute. See, however, Raymond 1973 and 1978.

11. Lipietz (1977) provides an approach to geography more or less along those lines.

12. On how Spinoza conceives the social, see Balibar 1985, Montag 1999, and Gavriilides 2000.

13. A perspective in this sense always involves both ideational and material components.

14. Elsewhere (Baltas 1997) I have argued in favor of this conception of science without bringing in the Spinozistic Attributes, although I admit they were hovering in my mind.

15. For the case of HM, see Baltas 1986 and 1988.

16. Compare Wittgenstein's (1984, 18e) comment to the effect that "what a Copernicus or a Darwin really achieved was not the discovery of a true theory but of a fertile new point of view." This fertile point of view translates readily into what I have been calling "scientific perspective." Jerry Massey elaborated on the same topic in a long conversation.

17. It is worthwhile stressing that indefinitely multiplying scientific perspectives opposes physicalism, which pushes toward reducing everything to bodies and their interactions, however microscopic or unobservable. Spinoza does not succumb to physicalism, for even independent of the way one could interpret his Attributes today, he considers Thought as perfectly equivalent to Extension and hence irreducible to it.

18. This seems to be the intuition guiding Aenishänslin (1993) to identify Wittgenstein's objects with Spinoza's Attributes. However, since he does not distinguish logic from grammar, he misses the mark.

19. Spinoza does not employ the term and for good reasons: the notion of parallelism implies lines never meeting, while Spinoza's Attributes are objective perspectives on Substance which do "meet" on it.

20. For the uses of the term "idea" in Spinoza's time and subsequent changes in focus, see Hacking 1982.

21. Curley (1969, 56–58), among others (e.g., Hampshire, whom Curley quotes), argues persuasively for this.

22. For a discussion of this cluster of propositions, see Russell's introduction to the *Tractatus* as well as Bogen 1996 and Diamond 2000a.

23. The battle cry of phenomenology "consciousness is always consciousness of" might find one of its historical roots here. From a different perspective, McDowell's "direct realism" can be grounded here as well.

24. All living organisms, which by being living encompass the striving—the conatus—for survival, might be considered to possess some kind of mind that, *mutatis mutandis,* "makes to itself pictures of the facts." Such pictures, particular as they are to the biological specificity of the organism, cannot fail to be *of* the facts as well as *objective* with respect to those facts (in the ways proper to the living organism), for their function is precisely that of *objectively* assisting the organism to survive. Nonetheless, we should distinguish the biologically determined *capacity* for thinking with which humans are congenitally endowed from the possible actualizations of that capacity. This capacity can be become actualized in properly human fashion only

through a biologically human individual's coming to participate in the human form of life, that is, of language and of the gendered human social order. But if, say, one secures one's biological survival though being "adopted" by some other kind of animal, then the capacity in question becomes compelled to conform to the corresponding animal form of life with the "pictures of the facts" peculiar to it. Various cases attesting such plasticity have been reported in the literature. See also Finkelstein 2007.

25. The following chapter discusses why Wittgenstein refuses to simply say that a proposition is the picture of *a fact* and the sense in which this refusal relates to what Spinoza has to say on the matter.

26. Another way to say this is to assert that truth and falsity are inextricably intertwined with Spinoza's understanding of knowledge. See Parkinson 1973; for a succinct summary, see Parkinson 2000.

Chapter Eight. Matching Form

1. Guéroult (1974, 101–2) and Della Rocca (1996, 94–106) offer discussions along this line.

2. Recall that the manifold of logical possibilities concerning the spatial location of facts is, in a sense, included among Wittgensteinian objects. Singling out this manifold would not serve the purposes of the *Tractatus*.

3. Wittgenstein calls this his "fundamental thought" (TLP 4.0231). See McDonough (1986) for a reading of the *Tractatus* based precisely on this.

4. For a detailed analysis of the issue, see Della Rocca 1996.

5. I have tried to "experiment" with this idea elsewhere (Baltas 2002b).

6. The "I" overarching all ideas in my mind thus appears to be structurally similar to the "I" issuing the discourse of solipsism. For this reason, solipsism threatens even for the most mundane uses of the "I."

7. I use cautionary quotation marks to stress that I want to be as vague as necessary concerning the way a mental "picture" of a fact comes to have some kind of bodily correlate while accepting that such bodily correlates should somehow exist: it is difficult to see how having different thoughts in my mind can leave my body *absolutely* unchanged.

8. That is, state S may be electrical, chemical, neurophysiological, or whatever, depending on how it could be eventually described by the relevant scientific discipline. Della Rocca (1996, 152–56) argues that Davidson, for example, would tend to accept this thesis, and Davidson (1999) himself admits that much. See also Heidelberger 2003.

9. Damasio 1994 presents strong arguments and neurophysiological evidence to demonstrate that part of the brain is devoted to getting input from the various parts of the body indicating their states and that the mind somehow comes to sense these states, though not at a conscious level. Damasio does not mention Spinoza in that work but does so with a vengeance elsewhere (Damasio 2003).

10. I owe the idea to Gideon Freudenthal, who proposed it to me in a very different context. The full relevance of the concept—homomorphism at every level of iteration—will appear later on.

11. Contemporary physical chemistry offers a good illustration of this Spinozistic idea. A clathrate is a complex molecular structure that exhibits a form of plasticity. If some molecule attaches to the clathrate for a certain length of time, then the form the clathrate had assumed while connected to that molecule is retained for some time after that molecule is withdrawn. This "trace" is proper to that other molecule, and hence its identity can be recovered from the form of the clathrate alone. The clathrate thus retains a "memory" of that molecule. We might add that the existence of clathrates offers some chemical grounds to the basic principles of homeopathic medicine. See Franks and Reid 1993; for the relation of clathrates to homeopathic medicine, see Anagnostatos et al. 1998. I owe these clarifications to my physicist colleagues L. Apekis, C. Christodoulides, and P. Pissis.

12. This is the fundamental idea behind Hallett's (1973, 154–60) impressive formalization.

13. See Greene and Nails 1986, where, however, such distance is not clearly marked overall.

14. In Newton's words: "Absolute, true, and mathematical time, in and of itself and of its own nature, without reference to anything external, flows uniformly and by another name it is called duration. Relative, apparent, and common time is any sensible and external measure (precise or imprecise) of duration by means of motion; such a measure—for example, an hour, a day, a month, a year—is commonly used instead of true time" (Newton 1999, 408).

15. I discuss such issues elsewhere (see Baltas 1986b and 2007b), though independent of Spinoza.

16. Mason (1999, 61) puts it succinctly: "Laws are not rules governing how things act but *are* how things exist and act as they do."

17. See Sklar 1992. I am indebted to Aristidis Arageorgis for a discussion of the issue.

18. Perhaps like quarks, which are taken to compose some "elementary" particles but never appear on their own, for they seem to obey what physicists call the "confinement" hypothesis. In that case, the "simplest bodies" (the quarks) would never appear as such, while the "atoms" in the enclosure would be elementary particles.

19. It is noteworthy that work on the ergodic hypothesis was initiated by Boltzmann whom, all biographical accounts agree, Wittgenstein considered as one of his heroes in natural science.

20. For a thorough discussion of Spinoza's definitions, see A. Garrett 2003.

21. See Janik and Toulmin 1973. Hertz is one of the few people to whom Wittgenstein explicitly refers in the *Tractatus* (TLP 4.04 and 6.361).

22. Beyssade (1994), however, equates the infinite mediate mode in the Attribute of Thought with the intellectual love of God, and Nadler (2006, 97) equates it with the total of the actual minds. This proposal coheres well with my interpretation of the infinite mediate mode in the Attribute of Extension.

23. Curley (1988, 34–50) offers an interesting discussion and useful references.

24. The totality of extended modes should be correlative to the totality of minds. Hence, given that E I p17s asserts intellect as such to "pertain [not] to the nature of God" but only to modes, Nadler's proposal (Nadler 2006, 96–98) makes the infinite mediate mode of the Attribute of Thought correspond well to (my interpretation of) the infinite mediate mode in the Attribute of Extension. We might perhaps say that the infinite mediate mode in the Attribute of Thought has the "face" of the truth with respect to the universe constituting our enclosure.

25. Earman (1986) has produced an illuminating discussion of such matters.

26. In his third rule "for the study of natural philosophy," Newton formulates this idea as follows: "Those qualities of bodies which cannot be intended and remitted [i.e. qualities that cannot be increased and diminished] and that belong to all bodies on which experiments can be made should be taken as qualities of all bodies universally" (Newton 1999, 795).

27. As noted, the issue of theory comparison in Wittgenstein's time was restricted to invoking criteria such as simplicity.

28. Wittgenstein's "nets" or "networks" thus do leave room for scientific progress and for debates among scientists at the conceptual level.

29. Hence there is a match with the Spinozistic Attributes if these are considered (in the way we saw in the preceding chapter) as self-sufficient (scientific) perspectives on Substance, each of which exhibits the lawlike connections, "of the causality form," proper only to it.

Exodus: Toward History and Its Surprises

1. It seems natural to surmise that the pains Spinoza took in composing his *Theologico-Political Treatise* might have clarified things and thus helped him to ameliorate the *Ethics*; the work at our disposal might well be the result of such critical toil. As I noted, Negri (1982) oversteps the limits of this natural surmise and goes as far as to maintain that the last parts of the *Ethics*—those presumably written after the completion of the *Theologico-Political Treatise*—put

forth what he calls a "second foundation" of Spinoza's philosophy. This proposal has not met the assent of Spinoza scholars, and Negri himself all but withdraws it later (Negri 2004). At any rate, the discussion lies outside the confines of the present work.

2. Monk (1990, 260–61) highlights Wittgenstein's conversations with Sraffa as reported to Malcolm. Wittgenstein reported that these conversations made him come to realize the role of bodily gestures in communication. See also Eagleton 1986.

3. Bogen (1972) has written a book interesting in this respect.

4. Compare PI §119: "The results of philosophy are the uncovering of one or another piece of plain nonsense and of bumps that the understanding has got by running its head up against the limits of language."

5. Another way of saying the same thing might be to assert that the formal cannot be abstracted away from content perfectly. Whether a formal contradiction such as "p and $-p$" constitutes a logical standofff might depend on the content of p, which cannot be analyzed down to a "concatenation of names" even in principle. It is presumably in this sense that logic lacks "crystalline purity."

6. The eureka experience amounts, in the last analysis, to the relief felt by such reinstatement.

7. Since this "dialectic" recovers strictly the same logic, it exhibits no Hegelian *Aufhebung*.

8. The presently imaginable (what *can* be presently imagined) constitutes the horizon of the real presently accessible, while the presently accessible real and the presently imaginable entertain the same relation to logic: recall that, on the one hand, the *Tractatus* holds that the real and the imaginable share logical form (TLP 2.022), while on the other hand, the fact that Wittgenstein's imaginary and imaginative examples in his later work can be thought at all implies they are not illogical—in stretching the imagination to investigate the grammatical, they necessarily remain within the bounds of logic. Or to say it more correctly, the fact that logic becomes reinstated after radical grammatical change implies that the same logic underlies all *possible* thought (presently accessible thought, including the imaginable), having itself no bounds. It bounces only on a logical standoff, which in being a *logical* standoff, rules out what cannot be thought at all.

9. McDowell's (1996) "direct realism" can thus be vindicated even in cases of radical grammatical breaks.

10. Here is the place where age-old philosophical questions tend to emerge as a matter of course: is there reality *beyond* the presently accessible? Is there reality per se that we discover by our cognitive practices? Were the features or facets of reality brought forth by the break already "there," or are they constructs of the break? And if they are constructs, are they at bottom cognitive or social? And together with these questions arises the temptation to answer them by reengaging the discussion on idealism, Platonic realism, the interpretation of the Kantian "thing-in-itself" and the like or on scientific realism and antirealism, on social constructivism, and so forth. In an interesting reawakening of such issues, Psillos (1999) replies to van Fraassen (1980), while Stergiopoulos (2006) tries to subvert the basis of this controversy. As we saw, the *Tractatus* undercuts such temptation at its roots by, among other things, identifying realism with solipsism. The question now at issue is how Wittgenstein's later work, if brought to bear on these grammatical breaks, does the same.

11. Doubtless, electromagnetic waves or the whole quantum realm have enriched the real we confront in our practices.

12. Kuhn (1962) has remarked that what I have been calling the postbreak real cannot be deemed *richer than* the prebreak real, for successes of the prebreak conceptual with respect to the prebreak real are sometimes lost by the postbreak conceptual in its relation to the postbreak real. The phenomenon has been dubbed "Kuhnian losses." And there cannot be any "transparadigmatic" criteria to assess a relation such as "richer/poorer."

13. This may provide the Wittgensteinian justification, with no idealism implied, of Kuhn's (1962) claim that we "live in different worlds" before and after the break. By the same token, no

enrichment of the kind being discussed can bring us closer to reality per se, which is only an illusion. It follows that no convergence to the real per se (to "The Truth") can make sense on the present account. Nonetheless, the asymmetry and incommensurability discussed in chapter 3 find their roots here.

14. This is perhaps the bottom reason why the young Wittgenstein never raised the issues of asymmetry, incommensurability, and the like. There is no distinction to be drawn between a pre- and a postbreak real at the level of logic, which is the only object of interest to him in the *Tractatus*. As was confirmed, the distinction between the two concerns only the conceptual (or grammatical) and thus only the disguise constituting colloquial language.

15. Scientists involved in the change go on working with the novel theory even as they develop intuitions on the basis of the widened grammatical space. Their philosophical worries in trying to domesticate—or philosophically ground—the novelties are appeased by the very developments to which their work gives rise. The Bohr-Einstein debate is famous in this respect. Laypersons, too, eventually come to feel at ease with the corresponding novelties, even if only indirectly, say, through the associated technological applications. The history of electromagnetic theory—its formulation, its development, its reception, its reinterpretation by the STR and quantum mechanics, and the technological achievements it allowed—is instructive in such respects. For an important facet of the history of electromagnetic theory considered under an interesting angle, see Arabatzis 2006.

16. This resistance is passive because the real has no way to voice its "dissatisfaction" with the conceptual.

17. Elsewhere (Baltas 1997 and 2004) I discuss the issue more thoroughly.

18. The fundamentals of this idea are already in place in the *Tractatus*. The difference between Wittgenstein's early and later work with respect to the issue consists in the modifications "really" and "effective," which I have emphasized to make this difference apparent.

19. Such predominance of the real over the conceptual (the ideal) is what any sober realism (or naturalism or materialism or . . .) should pursue.

20. Lacan equates what he calls "*le Réel*" with what resists symbolization, that is, with what does not conform to the conceptual. This is the Freudian "uncanny" catching us by total surprise. It becomes subjected to symbolization, domesticated by the conceptual, after the event.

21. Such reappraisal might either discard these relations or sanction some parts of them but under stringent conditions of translation or reinterpretation (Baltas 2004). Bachelard's (1975) distinction between *histoire sanctionnée* and *histoire périmée* (past scientific theories are either saved, while suitably transformed, by scientific development or are discarded from scientific practice, becoming objects of interest only for historians of science or of ideas) seems to formulate the issue under consideration here, namely, the necessary relations between the postbreak conceptual and the prebreak conceptual.

22. For example, we may still encounter difficulties in trying to make logical sense of some writings of Paracelsus.

23. Between grammatical breaks, the grammatical space on whose basis the inquiry unfolds remains at peace. This is the grammatical, Wittgensteinian rendering of Kuhn's distinction between normal and extraordinary science.

24. I more fully discuss the Whig conception of history in relation to historiography of science elsewhere (Baltas 1994).

25. Such work might come to show that a conceptual/experimental impossibility encountered in scientific practice was not a logical standoff whose overcoming amounts to a radical grammatical break. Working out the implications of the conceptual already in place might prove it capable of accommodating what had appeared as impossibility. The case of superconductivity and superfluidity is instructive in this respect. See Gavroglu and Goudaroulis 1989.

26. In this respect, see Fourtounis 2007 and 2011.

27. For an interpretation of Benjamin's "Theses on the Philosophy of History" along such

lines, see Baltas and Athanassakis 2007. The relation to Marxism of all the authors invoked might nevertheless offer grounds for justifying Wittgenstein's flirtation with events in Soviet Union. See Eagleton 1986, Monk 1990, and Crary 2000b.

28. This is no sacrifice of history overall but a sacrifice only in the sense that history is taken as subsumed to the ultimate per se. For Spinoza's conception of history, see Stylianou 1994.

29. See, however, Mason 2000.

30. Monk (1990) examines the reasons that led Wittgenstein to change his static outlook and perform the rotation at issue. My question—why he had adopted this outlook in the first place—is obviously different.

31. Negri (1982) and Althusser (1994) develop different arguments to claim that the theoretical lineage at work connects Marx to Machiavelli and Spinoza rather than to Rousseau and Hegel.

32. I am referring to Husserl's work of this title. See Theodorou 2000.

33. Interesting debates between those considered to be founders of the two traditions—e.g., the debate on the foundations of mathematics, involving Frege, Hilbert, and Husserl, or that on positivism, involving members or "relatives" of both the Frankfurt school and the Vienna Circle—attest that the unbridgeable gap between the two traditions has been very much a retroactive construct.

34. Russell's abandonment of Hegel and his subsequent intellectual life (see Monk 1996 and 2000) mark how these two opposing poles exerted unequal forces on professional philosophy.

References

Aenishänslin, M. 1993. *"Le Tractatus" de Wittgenstein et "l'Éthique" de Spinoza: étude de comparaison structurale*. Basel, Switzerland: Birkhäuser.

Althusser, L. 1966. "Du *Capital* à la philosophie de Marx." In *Lire le "Capital,"* by L. Althusser, J. Rancière, and P. Machery, 1:11–89. Paris: Maspero.

———. 1972. *Lénine et la philosophie, suivi de Marx et Lénine devant Hegel*. Paris: Maspero.

———. 1974a. *Eléments d'autocritique*. Paris: Hachette.

———. 1974b. *Philosophie et philosophie spontanée des savants*. Paris: Maspero.

———. 1976. "Freud et Lacan." In *Positions,* by Althusser, 9–34. Paris: Éditons Sociales.

———. 1990. "The Transformation of Philosophy." In *Philosophy and the Spontaneous Philosophy of the Scientists*, by Althusser, edited by G. Elliott, 241–66. London: Verso.

———. 1994. *Sur la philosophie*. Paris: Gallimard.

. 1996a. "Contradiction et surdétermination." In Althusser 1996c, 87–116.

———. 1996b. "Marxisme et humanisme." In Althusser 1996c, 227–58.

———. 1996c. *Pour Marx*. Paris: La Découverte.

Anagnostatos, G. S., P. Pissis, K. Viras, and M. Soutzidou. 1998. In *Homeopathy: A Critical Appraisal,* edited by E. Ernst and E. G. Hahn, 153 66. Oxford: Butterworth Heinemann.

Anapolitanos, D. 1999. *Leibniz: Representation, Continuity, and the Spatiotemporal*. Dordrecht, the Netherlands: Kluwer.

Anscombe, G. E. M. 1971. *An Introduction to Wittgenstein's* Tractatus. Philadelphia: University of Pennsylvania Press.

Arabatzis, T. 2006. *Representing Electrons: A Biographical Approach to Theoretical Entities*. Chicago: University of Chicago Press.

Bachelard, G. 1975. *Le nouvel esprit scientifique*. Paris: Presses Universitaires de France.

Balibar, É. 1985. *Spinoza et la politique*. Paris: Presses Universitaires de France.

Baltas, A. 1986a. "La confutabilità del materialismo storico e la struttura della pratica politica." *Nuova Civiltà delle Machine* 3–4:21–35.

———. 1986b. "Ideological 'Assumptions' in Physics: Social Determinations of Internal Structures." *PSA: Proceedings of the Biennial Meeting of the Philosophy of Science Association, 1986* 2:130–51.

———. 1988. "On the Concept of Historical Time—Physics and Historical Materialism." *O Politis* 94:59–70; repr., *Almanac of Sociology '88,* edited by G. Stamatis and J. Milios, 341–80. Athens: Exantas. In Greek.

———. 1991. "Physics as a Mode of Production." *Science in Context* 89:299–320.

———. 1994. "On the Harmful Effects of Excessive Antiwhiggism." In *Trends in the Historiography of Science,* edited by K. Gavroglu, J. Christianidis, and E. Nicolaidis, 107–20. Dordrecht, the Netherlands: Kluwer.

———. 1995. "Louis Althusser: The Dialectics of Erasure and the Materialism of Silence." *Strategies* 9–10: 152–94 (published by the Strategies Collective of UCLA).

———. 1997. "Constraints and Resistance: Stating a Case for Negative Realism." In *Realism and Quantum Physics,* edited by E. Agazzi, 74–96. Amsterdam: Rodopi.

———. 2000. "Classifying Scientific Controversies." In *Scientific Controversies: Philosophical and Historical Perspectives,* edited by A. Baltas, P. K. Machamer, and M. Pera, 40–49. Oxford: Oxford University Press.

———. 2002a. *Newton's Unconscious and Freud's Apple.* Athens: Exantas. In Greek.

———. 2002b. *Objects and Aspects of Self.* Athens: Hestia. In Greek.

———. 2004. "On the Grammatical Aspects of Radical Scientific Discovery." *Philosophia Scientia* 8.1:169–201.

———. 2007a. "Background 'Assumptions' and the Grammar of Conceptual Change: Rescuing Kuhn by Means of Wittgenstein." In *Reframing the Conceptual Change Approach in Learning and Instruction,* edited by S. Vosniadou, A. Baltas, and X. Vamvakoussi, 63–80. Amsterdam: Elsevier.

———. 2007b. "Physics as Self Historiography *in Actu*: Identity Conditions for the Discipline." Manuscript available on request.

———. 2009. "Nonsense and Paradigm Change." In *Rethinking Scientific Change and Theory Comparison: Stabilities, Ruptures and Incommensurabilities,* edited by L. Soler, H. Sankey, and P. Hoyningen-Huene, 47–68. New York: Springer.

Baltas, A., and M. Athanassakis. 2007. "History, Historiography, and Political Practice: A Comment on Walter Benjamin's *Theses on the Philosophy of History*." In *Walter Benjamin: Images and Myths of Modernity,* edited by A. Spyropoulou, 292–313. Athens: Alexandria. In Greek.

Benjamin, W. 1969. "Theses on the Philosophy of History." In *Illuminations,* by Benjamin, edited by H. Arendt, 253–64. New York: Shocken.

Bennett, J. 1984. *A Study of Spinoza's Ethics.* Indianapolis: Hackett.

Beyssade, J.-M. "Sur le mode infini médiat dans l'Attribut de la Pensée." *Revue Philosophique de la France et de l'Etranger* 1:23–26.

Black, M. 1964. *A Companion to Wittgenstein's Tractatus.* Ithaca, N.Y.: Cornell University Press.

Bloch, O., ed. 1993. *Spinoza au XXᵉ siècle.* Paris: Presses Universitaires de France.

Bogen, J. 1972. *Wittgenstein's Philosophy of Language.* New York: Humanities.

———. 1996. "Wittgenstein's *Tractatus*." In *Philosophy of Science, Logic, and Mathematics in the Twentieth Century,* edited by S. G. Shanker, 157–92. Routledge History of Philosophy, vol. 3. London: Routledge.

Boghosian, P., and S. Peacocke, eds. 2000. *New Essays on the A Priori.* Oxford: Oxford University Press.

Borges, J. L. 1998. *Collected Fictions.* Harmondsworth, U.K.: Penguin.

Bourlakis, P. 2006. "Wittgenstein and the Sceptical Stance in Philosophy." PhD dissertation, University of Athens and National Technical University of Athens. In Greek.

Bouveresse, J. 1995. *Wittgenstein Reads Freud: The Myth of the Unconscious.* Princeton, N.J.: Princeton University Press.

Brockhaus, R. D. 1991. *Pulling Up the Ladder: The Metaphysical Roots of Wittgenstein's Tractatus Logico-Philosophicus.* LaSalle, Ill.: Open Court.

Butler, J. 2005. *Giving an Account of Oneself.* New York: Fordham University Press.

Butterfield, H. 1982. *The Origins of Modern Science.* London: Bell and Hyman.

Camp, J. L. 2002. *Confusion: A Study in the Theory of Knowledge.* Cambridge, Mass.: Harvard University Press.

Carnap, R. 1978. "The Overcoming of Metaphysics through the Logical Analysis of Language." In *Heidegger and Modern Philosophy*, edited by M. Murray, 23–34. New Haven, Conn.: Yale University Press.

Carriero, J. 2005. "Spinoza on Final Causality." In *Oxford Studies in Early Modern Philosophy*, edited by D. Garber and S. Nadler, 2:105–148. Oxford: Clarendon.

Carroll, L. 1974. *Alice's Adventures in Wonderland* and *Through the Looking Glass*. In *The Philosopher's Alice*, by Carroll, introduction and notes by P. Heath. New York: St. Martin's.

Cavell, S. 1989. *Must We Mean What We Say?* Cambridge: Cambridge University Press.

———. 1999. *The Claim of Reason*. Oxford: Oxford University Press.

———. 2010. *Little Did I Know: Excerpts from Memory*. Stanford, Calif.: Stanford University Press.

Cavell, S., C. Diamond, J. McDowell, I. Hacking, and C. Wolfe. 2008. *Philosophy and Animal Life*. New York: Columbia University Press.

Coffa, J. A. 1993. *The Semantic Tradition, from Kant to Carnap to the Vienna Station*. Cambridge: Cambridge University Press.

Cohen, H. 1972 [1919]. *Religion of Reason, out of the Sources of Judaism*. Translated by S. Kaplan. New York: Ungar.

Cohen, R. S. 1956. "Introductory Essay." In Hertz 1956, i–xx.

Conant, J. 1991a. "The Search for Logically Alien Thought: Descartes, Kant, Frege, and the *Tractatus*." *Philosophical Topics* 20:115–80.

———. 1991b. "Throwing Away the Top of the Ladder." *Yale Review* 79:328–64.

———. 1995. "Putting Two and Two Together: Kierkegaard, Wittgenstein and the Point of View for Their Work as Authors." In S. Tessin and S. von der Ruhr 1995, 248–331.

———. 1997. "How to Pass from Latent to Patent Nonsense: Kierkegaard's *Postscript* and Wittgenstein's *Tractatus*." *Wittgenstein Studies* 2 (1997).

———. 2000a. "Elucidation and Nonsense in Frege and Early Wittgenstein." In Crary and Read 2000, 174–217.

———. 2000b. "Nietzsche's Perfectionism: A Reading of *Schopenhauer as Educator*." In *Nietzsche's Postmoralism*, edited by R. Schacht, 181–257. Cambridge: Cambridge University Press.

———. 2001. "Philosophy and Biography." In Klagge 2001, 16–50.

———. 2002. "The Method of the *Tractatus*." In Reck 2000, 374–462.

———. 2004. "Varieties of Scepticism." In McManus 2004, 97–136.

———. 2005. "The Dialectic of Perspectivism, I." *SATS: Nordic Journal of Philosophy* 6.2:7–50.

———. 2006. "The Dialectic of Perspectivism, II." *SATS: Nordic Journal of Philosophy* 7.1:7–57.

———. 2007. "Mild Mono-Wittgensteinianism." In Crary 2007b, 28–142.

Conant, J., and C. Diamond. 2004. "On Reading the *Tractatus* Resolutely." In *The Lasting Significance of Wittgenstein's Philosophy*, edited by M. Max Kölbel and B. Weiss, 42–97. London: Routledge.

Cornford, F. M. 1970. *Plato's Theory of Knowledge: The Theaetetus and the Sophist*. London: Routledge and Kegan Paul.

Crary, A. 2000a. "Introduction." In Crary and Read 2000, 1–18.

———. 2000b. "Wittgenstein's Philosophy in Relation to Political Thought." In Crary and Read 2000, 118–45.

———. 2007a. *Beyond Moral Judgment*. Cambridge, Mass.: Harvard University Press.

———, ed. 2007b. *Wittgenstein and the Moral Life: Essays in Honor of Cora Diamond*. Cambridge, Mass.: MIT Press.

Crary, A., and R. Read, eds. 2000. *The New Wittgenstein*. London: Routledge.

Curley, E. 1969. *Spinoza's Metaphysics: An Essay in Interpretation*. Cambridge, Mass.: Harvard University Press.

———. 1988. *Behind the Geometrical Method*. Princeton, N.J.: Princeton University Press.

Damasio, A. 1994. *Descartes' Error.* New York: Putnam's.

———. 2003. *Looking for Spinoza: Joy, Sorrow, and the Feeling Brain.* Orlando, Fla.: Harcourt.

Damianos, P. 2007. "Realism in Ecosystems: Scientific Realism and Non-Conceptual Content." PhD dissertation, University of Athens and National Technical University of Athens. In Greek.

Davidson, D. 1999. "Spinoza's Causal Theory of the Affects." In Yovel 1999, 95–111.

De Deugt, C. 2004. "Spinoza and Freud: An Old Myth Revisited." In Yovel and Segal 2004, 227–52.

De Dijn, H. 2004. "The Ladder, Not the Top: The Provisional Morals of the Philosopher." In Yovel and Segal 2004, 37–56.

Deleuze, G. 1968. *Spinoza et le problème de l'expression.* Paris: Minuit.

———. 1981. *Spinoza: philosophie pratique.* Paris: Minuit.

———. 2001. "Confrontation avec le commentaire de Guéroult." Cours Vincennes: Infini actuel—éternité, logique des relations. Internet file, http://www.webdeleuze.com/php/texte.php?cle=40&groupe=Spinoza&langue=1.

———. 2005. *Pure Immanence.* New York: Zone.

Della Rocca, M. 1996. *Representation and the Mind Body Problem in Spinoza.* Oxford: Oxford University Press.

Derrida, J. 1976. *Of Grammatology.* Translated by G. C. Spivak. Baltimore, Md.: Johns Hopkins University Press.

———. 1981. "Plato's Pharmacy." In *Dissemination,* by Derrida, translated and with an introduction and additional notes by B. Johnson, 61–172. Chicago: University of Chicago Press.

Diamond, C. 1995a. *The Realistic Spirit: Wittgenstein, Philosophy, and the Mind.* Cambridge, Mass.: MIT Press.

———. 1995b. "Throwing away the Ladder: How to Read the *Tractatus.*" In Diamond 1995a, 179–204.

———. 2000a. "Does Bismarck Have a Beetle in His Box? The Private Language Argument in the *Tractatus.*" In Crary and Read 2000, 262–92.

———. 2000b. "Ethics, Imagination, and the Method in Wittgenstein's *Tractatus.*" In Crary and Read 2000, 149–73.

Di Poppa, F. 2006. "God Acts from the Laws of His Nature Alone." PhD dissertation, University of Pittsburgh.

Dreyfus, H. L. 1992. *What Computers Still Can't Do.* Cambridge, Mass.: MIT Press.

Dreyfus, H. L., and S. E. Dreyfus. 1988. *Mind over Machine: The Power of Human Intuition and Expertise in the Era of the Computer.* New York: Free Press.

Dummett, M. 1993. *Origins of Analytic Philosophy.* Cambridge, Mass.: Harvard University Press.

Eagleton, T. 1986. "Wittgenstein's Friends." In *Against the Grain,* by Eagleton, 99–130. London: Verso.

Earman, J. 1986. *A Primer on Determinism.* Dordrecht, the Netherlands: Reidel.

Einstein, A. 1923. "On the Electrodynamics of Moving Bodies." In Einstein, A., et al., *The Principle of Relativity.* New York: Dover.

Engels, F. 1974. "Socialisme utopique et socialisme scientifique." Extract. In Marx and Engels 1974, 226–36.

Euclid. 1956. *The Thirteen Books of the Elements.* Translated and with an introduction and commentary by T. Heath. New York: Dover.

Fink, B. 1995. *The Lacanian Subject: Between Language and Jouissance.* Princeton, N.J.: Princeton University Press.

Finkelstein, D. 2007. "Holism and Animal Minds." In Crary 2007b, 251–78.

Floyd, J. 2000. "Number and Ascription of Number in Wittgenstein's *Tractatus.*" In Reck 2000, 308–52.

———. 2005. "Wittgenstein on Philosophy of Logic and Mathematics." In Shapiro 2005, 75–128.

Floyd, J., and S. Shieh, eds. 1999. *Future Pasts: Perspectives on the Place of the Analytic Tradition in Twentieth Century Philosophy.* Cambridge, Mass: Harvard University Press.

Fourtounis, G. 2002. "Immanence and Structure." In V. Grigoropoulou and A. Stylianou 2002, 191–228. In Greek.

———. 2005. "On Althusser's Immanent Structuralism: Reading Montag Reading Althusser Reading Spinoza." *Rethinking Marxism* 17.1:101–18.

———. 2007. "Althusser's Late Materialism and the Epistemological Break." In Turchetto 2007, 81–92.

———. 2011, forthcoming. "'An Immense Aspiration to Being': The Causality and Temporality of the Aleatory."

Franks, F., and D. S. Reid. 1993. In *Water: A Comprehensive Treatise,* edited by F. Franks, vol. 2, ch. 5. New York: Plenum.

Friedlander, E. 2001. *Signs of Sense: Reading Wittgenstein's* Tractatus. Cambridge, Mass.: Harvard University Press.

Garcia, J. J. E. 1992. *Philosophy and Its History.* Albany, N.Y.: SUNY Press.

Garrett, A. V. 2003. *Meaning in Spinoza's Method.* Cambridge: Cambridge University Press.

Garrett, D., ed. 1996. *The Cambridge Companion to Spinoza.* Cambridge: Cambridge University Press.

Garver, N. 1994. *This Complicated Form of Life.* Chicago: Open Court.

Garver, N., and S. Lee. 1994. *Derrida and Wittgenstein.* Philadelphia: Temple University Press.

Gavriilides, A. 2000. *Democracy against Liberalism.* Athens: Ellinika Grammata. In Greek.

Gavroglu, K., and Y. Goudaroulis. 1989. *Methodological Aspects of the Development of Low Temperature Physics 1881–1956: Concepts out of Contexts.* Dordrecht, the Netherlands: Kluwer.

Gibson, J. J. 1979. *The Ecological Approach to Visual Perception.* Boston: Houghton Mifflin.

Gier, N. F. 1981. *Wittgenstein and Phenomenology: A Comparative Study of the Later Wittgenstein, Husserl, Heidegger, and Merleau-Ponty.* Albany, N.Y.: SUNY Press.

Glock, H.-J. 1996. *A Wittgenstein Dictionary.* Oxford: Blackwell.

Goldstein, H. 1959. *Classical Mechanics.* San Francisco: Addison-Wesley.

Goldstein, R. N. 2006. *Betraying Spinoza: The Renegade Jew Who Gave Us Modernity.* New York: Schoken.

Gourgouris, S. ed. 2010. *Freud and Fundamentalism.* New York: Fordham University Press.

Greene, M., ed. 1973. *Spinoza: A Collection of Critical Essays.* Notre Dame, Ind.: University of Notre Dame Press.

Greene, M., and D. Nails, eds. 1986. *Spinoza and the Sciences.* Dordrecht, the Netherlands: Kluwer.

Grigoropoulou, V., and A. Stylianou, eds. 2002. *Spinoza: Toward Freedom, Ten Contemporary Greek Studies.* Special issue 2 of *Axiologica.* In Greek.

Guéroult, M. 1968. *Spinoza I.* Hildesheim, Germany: Georg Olms Verlagsbuchhandlung.

———. 1974. *Spinoza II.* Hildesheim, Germany: Georg Olms Verlagsbuchhandlung.

Haack, S. 1978. *Philosophy of Logics.* Cambridge: Cambridge University Press.

Hacker, P. M. S. 1986. *Insight and Illusion.* Oxford: Oxford University Press.

Hacking, I. 1982. *Why Does Language Matter to Philosophy?* Cambridge: Cambridge University Press.

Hallett, H. F. 1973. "Substance and Its Modes." In Greene 1973, 131–63.

Hampshire, S. 1987. *Spinoza: An Introduction to His Philosophical Thought.* Harmondsworth, U.K.: Penguin.

Hanson, N. R. 1958. *Patterns of Discovery.* Cambridge: Cambridge University Press.

Harnecker, M. 1974. *Les concepts élémentaires du matérialisme historique.* Brussels: Contradictions.

Harris, R. 1988. *Language, Saussure, and Wittgenstein.* London: Routledge.

Haugeland, J. 2000. *Having Thought: Essays in the Metaphysics of Mind*. Cambridge, Mass.: Harvard University Press.

Heaton, J. M. 2000. *Wittgenstein and Psychoanalysis*. Cambridge: Icon.

Heidegger, M. 1962. *Being and Time*. Translated by J. Macquarrie and E. Robinson. New York: Harper and Row.

Heidelberger, M. 2003. "The Mind-Body Problem in the Origins of Logical Empiricism: Herbert Feigl and Psychophysical Parallelism." In Parrini et al. 2003, 233–62.

Hertz, H. 1956. *The Principles of Mechanics*. Translated by D. E. Jones and J. T. Walley. New York: Dover.

Hutto, D. D. 2003. *Wittgenstein and the End of Philosophy: Neither Theory nor Therapy*. New York: Palgrave Mcmillan.

Israel, J. I. 2001. *Radical Enlightenment*. Oxford: Oxford University Press.

Janik, A., and S. Toulmin. 1973. *Wittgenstein's Vienna*. New York: Simon and Schuster.

Kashap, P., ed. 1972. *Studies in Spinoza: Critical and Interpretative Essays*. Berkeley: University of California Press.

Kindi, V. 1995. *Kuhn and Wittgenstein: Philosophical Investigations on the Structure of Scientific Revolutions*. Athens: Smili. In Greek.

Klagge J. C., ed. 2001. *Wittgenstein: Biography and Philosophy*. Cambridge: Cambridge University Press.

Kline, M. 1980. *Mathematics: The Loss of Certainty*. Oxford: Oxford University Press.

Koistinnen, O., and J. Biro, eds. 2002. *Spinoza: Metaphysical Themes*. Oxford: Oxford University Press.

Kordela, K. A. 2007. *Surplus: Spinoza, Lacan*. Albany, N.Y.: SUNY Press.

Koyré A. 1957. *From the Closed World to the Infinite Universe*. Baltimore, Md.: Johns Hopkins University Press.

Kuhn, T. S. 1962. *The Structure of Scientific Revolutions*. Chicago: Chicago University Press.

———. 2000. *The Road since "Structure": Philosophical Essays 1970–1993, with a Autobiographical Interview*. Edited by J. Conant and J. Haugeland. Chicago: Chicago University Press.

Lacan, J. 2006a. *Écrits*. translated by B. Fink. New York: Norton.

———. 2006b. "The Function of and Field of Speech and Language in Psychoanalysis." In Lacan 2006a, 237–68.

———. 2006c. "The Mirror Stage as Formative of the I Function as Revealed in Psychoanalytic Experience." In Lacan 2006a, 75–81.

———. 2006d. "On the Subject Who Is Finally in Question." In Lacan 2006a, 229–37.

———. 2006e. "Position of the Unconscious." In Lacan 2006a, 701–21.

———. 2006f. "Presentation on Psychical Causality." In Lacan 2006a, 123–58.

Lakatos I. 1980. "History of Science and Its Rational Reconstructions." In *Imre Lakatos: Philosophical Papers*. Edited by J. Worrall and G. Currie. Vol. 1, *The Methodology of Scientific Research Programmes*. Cambridge: Cambridge University Press.

Laplanche, J., and J.-B. Pontalis. 1973. *Vocabulaire de la psychanalyse*. Paris: Presses Universitaires de France.

Lear, J. 1999. *Open Minded: Working out the Logic of the Soul*. Cambridge, Mass.: Harvard University Press.

Lecourt, D. 1981. *L'Ordre et les jeux*. Paris: Grasset.

———. 1982. *La philosophie sans feinte*. Paris: J.-E. Hallier/Albin Michel.

Lénine V. 1962. *Matérialisme et empiriocriticisme*. In *Œuvres*, vol. 14. Paris: Éditions Sociales.

Lennox, J. G. 1976. "The Causality of Finite Modes in Spinoza's *Ethics*." *Canadian Journal of Philosophy* 3:479–500.

Lipietz, A. 1977. *Le capital et son espace*. Paris: Maspero.

Lock, G. 1992. *Wittgenstein: philosophie, logique, thérapeutique*. Paris: Presses Universitaires de France.

Loyd, G. 1994. *Part of Nature: Self-Knowledge in Spinoza's* Ethics. Ithaca, N.Y.: Cornell University Press.

———. 1996. *Spinoza and "the Ethics."* London: Routledge.

Machamer, P. K., and J. E. McGuire. 2009. *Descartes's Changing Mind.* Princeton, N.J.: Princeton University Press.

Macherey, P. 1979. *Hegel ou Spinoza.* Paris: Maspero.

———. 1993. "Spinoza est-il moniste?" In Revault d'Allones 1993, 39–53.

———. 1997. *Introduction à "l'Éthique" de Spinoza; la seconde partie: la réalité mentale.* Paris: Presses Universitaires de France.

———. 1998. *Introduction à "l'Éthique" de Spinoza; la première partie: la nature des choses.* Paris: Presses Universitaires de France.

Mandelbaum, M., and E. Freeman, eds. 1975. *Spinoza: Essays in Interpretation.* Chicago: Open Court.

Marx K., and F. Engels. 1974. *Études philosophiques.* Introduction by Guy Besse. Paris: Éditions Sociales.

Maslow, A. 1997. *A Study in Wittgenstein's* Tractatus. Bristol, U.K.: Thoemmes.

Mason, R. 1999. *The God of Spinoza.* Cambridge: Cambridge University Press.

———. 2000. *Before Logic.* Albany, N.Y.: SUNY Press.

———. 2003. *Understanding Understanding.* Albany, N.Y.: SUNY Press.

———. 2007. *Spinoza: Logic, Knowledge, and Religion.* Aldershot, U.K.: Ashgate.

Matheron, A. 1988. *Individu et communauté chez Spinoza.* Paris: Minuit.

Mathews, M. R. 2000. *Time for Science Education.* Dordrecht, the Netherlands: Kluwer.

McDonough. R. 1986. *The Argument of the "Tractatus."* Albany, N.Y.: SUNY Press.

McDowell, J. 1996. *Mind and World.* Cambridge: Harvard University Press.

McGuinness, B. 2005. *Young Ludwig: Wittgenstein's Life 1889–1921.* Oxford: Clarendon.

McManus, D. 2004. *Wittgenstein and Skepticism.* London: Routledge.

Medina, J. 2002. *The Unity of Wittgenstein's Philosophy: Necessity, Intelligibility, and Normativity.* Albany, N.Y.: SUNY Press.

Monk, R. 1990. *Ludwig Wittgenstein: The Duty of Genius.* London: Penguin.

———. 1996. *Bertrand Russell: The Spirit of Solitude.* London: Vintage.

———. 2000. *Bertrand Russell: The Ghost of Madness 1921–1970.* London: Vintage.

———. 2001. "Philosophical Biography: The Very Idea." In Klagge 2001, 3–15.

———. 2005. *How to Read Wittgenstein.* New York: Norton.

Montag, W. 1999. *Bodies, Masses, Power: Spinoza and His Contemporaries.* London: Verso.

Montag, W., and T. Stolze, eds. 1997. *The New Spinoza.* Minneapolis: University of Minnesota Press.

Morawetz, T. 1978. *Wittgenstein and Knowledge: The Importance of "On Certainty."* London: Harvester.

Moreau, P. F. 1994. *Spinoza: l'expérience et l'éternité.* Paris: Presses Universitaires de France.

Morris, M. 2008. *Wittgenstein and the "Tractatus."* London: Routledge.

Mounce, H. O. 1985. *Wittgenstein's "Tractatus": An Introduction.* Chicago: University of Chicago Press.

Mulhall, S. 1990. *On Being in the World: Wittgenstein and Heidegger on Seeing Aspects.* London: Routledge.

———. 2001. *Inheritance and Originality: Wittgenstein, Heidegger, Kierkegaard.* Oxford: Clarendon.

———. 2005. *Philosophical Myths of the Fall.* Princeton, N.J.: Princeton University Press.

Nadler, S. 1999. *Spinoza: A Life.* Cambridge: Cambridge University Press.

———. 2004. *Spinoza's Heresy: Immortality and the Jewish Mind.* Oxford: Oxford University Press.

———. 2006. *Spinoza's* Ethics: *An Introduction.* Cambridge: Cambridge University Press.

Narboux, J.-P. 2001. "La logique peut-elle prendre soin d'elle-même?" *Critique* 654:830–46.

Nasio, J.-D. 1998. *Five Lessons on the Psychoanalytic Theory of Jacques Lacan.* Translated by D. Pettigrew and F. Raffoul. Albany, N.Y.: SUNY Press, 1998. Originally published as *Cinq leçons sur la théorie de Jacques Lacan.* Paris: Rivages.

Negri, A. 1982. *L'anomalie sauvage.* Translated by F. Matheron. Paris: Presses Universitaires de France.

———. 2004. *Subversive Spinoza.* Edited by T. S. Murphy. Manchester, U.K.: Manchester University Press.

Nehamas, A. 1985. *Nietzsche: Life as Literature.* Cambridge, Mass: Harvard University Press.

———. 1998. *The Art of Living.* Berkeley: University of California Press.

Newton, I. 1999. *The Principia.* Translated by I. B. Cohen and A. Whitman. Berkeley: University of California Press.

Nietzsche, F. 1982. *The Portable Nietzsche.* Translated and edited by W. Kaufmann. New York: Viking.

Nordmann, A. 2005. *Wittgenstein's* Tractatus: *An Introduction.* Cambridge: Cambridge University Press.

Norris, S. 1991. *Spinoza and the Origins of Modern Critical Theory.* Oxford: Basil Blackwell.

Ostrow, M. B. 2002. *Wittgenstein's "Tractatus": A Dialectical Interpretation.* Cambridge: Cambridge University Press.

Pagondiotis, K. 2005. "Can the Perceptual Be Conceptual and Non-Theory-Laden?" In *Cognitive Penetrability of Perception: An Interdisciplinary Approach,* edited by A. Raftopoulos, 165–80. New York: Nova Scotia.

Parkinson, G. H. R. 1973. "Language and Knowledge in Spinoza." In Greene 1973, 73–100.

———. 2000. "Introduction" to Spinoza 2000a, 5–54.

Parrini, P., W. C. Salmon, and M. H. Salmon, eds. 2003. *Logical Empiricism: Historical and Contemporary Perspectives.* Pittsburgh, Pa.: University of Pittsburgh Press.

Pissis, G. 2002. *The Philosophy of Mathematics of Ludwig Wittgenstein.* Master's thesis, University of Athens and National Technical University of Athens. In Greek.

Popkin, R. H. 1979. *The History of Skepticism from Erasmus to Spinoza.* Berkeley: University of California Press.

Psillos, S. 1999. *Scientific Realism: How Science Tracks Truth.* London: Routledge.

Putnam, H. 2000. "Rethinking Mathematical Necessity." In Crary and Read 2000, 218–31.

Ramsey, F. P. 1931. "General Propositions and Causality." In *The Foundations of Mathematics,* edited by R. B. Braithwaite, 238. London: Kegan Paul, Trench, and Trubner.

Rancière, J. 1988. *Le maître ignorant.* Paris: Arthème Fayard.

Raymond, P. 1973. *Le passage au matérialisme.* Paris: Maspero.

———. 1978. *L'histoire et les sciences.* Paris: Maspero.

Reck, E. H., ed. 2002. *From Frege to Wittgenstein.* Oxford: Oxford University Press.

Reid, L. 1998. "Wittgenstein's Ladder: The *Tractatus* and Nonsense." *Philosophical Investigations* 21:97–151.

Revault d'Allones, M. 1993. *Spinoza: puissance et ontologie.* Paris: Éditions Kimé.

———. 2002. *Fragile humanité.* Paris: Aubier.

Rhees, R. 2005. *Wittgenstein's "On Certainty."* Edited by D. Z. Phillips. Oxford: Blackwell.

Ricketts, T. 1985. "Frege, the *Tractatus,* and the Logocentric Predicament." *Nous* 19:3–15.

———. 1996. "Pictures, Logic, and the Limits of Sense in the *Tractatus.*" In Sluga and Stern 1996, 59–99.

Rorty, R., ed. 1988. *The Linguistic Turn: Recent Essays in Philosophical Method.* Chicago: University of Chicago Press.

Rorty, R., J. B. Shneewind, and Q. Skinner. 1984. *Philosophy in History.* Cambridge: Cambridge University Press.

Saussure, F. 1968. *Cours de linguistique générale.* Paris: Payot.

Savan, D. 1958. "Spinoza and Language." In Greene 1973, 60–72.

Schacht, R. 1999. "The Spinoza-Nietzsche Problem." In Yovel 1999, 211–32.

Schipper, L. 1993. *Spinoza: The View from Within.* New York: Peter Lang.

Schouls, Peter A. 1980. *The Imposition of Method: A Study of Descartes and Locke.* Oxford: Clarendon.

Sellars, W. 1991. *Science, Perception and Reality.* Atascadero, Calif.: Ridgeview.

Shahan, R. S., and Biro, J. I. 1978. *Spinoza: New Perspectives.* Norman: University of Oklahoma Press.

Shapiro, S., ed. 2005. *The Oxford Handbook of Philosophy of Mathematics and Logic.* Oxford: Oxford University Press.

Shields, P. R. 1993. *Logic and Sin in the Writings of Ludwig Wittgenstein.* Chicago: Chicago University Press.

Shirley, S. 1982. "Introduction." In Spinoza 1982, 1–20.

Sklar, L. 1992. *Philosophy of Physics.* Boulder, Colo.: Westview.

Sluga H., and D. Stern. 1996. *The Cambridge Companion to Wittgenstein.* Cambridge: Cambridge University Press.

Smith, S. B. 2003. *Spinoza's Book of Life.* New Haven, Conn.: Yale University Press.

Spiliopoulos, D. 2011. "Responsibility: Before and beyond Duty." Master's thesis, University of Athens and National Technical University of Athens. In Greek.

Spinoza, B. 1982. *The Ethics and Selected Letters.* Translated by S. Shirley. Indianapolis: Hackett.

———. 1985a. *The Collected Works of Spinoza.* Vol. 1. Translated by E. Curley. Princeton, N.J.: Princeton University Press.

———. 1985b. *Ethics.* Translated by E. Curley. In Spinoza 1985a, 408–617.

———. 1985c. *Short Treatise on God, Man, and His Well-Being.* Translated by E. Curley. In Spinoza 1985a, 53–156.

———. 1985d. *Treatise on the Emendation of the Intellect.* Translated by E. Curley. In Spinoza 1985a, 7–45.

———. 1988. *Éthique.* Bilingual edition. Translated by B. Pautrat Paris: Éditions du Seuil.

———. 1995. *The Letters.* Translated by S. Shirley. Indianapolis: Hackett.

———. 2000a. *Ethics.* Translated by G. H. R. Parkinson. Oxford: Oxford University Press.

———. 2000b. *Treatise on the Emendation of the Intellect.* Translated by B. Jaquemart and V. Grigoropoulou, edited by V. Grigoropoulou. Athens: Polis. In Greek.

———. 2009. *Ethics.* Translated by E. Vandarakis. Athens: Ekkremes. In Greek.

Staten, H. 1984. *Wittgenstein and Derrida.* Lincoln: University of Nebraska Press.

Stenius, E. 1996. *Wittgenstein's "Tractatus": A Critical Exposition of the Main Lines of Thought.* Bristol, U.K.: Thoemmes.

Stergiopoulos, K. 2006. "Empiricism, Science, and Metaphysics: Criticizing the Controversy between Scientific Realism and Constructive Empiricism." PhD dissertation, University of Athens and National Technical University of Athens. In Greek.

Stern, D. G. 2003. "The Methods of the *Tractatus*: Beyond Positivism and Metaphysics?" In Parrini et at. 2003, 125–56.

Stone, M. 2000. "Wittgenstein on Deconstruction." In Crary and Read 2000, 83–117.

Stylianou, A. 1994. "Histoire et politique chez Spinoza." PhD dissertation, University of Paris I.

Tait, W. W., ed. 1997. *Early Analytic Philosophy.* Chicago: Open Court.

Tessin, S., and S. von der Ruhr, eds. 1995. *Philosophy and the Grammar of Religious Belief.* London: Macmillan.

Theodorou, P. 2000. "Thought Experiments: A Husserlian Analysis of Their Structure." PhD dissertation, National Technical University of Athens. In Greek.

Turchetto, M., ed. 2007. *Rileggere il "Capitale": la lezione di Louis Althusser.* Milan: Mimesis Edizioni.

Van Fraassen, B. C. 1968. "Presupposition, Implication, and Self-Reference." *Journal of Philosophy* 65:132–56.

———. 1980. *The Scientific Image.* Oxford: Clarendon.

Vokos, G. 2002. "*Treatise on the Emendation of the Intellect:* The Preface." In Grigoropoulou and Stylianou 2002, 15–34. In Greek.

Vosniadou, S. 2002. "On the Nature of Naïve Physics." In *Reconsidering the Processes of Conceptual Change,* edited by M. Limon and L. Mason, 61–76. Dordrecht, the Netherlands: Kluwer.

Waugh, A. 2008. *The House of Wittgenstein: A Family at War.* London: Bloomsbury.

Weinberg, S. 1977. "The Search for Unity: Notes for a History of Quantum Field Theory." *Daedalus* 107:17–35.

Weiner, D. 1992. *Genius and Talent: Schopenhauer's Influence on Wittgenstein's Early Philosophy.* London and Toronto: Associated University Presses.

Weiner, J. 1990. *Frege in Perspective.* Ithaca, N. Y.: Cornell University Press.

White, R. M. 2006. *Wittgenstein's* Tractatus Logico-Philosophicus. London: Continuum.

Wittgenstein L. 1968. *Philosophical Investigations.* Translated by G. E. M. Anscombe. New York: Macmillan.

———. 1969. *On Certainty.* Edited by G. E. M. Anscombe and G. H. von Wright. New York: Harper and Row.

———. 1971. *Tractatus Logico-Philosophicus.* Translated by Th. Kitsopoulos. Special issue, *Deucalion* 7—8 (1971). In Greek.

———. 1973. *Letters to C. K. Ogden.* Translated by G. H. von Wright. Oxford: Basil Blackwell.

———. 1974. *Tractatus Logico-Philosophicus.* Translated by D. F. Pears and B. F. McGuinness. London: Routledge.

———. 1984. *Culture and Values.* Translated by P. Winch. Chicago: University of Chicago Press.

———. 1986. *Tractatus Logico-Philosophicus.* Translated by C. K. Ogden. London: Routledge and Kegan Paul.

———. 1993. *Tractatus Logico-Philosophicus.* Translated by G.-G. Granger. Paris: Gallimard.

———. 2009. *Philosophical Investigations.* Translated by G. E. M. Anscombe, P. M. S. Hacker, and J. Schulte; edited by P. M. S. Hacker and J. Schulte. 4th ed., rev. Chichester, U.K.: Wiley/ Blackwell.

Wolfson, H. A. 1983. *The Philosophy of Spinoza.* Cambridge, Mass.: Harvard University Press.

Yovel, Y. 1989. *Spinoza and Other Heretics: The Adventures of Immanence.* Princeton, N.J.: Princeton University Press.

———. ed. 1999. *Desire and Affect: Spinoza as Psychologist.* New York: Little Room.

Yovel, Y., and G. Segal, eds. 2004. *Spinoza on Reason and the Free Man.* New York: Little Room.

Index